Study Guide

Chemistry

TENTH EDITION

Kenneth W. Whitten
University of Georgia, Athens

Raymond E. Davis
University of Texas at Austin

M. Larry Peck
Texas A&M University

George G. Stanley
Louisiana State University

Prepared by

James Petrich
San Antonio College

BROOKS/COLE
CENGAGE Learning

Australia • Brazil • Japan • Korea • Mexico • Singapore • Spain • United Kingdom • United States

For product information and technology assistance, contact us at **Cengage Learning Customer & Sales Support, 1-800-354-9706**

For permission to use material from this text or product, submit all requests online at **www.cengage.com/permissions** Further permissions questions can be emailed to **permissionrequest@cengage.com**

ISBN-13: 978-1-133-93354-0
ISBN-10: 1-133-93354-8

Brooks/Cole
20 Davis Drive
Belmont, CA 94002-3098
USA

Cengage Learning is a leading provider of customized learning solutions with office locations around the globe, including Singapore, the United Kingdom, Australia, Mexico, Brazil, and Japan. Locate your local office at: **www.cengage.com/global**

Cengage Learning products are represented in Canada by Nelson Education, Ltd.

To learn more about Brooks/Cole, visit **www.cengage.com/brookscole**

Purchase any of our products at your local college store or at our preferred online store **www.cengagebrain.com**

Printed in the United States of America
1 2 3 4 5 6 7 17 16 15 14 13

PREFACE

The science of chemistry is a tremendously broad subject of study, with implications ranging from applications in the research laboratory, industry, medicine, and agriculture, through innumerable uses in our daily lives, to the challenging intellectual exercise of making sense of such a vast subject. The only way to make all of this a manageable study is to organize it. Therefore, we systematize our observations, trying to see common features in many different experimental results; we use the resulting summaries of observed behavior, called scientific laws, to help us predict chemical and physical behavior in unknown or untested cases; and we try to understand our observations and their summaries in terms of broad concepts such as the atomic theory. The merging of literally millions of chemical and physical observations, along with theories to explain them, into the ever-growing science of chemistry is one of the grandest intellectual achievements of the human mind.

The biggest challenge for many students undertaking their first college-level study of chemistry is usually determining what, in the wealth of detailed information in a chapter, comprises the key central themes of that chapter, and what other material merely supports, explains, or exemplifies these main ideas. This Study Guide was prepared to assist students in their study of chemistry from the text *Chemistry, Tenth Edition*, by Whitten, Davis, Peck, and Stanley. It is intended to serve as a supplement to lectures and text reading. In preparing this Study Guide, we have been guided by the belief that the primary functions of such a guide are similar to those of an effective teacher—to summarize, to focus study toward particular goals, to stimulate practice at applying concepts and sharpening skills, and to provide an assessment of how the study is progressing.

Each chapter contains five parts:

1. A **Chapter Summary**, highlighting the main themes of the chapter, tying together the various key ideas of the chapters, and relating them to previous study and to topics to be encountered later.
2. Explicit **Study Goals,** listing each by the appropriate sections in the text and including references to text exercises related to each goal.
3. Some **Important Terms** from the chapter, including many of the Key Terms listed in the text chapter, plus other important terms that first appeared in an earlier chapter. The list is followed by paragraphs containing blanks that are to be filled with the appropriate terms. This kind of procedure develops comprehension skills and vocabulary through structured and contextual analysis strategies. Some chapters also include Quotefalls puzzles, using the scientific laws introduced in the chapter.
4. A **Preliminary Test**, consisting of many questions (averaging more than 90 per chapter) of easy to moderate difficulty. These supply extensive practice at applying the terminology, basic concepts, and fundamental calculations of the chapter. They lay the necessary groundwork for practice with more difficult exercises from the textbook. Many students find this the most useful aspect of the Study Guide.

iii

5. The **Answers to the Preliminary Test**, containing answers to *all* Preliminary Test questions, most with additional comments, reasoning, or stepwise solutions presented.

The following section, **TO THE STUDENT**, presents suggestions to guide students in developing systematic, productive study habits, and in coordinating material from classroom, text reading, Study Guide, and homework.

My thanks to my family for their support:

- to my wife Janis for enabling me to have the time to create, for developing graphics, for providing many valuable suggestions, and for proofreading;
- to my son Matthew, who wrote and ran the Java program that produced the Quotefalls puzzles; and
- to my daughter Christine, another Dr. Petrich and mother of my two precious grandsons.

Many thanks also to my students, friends, and associates:

- to my students over many years of teaching, who didn't know that their questions would be so helpful to me and
- to my colleagues in the Chemistry Department at San Antonio College, who have helped me with their suggestions and support.

James A. Petrich

My gratitude and my great respect are due to the late John Vondeling of Saunders College Publishing, who suggested to me the writing of the first edition of this Study Guide, and who has encouraged and guided me in all aspects of my involvement in the project. I am especially indebted to my friends, Professor Ken Whitten, Professor Larry Peck, and the late Professor Ken Gailey, for their cooperation, their enthusiasm, and their many helpful suggestions at all stages of previous editions. I also thank the students in my many introductory chemistry classes; our study of chemistry together, their difficulties and successes, and their many questions and discussions have helped to provide the point of view from which the guide was written; their welcome comments have aided in developing the guide into its present form. We would appreciate further comments and suggestions from readers of this guide. I thank my wife, Sharon, for cheerfully typing major portions of this guide. Most of all, I again express my deepest appreciation to my family—Angela, Laura, and Brian, and especially Sharon—for the continued love, understanding, encouragement, and patience they have given me through yet another writing project.

Raymond E. Davis

Dedicated with love to our families.

TO THE STUDENT

The true understanding of chemistry is not just memorizing material, and you will not perform adequately in your chemistry course if you try just to *remember* everything that you hear or read. To be successful in your study of chemistry you must (1) *organize* the material as you study and (2) *practice applying* the concepts and skills you learn to particular problems or experiments. You should always keep these two goals in mind in any study session. You will find that your performance on examinations will be directly related to the amount of real practice that *you* have devoted to thinking about and applying the concepts, and to solving problems and answering questions. You will find that almost every stage of your study of chemistry will depend on a firm working knowledge of concepts that you have already studied. Do not fall behind in your study, and do not neglect a topic just because you have already taken the exam that covers it!

The most helpful way to study is to take the most active role you can. Just reading (your class notes, the text, examples, …) rarely does as much good as getting involved—for instance, working the examples out yourself, or outlining a text chapter. Students often underestimate the great help that the act of *writing* something down can be in the learning process. This forces you to pay more attention than if you were just reading, and will help you to remember what you have written down.

How to Learn Chemistry

1. Know the *vocabulary*. Learning chemistry is a lot like learning a foreign language. In fact, there are as many new words in a beginning chemistry course are there are in a first-year language course. At the back of your notebook begin your own personal glossary. As you read the textbook, you will encounter new terms. (They're often in **bold** or *italics*.) Write each of these terms, with a definition and/or examples, in your personal glossary.

   As early as possible in your study of each chapter, you must get a working knowledge of the important new terms as they arise and review terms from earlier chapter. Each chapter of this Study Guide contains a brief section entitled **Some Important Terms in This Chapter**. You should try to write down, in your own words, what these terms mean to you. Then look them up in the textbook. Many of them appear in the Key Terms list in the text, while you will have to find others in the chapter (or in preceding chapters). Try to improve your own definitions. Again *do not* just read the definitions or copy them from the text. Putting them into your own words, in your own writing, makes you think.

2. Give it a *try*. Then … try *again*. Chemistry deserves a chance. *You* deserve a chance to learn chemistry. *Decide* that you'll give it your best effort.

3. Read each assignment *before* the lecture. (What a concept!) You should always read the material that will be covered in class a little ahead of time. Your instructor will probably not assume that you have completely understood the text material before class, but it is always helpful if you have read ahead at least lightly so you have some idea what is

v

coming. This Study Guide contains a **Chapter Summary** of the main ideas and a list of definite **Study Goals** for each chapter. Even though you may not understand these in much detail before you hear lectures over the chapter, read them before class anyway. Then some of the words will begin to be familiar to you, and you can recognize key ideas when they come up in class. For the same reasons, look in advance at the list of **Important Terms** in this guide so you can pay special attention when these come up in class.

4. Chemistry requires *practice*. As you read the textbook:
   a. Study the examples.
   b. Work the suggested exercises.
   c. Check the answers, when available, at the back of the book.
   d. Work other exercises at the end of the chapter.
   e. Check the answers, when available, at the back of the book.

5. All of this study will take *time*. Spend an hour or two *each day* studying chemistry. Don't expect to learn chemistry by studying several hours at a time just once or twice a week. Take a look at your schedule. Designate a time, or times, each day when you can say, "It's time to study chemistry."

6. *Attend* all class meetings. You miss a lot when you're absent even once.

7. Take *good notes* in class. What *are* good notes? Try to take class notes that are sufficient to remind you of the general development being presented in class. You will not be able, nor should you try, to write down everything that is said or even everything that appears on the board or the screen. It is much more important to pay attention and think about the reasoning being presented. (See the next section for more information on note-taking.)

8. Very soon after each class (the *same day* if possible) rewrite your class notes. This should be done not just to improve their legibility, but to expand them by adding material mentioned in class that you did not have time to write down fully. As you rework your class notes, read the appropriate sections of the text (perhaps your instructor has suggested special pages, sections, or examples to supplement the class discussion). But do not *just* read—incorporate this material to improve your notes. As you do this, you should think carefully about what you are writing—do not just copy. Thus, you will be organizing the material in your mind, relating the various ways the same topics are explained in the lecture, in the text, and in the Study Guide, and incorporating more examples than those given in class. Note two important points about this reworking of your notes: (a) it should be done as *soon as possible* after each class, while the material is still fresh in your mind, and (b) it involves *writing* about the material, which forces you to concentrate more effectively than if you just read.

At this stage, too, you should reread the **Chapter Summary** in the guide. The list of **Study Goals** that this guide contains will serve as a framework for organizing your study of the text and lecture notes, and will aid in pointing your study in the right direction. These goals emphasize both what are the central ideas of the chapter and what you should accomplish with these ideas. So some of the Study Goals will say that you should "Know what is meant by ...," "Relate ... to ...," or "Understand ...," while others emphasize that you should "Be able to calculate ...," or "Know how to ... ." Each goal is accompanied by lists of related text sections and suggested Exercises from the end of the text chapter.

A good technique for summarizing the material is to pretend that you have to teach it to someone else. Write out your own notes, and go over what you would say to help someone else understand the concept or the problem. "Explain" it to yourself in your own words.

9. Ask *questions*. Be as specific as possible.

10. It is essential that you *use* the material as early and as extensively as possible. The best way to do this is to answer questions and work problems that require the concepts you are learning. One of the most dangerous mistakes you can make when you study is to read the question, then read the answer or solution, and then say to yourself, "Yes, I can do that." You are only fooling yourself—and it is better to find out during your study than on an exam that you really did not know how to work that problem. It is crucial to practice using the material as soon as possible after each class—do not wait until a day or two before the examination, or you will be missing something that you might need to know in order to understand the next class meeting! This important phase of your study should follow this general order:

   a. The **Preliminary Test** questions in this Study Guide. These short questions, averaging more than 90 per chapter, are designed to help you master the fundamental terminology, the basic concepts, and the main types of calculations from the chapter. At the end of the guide chapter, you will find the **Answers to Preliminary Test Questions**, containing answers to *all* these questions. Many of these answers include comments, reasoning, stepwise solutions, or text study suggestions. These questions are usually easier than homework or questions in the main text, and they help you learn to use terms and concepts a little at a time. And *do not* just read the questions— write out the answers. On multiple choice questions, do not just find the correct answer and go on. Rather, try to understand *why* each wrong answer is wrong; this helps you to think about the underlying concepts from different viewpoints, and make study of the question four or five times as useful. Do not cut short this preliminary portion of the study—it is usually best to "overlearn" these basic terms and operations, because you will be using them in many ways in more difficult questions. Be sure you can walk before you try to run!

   b. Examples from the textbook. Be sure to work these out yourself. It might help to cover up the solution to the example and then uncover it a line at a time, after you have already tried to figure out the next step of the reasoning or calculations. Write down, as fully as you can, the reasoning for each step.

   c. Homework that your instructor may assign. These may be as simple as the Study Guide questions, or they may involve using several ideas in the same question. In any case, be systematic in answering the questions.

   d. Exercises at the end of the chapter in the Whitten, Davis, Peck, and Stanley text. Some of these are quite simple, similar to the Study Guide or the homework; others may be moderately or very difficult. Many of these questions require you to combine and use several concepts in your reasoning, to see whether you really understand the material. Again, the **Study Goals** in this guide will point out related exercises from the text, helping you to focus your study. The answers to all even-numbered numerical problems are in the back of the text, and complete solutions to these are available in the *Student Solutions Manual*, by Keeney-Kennicutt, which your instructor may wish to make available to you. Work the even-numbered exercises before the other numerical problems, so that you know when you have gotten the correct answer (but do not just start with the known answer and then work backwards). Then work other problems that are similar to these. A most productive technique for working problems is to write out comments for each stage of the reasoning, "explaining" to yourself how and why you did that step. This helps you to focus on how the concepts of the chapter are being applied to the problem.

11. Concentrate more on *concepts* (trends and patterns) than on memorization.

12. Finally, when you are doing your final preparation for examinations or quizzes, the **Study Goals** and the **Preliminary Test** questions will point out topics that require further work.

Many students find it helpful to set aside a separate portion of their notebook for working problems and answering questions. At every stage of this problem working and question answering, be as thorough and systematic as possible. Write down *why* you are doing each step in solving a problem. This forces you into the discipline of thinking about what you are doing, so it will be easier to remember the next time even if the problem is worded differently. It this way, it will also be easier for you to review this material at exam time. As you do this, you will find that a wide variety of problems and questions actually involve only a few *central concepts*, but that these can be combined in many different ways. The more practice you have had at working and applying these concepts to specific situations, the better prepared you will be.

You may wish to modify this suggested study approach to suit your own learning style, and your instructor may have additional suggestions. Whatever you do, be systematic in your study, and take an active part in working problems—writing is always better than just reading. In this way, you can share in the excitement and enjoyment of the complex, useful, and fascinating subject of chemistry.

<u>Notes on Note-Taking</u>

1. "Why should I develop good note-taking skills?"
   There are several purposes for developing good note-taking skills:
   a. Good notes organize the lecture.
   b. Good notes provide a record of information and announcements.
   c. Taking good notes helps maintain attention in class.
   d. Good notes provide information supplemental to the textbook.
   e. Good notes record questions occurring during the lecture.

2. "Is there anything I need to do *before* the lecture?"
   Taking good notes does begin before the lecture:
   a. Be prepared intellectually—read the chapter.
   b. Have all needed materials (pen and paper, etc.)
   c. Sit close to the lecturer.
   d. Date and title the notebook page.

3. "What should I do during the lecture that's different from what I'm doing now?"
   You may already have many of these good note-taking skills:
   a. Write clearly on only *one* side of the paper.

© 2014 Cengage Learning. All Rights Reserved. May not be scanned, copied or duplicated, or posted to a publicly accessible website, in whole or in part.

b. Listen closely for main ideas.
c. Paraphrase—do not copy the instructor's words verbatim, *except when*
    i. A definition is given
    ii. A formula is given
d. Take notes in a semi-outline style.
e. Use the margin as an index to your notes.
f. Leave generous space between main ideas and subtopics.
g. Write examples given by the instructor. (Treat them as precious jewels.)
h. Watch for cues that important information is being given. ("This will be on the test.")
i. Write down connections between points.
j. Keep taking notes during discussions.
k. Pay particular attention to the last ten or fifteen minutes of class.
l. Note questions, confusions, and things to look up that were generated by the lecture.
m. Number points if the professor is making a number of points.
n. Let go of judgments about lecture styles.
o. Use graphics.

4. "That's it, right?"
   No, good note-taking skills continue *after* the lecture:
   a. Immediately after the lecture, look over your notes to fill in missing information, expand abbreviations, etc.
   b. Within 24 hours read through your notes. Fill in gaps and review.
   c. Index your notes.
   d. Write comments, elaborations, questions, etc. in your index.
   e. Create "mind-maps" or "networks" as summaries

5. "You know, some instructors know their stuff, but they're really bad lecturers. What can I do about that?"
   There are some strategies that you can use when the note-taking situation is not ideal:
   a. Ask the instructor to slow down or repeat a point.
   b. Take "telegraphic notes"—nouns and verbs.
   c. Ask questions.
   d. Be persistent.
   e. Create and use a "lost" signal.

SUMMARY OF SUGGESTED STUDY STRATEGY

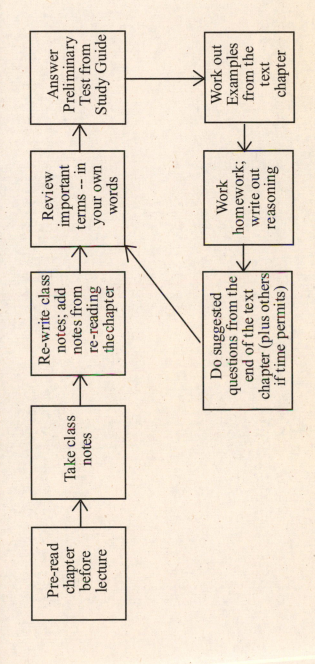

Pre-read chapter before lecture → **Take class notes** → **Re-write class notes; add notes from re-reading the chapter** → **Review important terms -- in your own words** → **Answer Preliminary Test from Study Guide** → **Work out Examples from the text chapter** → **Work homework; write out reasoning** → **Do suggested questions from the end of the text chapter (plus others if time permits)** → (back to Review important terms)

WRITE AT EVERY STAGE OF YOUR STUDY

CONTENTS

1 The Foundations of Chemistry

Chapter Summary

In your study of chemistry, you must manipulate numbers, quantitative ideas, words, and concepts. Yes, it's complex, but we've broken it up into "bite-sized chunks," and if you "eat" one bit at a time, by the time you reach the end of the course, you will have finished the gourmet meal.

The main goals of this chapter are (1) to begin to learn what chemistry is about and the ways in which the science of chemistry views the world, and (2) to acquire some skill in manipulations that are useful and necessary in the understanding of chemistry. We say *begin* because your progress in understanding chemistry, like any other complex subject, does not go in a single straight line. Rather we start with a few main ideas, perhaps oversimplified or understated at this stage. As we develop these ideas and expand them into others, you will find that you must keep coming back to rethink these ideas. This sharpens your thinking and your understanding of what the ideas mean.

For example, very early in this chapter you will encounter the idea of **energy**. This concept has a central role in your understanding of chemistry. Yet the first introduction to it (Section 1-1) seems rather formal and perhaps not too useful. If the authors told you now everything you will need to know about energy for your study of chemistry, it would be a very long and confusing chapter. Instead, the authors tell you enough to get started. Your ideas of energy then develop further as you use them at many stages throughout the book:

- later in Chapter 1 (Section 1-12), in the first introduction to heat as a form of energy;
- in Chapter 4, where you will learn about light as a form of energy as you study atomic structure;
- in Chapters 12 and 13, to help explain the properties of gases, liquids, and solids;
- in Chapter 15, where study of energy changes helps predict whether a reaction can occur; and
- many other places.

At each stage, you learn more about energy and more about chemistry.

Be willing to have your interest aroused in many of the topics that appear early in this first chapter—matter, energy, physical changes, chemical changes, and so on. Do not insist on a complete definition or understanding of all of these new concepts at the first encounter.

In the later sections of the chapter, 1-8 through 1-13, you will learn some of the very important quantitative skills that you will use throughout your study of chemistry. Your progress in studying chemistry will depend on how well you can perform the calculations you learn here. Developing these skills takes work and *personal* practice. It is easy to watch your instructor or a study partner work some problems and then to nod and say, "Yes, I can work that kind of problem." You probably cannot, however, until after *you* have worked such problems, and many

1

of them, with less and less reliance on text, instructor, or study partner. Only then will you have the skill at problem solving that you will need in your further study of chemistry. And then you need to keep that skill sharp by frequent review and practice, just as you could keep a hard-earned ability to play a musical instrument or to ride a unicycle only by further practice.

The chapter opens with some questions that indicate the vast scope of chemistry, a science that touches every aspect of our lives. Section 1-1 then introduces the two central concerns of chemistry—matter and energy. Early in Chapter 1 you read that chemistry is the science that describes matter—its chemical and physical properties, the chemical and physical changes that it undergoes, and the energy changes that accompany those processes. You need to learn what such words mean. We (scientists) summarize our experience in various statements known as **laws**. It is important to realize that in science, a law is not an arbitrary rule. Rather, it is a statement that summarizes what is common among a large number of observations. Then we use that generalization to predict the outcome of further experiments. You learn here some important ideas that you will use often throughout your study of chemistry—the Law of Conservation of Matter (to be used, for example, in Chapters 2 and 3) and the Law of Conservation of Energy (a central idea in Chapter 15).

Chemistry has at least three levels to it: (1) the **macroscopic**, working with test tubes, beakers, and chemicals, (2) the **nanoscale**, the individual atoms and molecules, and (3) the symbolic, representing atoms, molecules, and the changes they undergo by chemical symbols, formulas, and equations. Linking these three levels—learning to work skillfully with substances, to imagine the behavior of their component particles, and to represent this behavior symbolically—is a key to success with chemistry. We begin with the nanoscale level. Section 1-2 describes **Dalton's Atomic Theory**. According to this idea, matter is composed of extremely small particles called **atoms**. **Compounds** represent the combination of these atoms in definite ratios. A **molecule** is the smallest particle of an element or compound that can have a stable independent existence.

Matter commonly exists in one of three **states**—**gas**, **liquid**, and **solid** (Section 1-3). You will study this topic in much more detail in Chapters 12 and 13. We describe any sample of matter in terms of its **properties** (Sections 1-4 and 1-5). These may be **chemical properties**, involving the change of the substance into a different substance (a **chemical reaction**), or **physical properties**, not involving such a change. You learn (Section 1-6) that any sample of matter may be either a **substance** or a **mixture**. Any specimen of a substance has identical properties to those of any other sample of the same substance. A mixture has properties that can vary with gradually differing composition. (In Chapter 14, you will study the composition and properties of a **solution**, which is a particular kind of mixture.) Further, a substance may be either an **element** or a **compound**. The elements are the more than 100 substances that do not decompose into simpler stable substances. A compound is composed of two or more different elements in a **fixed ratio**.

The rest of the chapter concerns some basic skills that you must master before further study. Sections 1-7 and 1-8 deal with the systems of units used to express the results of measurements—the **metric system** and the related **SI system** of units. Each system of units arbitrarily defines a certain amount of the quantity we wish to describe (distance, mass, volume, time, energy, etc.) to correspond to a particular unit. Once you accept the meaning of each of these units, you may still wish to alter the size of units to keep numbers to a manageable size. In our English system of measurements, we prefer to describe some lengths or distances in inches whereas others are more conveniently (or more conventionally) described in feet, yards, miles, or light-years. In the metric or SI systems, we also use related units of different magnitudes, but one set of prefixes, applied to any of the basic units, determines the relationships among these units. Each prefix alters the value of the basic unit by multiplying or dividing it by some power of ten. This makes conversion within the system very easy—just a matter of shifting the decimal point.

See Appendix A-4 for a review of two aspects of dealing with numbers. **Scientific** or **exponential notation** helps us to represent large or small numbers in a convenient way. **Significant figures** indicate how well we know a quantity we have measured or one we have calculated from a measured quantity. The **unit factor method** (Section 1-9) is a simple method for converting from any unit to any other related one. Examine carefully the explanation of this method and the numerous examples. Practice this method until it is an easily used and reliable tool. Section 1-10 presents **percentage** as a unit factor calculation.

The remainder of the chapter (Sections 1-11 through 1-13) introduces you to other important physical properties, such as **density**, **specific gravity**, **heat**, and **temperature**. Here you will learn some calculations related to these quantities.

IMPORTANT. In your study of Sections 1-8 through 1-13, you develop and sharpen some tools and skills that you will need for your further understanding of chemistry. Do not just learn to go through the motions in a prescribed way to work or answer a certain kind of question. Instead, keep in mind *why* you (or the textbook or your instructor) are approaching a problem in a particular way. In each example throughout the text, the reasoning or strategy is given first— the **Plan**; then the detailed **Solution** is shown. Study the plan carefully so you understand how to apply the general concept to the specific question or problem.

Look at each question first to see what is *given* and what is *asked for*. This will often help you start on a question even if it is not a numerical problem. Writing down a list of these quantities, with units, is often helpful. Then try to recall a relationship you can use to link the *given* quantities to the *unknown* quantity—that is, develop *your own plan* for answering the question. Sometimes this will require using two or more relationships in sequence. Several examples in Sections 1-8 through 1-13 illustrate this. Once you have found the right relationship, you might need to manipulate it before you carry out the calculations required. Sometimes you need to rearrange the relationship algebraically, to isolate the unknown quantity in terms of the known ones. It is always helpful to include the units for all quantities. Pay special attention to the *Problem-Solving Tip* boxes that you find frequently in the text. They will not only alert you to common errors in problem solving, but they will often give you a new insight into some aspect of problem solving.

Finally, it is always a good idea to *think* about whether the answer you obtained makes sense. Suppose we had used conversion factors to calculate that 1 ft = 4.72 cm. Recall that 1 cm is about the width of a little fingernail. It would be clear that the numerical result *must* be wrong— it obviously takes more than 4.72 human-fingernail widths to span 1 ft. (When we go back over our calculation, we find that we divided when we should have multiplied. We could have easily avoided this error if we had written units throughout the calculation.) Checking to see whether the answer is reasonable is usually not a guarantee that you have the correct answer. But frequently it can tell you that your answer is certainly an *incorrect* one. Then you can begin to search for your error.

Above all, you should discipline yourself to think about and work problems systematically. Do not just turn to your calculator when you see that a problem involves numbers. Think your way through the problem, at least in broad terms, to develop a plan before you begin solving the

3

problem. Then write down the steps by which you will arrive at your answer before you start the actual computations, being careful to include units for each term. In this way, if you arrive at an incorrect answer, you can review your solution to the problem and perhaps find the mistake. If you get the correct answer, it will be easier for you to review and check the result, and then to review your plan in a later study session. And when you are finished, think about the answer.

> The section entitled "To the Student" at the beginning of this Study Guide suggests how you might organize your study of chemistry. Begin that systematic approach with this important chapter.

Study Goals

These study goals will help you determine specific directions your study will take. Use them to help organize your class notes and text study. Each study goal refers you to related sections in *Chemistry*, *Tenth Edition*, by Whitten, Davis, Peck, and Stanley. When appropriate, some related exercises at the end of the main text chapter are suggested. In addition, *always* work some of the "Mixed Exercises" for each chapter, so you learn to recognize additional types of questions.

Section 1-1
1. Define, distinguish among, and give examples of (a) matter, (b) mass, (c) energy, (d) kinetic energy, (e) potential energy, (f) exothermic, and (g) endothermic. *Work Exercises 3, 4, 8, and 9.*
2. State and use (a) the Law of Conservation of Matter, (b) the Law of Conservation of Energy, and (c) the Law of Conservation of Matter and Energy. Express these in words other than those used in the text, and give examples of each. *Work Exercises 5 and 10 through 13.*

Section 1-2
3. Understand the postulates of Dalton's Atomic Theory. Distinguish between atoms and molecules. *Work Exercise 17.*

Section 1-3
4. Describe and distinguish among the general properties of gases, liquids, and solids. *Work Exercises 14 and 76.*

Sections 1-4 and 1-5
5. Define, distinguish among, and give examples of (a) a chemical change, (b) a physical change, (c) a chemical property, (d) a physical property, (e) an intensive property, and (f) an extensive property. *Work Exercises 3, 4, and 21 through 28.*

Section 1-6
6. Define, distinguish among, and give examples of (a) substances, (b) heterogeneous mixtures, (c) homogeneous mixtures, (d) elements, and (e) compounds. *Work Exercises 15 through 20.*
7. Write proper symbols for most common elements (Table 1-3), and write the name of one of these elements, given its symbol.

Sections 1-7 and 1-8

8. Know the fundamental SI units in Table 1-5 and the meanings of the prefixes listed in Table 1-6. Know at least one conversion factor (Table 1-8) relating metric and English units of (a) mass, (b) length, and (c) volume. *Work Exercises 35 and 36.*

Appendix A

9. Be familiar with conventions regarding exponential notation and significant figures. Apply them properly when doing mathematical operations. *Work Exercises 29 through 34 and 42 through 44.*

Sections 1-9 and 1-10

10. Know how to construct unit factors from equalities. Use these in calculations that involve conversions from one set of units to another (dimensional analysis). *Work Exercises 37 through 41.*

Sections 1-8, 1-11, and 1-12

11. Distinguish between (a) mass and weight; (b) accuracy and precision; (c) density and specific gravity; and (d) heat and temperature. *Work Exercises 3 and 4.*

Section 1-11

12. Carry out calculations relating density, specific gravity, mass, and volume. *Work Exercises 45 through 52.*

Section 1-12

13. Relate the Celsius, Fahrenheit, and Kelvin temperature scales. Convert a specified temperature on one scale to the corresponding temperature on the other two scales. *Work Exercises 53 through 60.*

Section 1-13

14. Distinguish between endothermic and exothermic changes. Be able to carry out calculations that relate heat capacity or specific heat to the heat transfer that accompanies temperature changes. *Work Exercises 6 through 9 and 61 through 66.*

General

15. Use your understanding of this chapter to recognize and solve a variety of types of questions. *Work **Mixed Exercises** 67 through 74.*

16. Use your understanding of this chapter to answer conceptual questions, which often do not involve calculations. *Work **Conceptual Exercises** 75 through 98.*

17. The Internet is an increasingly important source of many kinds of information. Exercises at the end of chapter direct you to sources outside the textbook, such as websites, for information to use in solving them. Work **Beyond the Textbook** *exercises 99 through 104.*

Some Important Terms in This Chapter

IMPORTANT. Each new topic that you encounter in your study of chemistry will include some new terms. You must understand clearly what these new terms mean in order to understand explanations that use them. You will need these terms to communicate clearly, when you discuss the subject matter, when you ask your instructor questions, and when you give your answers on examinations. Some important new terms from Chapter 1 are listed below. These are only some of the new terms in the chapter—pay attention to others you encounter in the Key Terms list or in your reading. Fill the blanks in the following paragraphs with terms from the list. Use each term only once. The answers are on page 12.

atom	heat capacity	physical properties
chemical change	heterogeneous mixture	potential energy
chemical properties	homogeneous mixture	properties
compounds	kinetic energy	specific gravity
density	mass	specific heat
elements	matter	temperature
energy	molecule	unit factor
heat	physical change	

The universe is composed of [1]_____, which has mass and occupies space, and [2]_____, the capacity to do work or to transfer heat. An object, such as a bowling ball, will resist a change in motion because of its [3]_____. Held high overhead, the bowling ball possesses [4]_____, but when it is rolling toward the pins (or down the gutter), it has [5]_____.

The smallest particle of an element that is that element is an [6]_____. Similarly, the smallest particle of a molecular substance that is that compound is a [7]_____.

Various characteristics of matter can be observed or measured. These [8]_____ can be divided into two types: [9]_____ can be observed in the absence of any change in the identity of the matter, but [10]_____ are exhibited by matter as it undergoes changes in identity. When matter is heated, illuminated, or exposed to other matter, it will either: (a) do nothing (no change), (b) change *without* becoming a different substance (a [11]_____), or (c) change into a *different* substance (a [12]_____). Any change, physical or chemical, may be detected by a change in one or more of the physical properties of the substance.

We experience mixtures every moment of our lives. A stew, which has chunks of meat, potatoes, and other vegetables in broth, is clearly a mixture, specifically a [13]_____. One has to look more closely to determine that air is a mixture because it is a [14]_____. Mixtures can be separated into their components by physical changes. Some of these components may be [15]_____, pure substances that can be decomposed into simpler substances by chemical changes. Others may be [16]_____, which do not decompose into simpler substances by either physical or chemical changes.

Many chemical calculations can be accomplished by the technique of dimensional analysis, in which the original quantity is multiplied by a ratio equal to unity (the number one), called a [17]_____.

The ratio of the mass of an object to its volume is its [18]_____. When this value is divided by the density of water at the same temperature, the result is the [19]_____ _____, a number without units.

Four characteristics of matter are often confused because they are all related to the motion of the tiny particles that compose matter, and three of them have the same word in common. [20]_____ is the intensity of that motion, which causes an object to feel "hot" or "cold." [21]_____ is the form of energy that allows these particles to move and flows from a hotter object to a colder one. When an object, a specified mass of a specified substance, is heated, its temperature will go up. The amount of heat required to raise its temperature 1°C is its [22]_____. However, the amount of heat required to raise the temperature of specifically 1g of this specific substance is a very specific number, called the [23]_____.

Quotefalls

Each puzzle below contains an important quote from Chapter 1. A black square indicates the end of a word. Words starting at the end of a line may continue in the next line. Punctuation in the statement is included in the boxes. Above the boxes in each vertical column are the letters that belong in the boxes, in a randomized order. Place the letters into the boxes directly below them to form words across. The letters do not necessarily go into the squares in the order listed. Use each letter only once. When all the letters are in their correct boxes, you will be able to read the complete quotation across the diagram from left to right, line by line. **Hint**: It may be helpful at times to fill in small words, like "the," to use up some letters to help in determining other longer words. Some letters have been "seeded" to help you get started. The solutions are on page 13.

Sample:

Sample Completed:

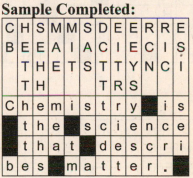

Puzzle #1:

Puzzle #2:

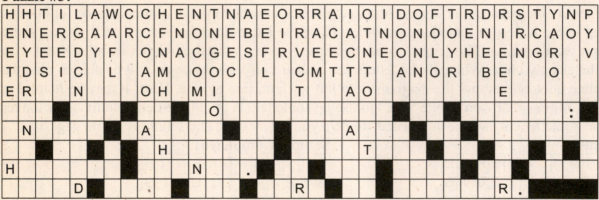

Puzzle #3:

Puzzle #4:

P	I	T	F	E	R	C	N	O	T	A	I	R	P	T	A	S	T	I	C	O	S	P	E	O	R	T	A	S	N	S	O	M	
N	R	E	H	D	W	W	A	M	E	S	A	M	O	T	E	I	T	S	F	P	M	O	Y	Y	I	T	E	N	S	S	C		I
O	O	F	N	E	A	A	E	S	C	A	P	E	N	I	R	O	A	O	M	N	O	S	E	M	I	O	N	S		C
D	H	U	L	A	L	E	S	O	F	C	P	E	N	S	L	H	E	E	O	N	A	B	L	P	U	R	O		
T			L				S	O	T		O	N	D	L	P	O	N	R		R		P	M	I	O			
								E										R									O		(
									T					C					O						)	:	
D			E											E									M				
P					N																A		.	
										P				R								

Preliminary Test

Each chapter of this study guide contains questions in the form of a preliminary test. The questions in these tests involve the fundamental terminology and basic concepts. Many questions involve only one idea or one skill at a time. Use the questions of these preliminary tests as an initial check of your understanding of fundamental terms, basic concepts, and main types of problems in the chapter. When you do well on the preliminary test for a chapter, you are ready to proceed with further detailed study, assigned homework, or problem solving from the text. Always work many problems and answer many questions at the end of the chapter. This practice will give you the skills and confidence that are necessary to understand and apply the material of the chapter. Refer to the textbook for additional data. Table 1-8 will be a useful source of conversion factors.

True-False

Mark each statement as true (T) or false (F).

_____ 1. The Law of Conservation of Matter probably does not apply outside our solar system.
_____ 2. Although a sample of matter can change its kinetic energy, its potential energy is always fixed unless it undergoes a chemical reaction.
_____ 3. If we have two samples of matter, the one that is traveling at the higher speed has the higher kinetic energy.
_____ 4. In writing a chemical symbol, we can use either capital or lower case letters, as we prefer.
_____ 5. The chemical symbol for mercury is Me.
_____ 6. The term "atom" can apply only to an element, not to a compound.
_____ 7. The term "molecule" can apply only to a compound, not to an element.
_____ 8. Molecules always consist of more than one atom.
_____ 9. Elements that do not exist in stable form as single atoms are always diatomic.
_____ 10. A molecule of a compound must consist of at least two atoms.
_____ 11. For most substances, the gaseous state is less dense than is the liquid state.
_____ 12. For most substances, the solid state is denser than is the liquid state.
_____ 13. When we do not stir a liquid, its molecules are motionless.
_____ 14. To observe the chemical properties of a substance we must convert at least some of the substance into other substances.
_____ 15. A substance whose melting point is −7.1°C and whose boiling point is 58.8°C is a liquid at room temperature (about 21°C).
_____ 16. The substance referred to in Question 15 must be bromine.
_____ 17. When a liquid on a surface evaporates, it removes some heat from the surface.
_____ 18. Most of the known elements actually occur in very small amounts on the earth.
_____ 19. Most of the naturally occurring elements occur in combination with other elements, rather than as free elements.
_____ 20. One of the elements that occurs in our atmosphere in considerable quantity is present in nature both as the free element and in compounds.
_____ 21. Different samples of a compound can have compositions that are different but only slightly so.
_____ 22. A mixture has properties that are similar to those of its component substances.
_____ 23. A compound has properties that are similar to those of its constituent elements.
_____ 24. Different parts of a solution have different properties.
_____ 25. A T-bone steak is an example of a heterogeneous mixture.
_____ 26. A cup of sweetened hot tea is an example of a heterogeneous mixture.
_____ 27. A glass of iced tea is an example of a homogeneous mixture.

_____ 28. To convert the mass of an object from grams to milligrams, we divide by 1000.

_____ 29. The weight of a body depends on where it is; its mass does not.

_____ 30. In the SI system of units, the meter (m) is more important than the kilogram (kg).

_____ 31. One gram is about the same mass as one ounce.

_____ 32. It is easy to tell, looking from across the room, the difference between a 2-L bottle and a 2-qt bottle.

_____ 33. Because 1 m = 39.37 in, the number of inches in 13.83 m is correctly described as 13.83 × 39.37 = 544.4871 in.

_____ 34. Because one dozen is 12 of the objects in question, the number of eggs in 16 dozen is correctly calculated as 16 × 12 = 192 eggs.

_____ 35. Density and specific gravity are the same thing.

_____ 36. The mass of a 1.25-L sample of saturated brine solution, specific gravity 1.45, is 1.81 kg at 25°C.

_____ 37. When we add heat to an object, we always raise its temperature.

_____ 38. No matter what temperature scale we use, one degree is always the same increment in temperature.

_____ 39. In order to tell the specific heat of an object, it is sufficient to know what substance composes the object; in order to tell its heat capacity, it is necessary to also know how much of the substance we have.

_____ 40. When a piece of iron is allowed to rust, the resulting rust is observed to weigh more than the original piece of iron; this is an example of an exception to the Law of Conservation of Matter.

Short Answer

Answer with a word, a phrase, a formula, or a number with units as necessary.

1. An example of an exothermic process is _____.

2. The symbol for the reactive metallic element magnesium is _____.

3. The element whose symbol is Na is _____.

4. Three examples of mixtures are _____, _____, and _____.

5. Three examples of elements are _____, _____, and _____.

6. Three examples of compounds are _____, _____, and _____.

7. Arrange the following in order from smallest to largest: meter, kilometer, millimeter, centimeter, and micrometer.

8. Of the units in Question 7, the one most convenient for measuring the distance from Philadelphia to Paris would be the _____, whereas the most convenient unit for measuring the distance from home plate to first base on a baseball field would be the _____.

9. The distance from home plate to first base (90 ft), expressed in metric units, is _____. (Refer to Table 1-8 for needed conversion factors.)

10. The weight of my 2.8-kg dog, expressed in pounds, is _____.

11. The distance 221,463 miles, expressed in kilometers, is _____.
12. The mileage rating of an automobile that is capable of 27.5 miles per gallon is equal to _____ km/L.
13. The number 221,463, expressed in scientific notation, is _____.
14. The number 0.000473, expressed in scientific notation, is _____.
15. Another name for scientific notation is _____.
16. In order to comply with a request to keep a room at 68.0°F, we would have to set a thermostat calibrated on the Celsius scale at _____.
17. The heat capacity of a 250-g block of metal whose specific heat is 0.220 J/g·°C is _____. (Be sure to include the units.)
18. The amount of heat necessary to raise the temperature of N grams of water by M degrees Celsius is equal to the expression _____. (Be sure to include the units.)
19. In trying to identify a sample of a pure substance, we make the following observations:
    (a) Its mass is 142.6 g.
    (b) It is a shiny solid at room temperature.
    (c) It is easily etched by hydrochloric acid.
    (d) It melts when heated to 540°C.
    (e) It is 24.4 cm long.
    (f) It is a good conductor of electricity.
    (g) It burns in air.
    Of these observations, the ones that would be helpful in identifying the substance composing the sample are _____.
20. Of the properties listed in Question 19, _____ are chemical properties, whereas _____ are physical properties.
21. Milk of Magnesia contains magnesium hydroxide. Analysis of any pure sample of magnesium hydroxide shows that it consists of 41.7% Mg, 54.9% O and 3.4% H by mass. This information represents an example of the Law of _____.
22. The information given in Question 21 shows that magnesium hydroxide is a (an) _____.
23. The substance(s) listed in Table 1-2 of the text that is (are) gaseous at room temperature (about 21°C) is (are) _____.
24. Fill in the missing entries in the following table of elements.

Element Name	Element Symbol	Element Name	Element Symbol
hydrogen	_____	phosphorus	_____
_____	C	_____	K
_____	Ca	_____	S
fluorine	_____	zinc	_____
_____	O	_____	Hg

Multiple-Choice

Each multiple-choice question in this guide has only one best answer. Select the best answer. If you use them properly, multiple-choice questions that are not calculations can be especially valuable for study. Do not just look for the correct answer and then go on. Try to see what is wrong with each of the other responses or how the correct response is better than the others are. This enables you to look at the ideas of the question from several different viewpoints. Experience with this viewpoint will be especially useful if you encounter multiple-choice questions on an exam.

_____ 1. Based on all of the evidence available, we believe that the Law of Conservation of Matter and Energy is
(a) always true.　　(b) usually true.　　(c) always false.　　(d) usually false.

_____ 2. Which one of the following is *not* an example of a compound?
(a) table salt　　(b) baking soda　　(c) nitrogen　　(d)　　carbon monoxide

_____ 3. Which one of the following SI prefixes corresponds to a multiplication factor of 1000?
(a) milli-　　(b) kilo-　　(c) micro-　　(d) centi-　　(e) nano-

_____ 4. Which one of the following units of volume is the largest?
(a) milliliter　　(b) liter　　(c) centiliter　　(d) deciliter　　(e) quart

_____ 5. Refer to Table 1-8 of the text. A conversion factor by which we could multiply a volume in quarts to convert it into milliliters is

(a) $\dfrac{1.057 \text{ qt}}{1 \text{ L}} \times \dfrac{1 \text{ L}}{1000 \text{ mL}}$　　(b) $\dfrac{1.057 \text{ qt}}{1 \text{ L}} \times \dfrac{1000 \text{ mL}}{1 \text{ L}}$　　(c) $\dfrac{1 \text{ L}}{1.057 \text{ qt}} \times \dfrac{1 \text{ L}}{1000 \text{ mL}}$

(d) $\dfrac{1 \text{ L}}{1.057 \text{ qt}} \times \dfrac{1000 \text{ mL}}{1 \text{ L}}$

_____ 6. Refer to Table 1-8 of the text. A conversion factor by which we could multiply an area in square miles (mi^2) to convert it to square kilometers (km^2) is

(a) $\dfrac{2.588881 \text{ km}^2}{1 \text{ mi}^2}$　　(b) $\dfrac{2.589 \text{ km}^2}{1 \text{ mi}^2}$　　(c) $\dfrac{1 \text{ mi}^2}{2.588881 \text{ km}^2}$　　(d) $\dfrac{1 \text{ mi}^2}{2.589 \text{ km}^2}$

_____ 7. The prefix "centi-" means
(a) the same as the unit to which it is attached.
(b) 1/1000 of the unit to which it is attached.
(c) 1/100 of the unit to which it is attached.
(d) 100 of the unit to which it is attached.
(e) 1000 of the unit to which it is attached.

_____ 8. Two samples, each of them a pure substance, are found to have different melting points. Which one of the following statements about the substances is true?
(a) The two samples are certainly different pure substances.
(b) The two samples are probably different pure substances, but we need more information to tell for sure.
(c) The two substances are certain to have identical chemical formulas.
(d) Both substances are certain to be compounds and not elements.

(e) The two substances are certain to have different densities.

_____ 9. Which one of the following processes is an example of a chemical change?
(a) Fermentation of wine
(b) Formation of salt by evaporation of seawater
(c) Separation of a solid from a liquid by filtration
(d) Condensation of moisture on a cold surface in a humid room
(e) Writing on a piece of paper with a pencil

_____ 10. Which one of the following is the longest?
(a) 100.0 cm   (b) 0.0010 km   (c) 1.000 m   (d) 1000 mm
(e) All of the preceding quantities are equivalent

_____ 11. The melting point (freezing point) of mercury is –35°C. What is the temperature, in degrees Fahrenheit, below which a mercury thermometer would not be usable because the mercury in it would be frozen?
(a) 35°F   (b) –63°F   (c) –31°F   (d) –5.4°F

_____ 12. We know that air is a mixture and not a compound because
(a) It can be heated.   (b) It can be compressed.   (c) It is colorless.
(d) Its composition can vary.

_____ 13. Which one of the following is an extensive property?
(a) density (b) temperature (c) volume (d) color (e) reactivity with hydrochloric acid

Answers to Some Important Terms in This Chapter

1. matter
2. energy
3. mass
4. potential energy
5. kinetic energy
6. atom
7. molecule
8. properties
9. physical properties
10. chemical properties
11. physical change
12. chemical change
13. heterogeneous mixture
14. homogeneous mixture
15. compounds
16. elements
17. unit factor
18. density
19. specific gravity
20. Temperature
21. Heat
22. heat capacity
23. specific heat

Answers to Quotefalls

Puzzle #1:
The Law of Conservation of Matter and Energy: The combined amount of matter and energy available in the universe is fixed.

Puzzle #2:

The Law of Conservation of Matter: There is no observable change in the quantity of matter during a chemical reaction or during a physical change.

Puzzle #3:

The Law of Conservation of Energy: Energy cannot be created or destroyed in a chemical reaction or in a physical change. It can only be converted from one form to another.

Puzzle #4:

The Law of Definite Proportions (or Law of Constant Composition): Different Samples of any pure compound contain the same elements in the same proportions by mass.

Answers to Preliminary Test

Do not just look up answers. *Think about why the answers that appear here are correct.* Reread sections of the text, if necessary. Practice many of the suggested exercises in the textbook. Also, work several of the Mixed Exercises in the text, to gain practice at recognizing additional types of questions.

True-False

1. False. We have never observed any exceptions to this law, and we believe that exceptions do not occur.
2. False. Its gravitational potential energy, for instance, depends on its vertical position.
3. False. Kinetic energy depends on both mass and speed. Remember that energy is the capacity for doing work. It might be easier to break a window with a very heavy rock thrown at low speed than with a very lightweight rock thrown at a higher speed.
4. False. When a symbol consists of only one letter, it is always a capital letter. When it consists of two letters, the first is always capitalized and the second is always lower case. Even though this convention seems (and is!) arbitrary, like any convention of communication it must be applied in the same way by everyone who hopes to use it to communicate. For example, to a chemist CO cannot mean cobalt—it means a compound consisting of carbon and oxygen atoms in a 1:1 ratio.
5. False. Mercury is Hg. See Table 1-3.
6. True. See Section 1-2. The smallest particle of a compound that is the compound is a molecule or formula unit.
7. False. See Section 1-2. Many elements consist of molecules that have two or more atoms of the element joined together.
8. False. See the description of the noble gases in a marginal note on page 6.
9. False. P_4 and S_8 are two exceptions discussed in Section 1-2. See Figure 1-4.
10. True. A compound is, by definition, composed of two or more elements, and the atoms of each element are different from the atoms of any other element. See Section 1-1.
11. True. (In fact, this is true for all substances.)
12. True. Water is one of the very few exceptions to this generalization.

13. False. Think about the diffusion of a drop of red dye throughout an unstirred glass of water. Try to relate what you study to your own experience.

14. True. Chemical properties are exhibited by matter as it undergoes changes in composition.

15. True. Suppose we start with a sample of the substance at such a low temperature that it is a solid, and then warm it. When it reaches –7.1°C, it melts—i.e., the solid changes to liquid. Then we would need to heat the liquid to 58.8°C for it to boil. Thus, it is a liquid over the temperature range –7.1 to 58.8°C.

16. False. It may be (see Table 1-2), but we cannot be sure just from these two data. Perhaps another substance coincidentally has the same melting and boiling points as bromine but with other very different properties.

17. True. For example, recall the cooling effect of the evaporation of perspiration.

18. True. See Table 1-4 and the accompanying discussion in Section 1-6.

19. True. Recall (Section 1-6) that only about one-fourth of the elements occur as free elements.

20. True. Oxygen occurs both as a free element and in compounds such as carbon dioxide.

21. False. A given compound always has the same composition of its elements (see Section 1-6).

22. True. Each of the substances in a mixture retains its own identity and properties.

23. False. A compound is a different substance from its constituent elements, with its own unique properties.

24. False. A solution is a homogeneous mixture. Be sure you understand the difference between heterogeneous and homogeneous mixtures (Section 1-6).

25. True. Fat, bone, and lean meat are parts of a steak that obviously have different properties. See Section 1-6.

26. False. The sweetener and tea completely dissolve in the hot water; this is a homogeneous mixture.

27. False. The solid ice and the liquid tea have different properties.

28. False. A milligram (1/1000 of a gram) is *smaller* than a gram; even if we had not learned about conversion factors, common sense tells us that the number of milligrams in an object must be *greater* than the number of grams, so we must multiply by a conversion factor *greater* than one.

29. True. See the discussion in Section 1-8. But we should keep in mind that a chemical reaction usually takes place at constant gravity. Therefore, weight relationships are just as useful as mass relationships for most applications in chemistry.

30. False. One (the meter) is the fundamental unit for the measurement of length, whereas the other (the kilogram) is the base unit for the measurement of mass.

31. False. Perhaps you have read in the grocery store that a 1-lb (16-oz) can of food contains 454 grams.

32. False. Many soft drinks are now being packaged in 2-liter bottles, which contain only about 6% more than 2 quarts.

33. False. Significant figures! The answer would be correctly written as 544.5 inches. See Appendix A.4 in the text and pay attention to Exercises such as 29 through 34 and 42 through 44 at the end of the chapter.

34. True. These are all exact numbers, not obtained from measurement. See Section 1-9.

35. False. Density is the ratio of the mass of a sample of a substance to its volume. Thus, it has units expressed as mass/volume. It could have different values for the same object,

depending on the units used to describe mass and volume. Specific gravity is the ratio of the density of a substance to that of water at the same temperature and is thus dimensionless. See Section 1-11.

36. True. Remember that the density of water is 1.00 g/mL at 25°C, so at that temperature density and specific gravity are numerically equal. Thus, the density of the brine solution is 1.45 g/mL. Constructing the necessary unit factors,

$$\underline{?} \text{ g solution} = 1.25 \text{ L} \times \frac{1000 \text{ mL}}{1 \text{ L}} \times \frac{1.45 \text{ g solution}}{1 \text{ mL solution}} \times \frac{1 \text{ kg}}{1000 \text{ g}} = 1.81 \text{ kg}$$

The result of carrying out the computation on a calculator is 1.8125. Rounding to the correct number of significant figures gives the answer 1.81.

37. False. Consider, for example, that when we add some heat to ice at 0°C, we melt some of it to form water, but the temperature does not change. Read Section 1-5 and then study Section 1-12 carefully.

38. False. For instance, on the Celsius scale the difference between the freezing and boiling points of water is 100 – 0 = 100 degrees, whereas measured on the Fahrenheit scale this *same difference* in temperature is described as 212 – 32 = 180 degrees.

39. True. Specific heat is an intensive property, whereas heat capacity is an extensive property. Refer to Section 1-4 for a reminder of the terms "intensive" and "extensive", and then study Section 1-13 carefully.

40. False. The iron reacts with the moist air, and the increase in weight is due to something (oxygen) being added to the iron. There are no known violations of the Law of Conservation of Matter in ordinary chemical reactions. See Section 1-1.

Short Answer

1. Many valid examples come to mind. Any process in which energy (often heat) is given off to the surroundings (Section 1-13) could be cited. Burning a piece of metallic magnesium in air (Section 1-1), combustion of a fuel, conversion of matter into energy as in a nuclear reaction (Section 1-1), and freezing of a liquid (Section 1-5) are among the many adequate answers.

2. Mg

3. sodium

4. Again, many answers would suffice. Among the ones mentioned in Chapter 1 are a solution of copper(II) sulfate in water (Figure 1-9c), a mixture of salt and sugar, a mixture of sand and salt, a mixture of iron powder and powdered sulfur (Figure 1-11), air, and vegetable soup.

5. There are presently more than 100 known answers to this question. Some of the elements mentioned in this chapter are hydrogen, oxygen, fluorine (note the spelling!), krypton, silicon, aluminum, iron, sulfur, and chlorine. How many others can you find mentioned in the chapter? Of how many elements have you seen a sample? You should memorize the list of elements and symbols in Table 1-3.

6. Now there are several million correct answers! Among those mentioned in the chapter are copper(II) sulfate, sodium chloride, water, and carbon dioxide.

7. micrometer, millimeter, centimeter, meter, kilometer. You should know the prefixes given in Table 1-6, as well as their abbreviations and values (meanings).

8. kilometer, meter

9. 27.432 m. The conversion is: $\underline{?}$ m $= 90$ ft $\times \dfrac{12 \text{ in}}{1 \text{ ft}} \times \dfrac{2.54 \text{ cm}}{1 \text{ in}} \times \dfrac{1 \text{ m}}{100 \text{ cm}} = 27.432$ m

   Notice that the answer is an exact number since the original value and all the relationships are exact. By definition, the distance required is exactly 90 ft., one foot is exactly 12 inches, one inch is exactly 2.54 cm, and 1 m is exactly 100 cm.

10. 6.2 lb. The conversion is: $\underline{?}$ lb $= 2.8$ kg $\times \dfrac{1000 \text{ g}}{1 \text{ kg}} \times \dfrac{1 \text{ lb}}{453.6 \text{ g}} = 6.2$ lb.

    Watch the significant figures. The number on the calculator was 6.172839506, but I certainly do not know the weight of my dog that precisely.

11. 3.563×10^5 km. $\underline{?}$ km $= 221{,}463$ mi $\times \dfrac{1.609 \text{ km}}{1 \text{ mi}} = 3.563 \times 10^5$ km

    If you use 1 mi = 5280 ft., 1 ft. = 12 in., 1 in. = 2.54 cm, 1 cm = 0.01 m, and 1 m = 0.001 km, all of which are exact, the answer is 3.56410×10^5 km (6 significant figures)

12. 11.7 km/L. $\underline{?}$ km/L $= 27.5 \dfrac{\text{mi}}{\text{gal}} \times \dfrac{1.609 \text{ km}}{1 \text{ mi}} \times \dfrac{1 \text{ gal}}{4 \text{ qt}} \times \dfrac{1.057 \text{ qt}}{1 \text{ L}} = 11.7$ km/L

13. 2.21463×10^5

14. 4.73×10^{-4}

15. exponential notation

16. 20.0°C. $°C = \dfrac{1.0° \text{C}}{1.8° \text{F}} (68.0°F - 32°C) = \dfrac{1.0° \text{C}}{1.8° \text{F}} (36.0°C) = 20.0°C$

    See how the rules for significant figures are applied in this case. The values in the formula (1.0°C, 1.8°F, and 32°C) are exact numbers. See Section 1-12 and especially Example 1-16 for conversions of this type. Be sure to pay attention to problems such as Example 1-17, where two stages of conversion are necessary. Exercises 53 and 54 will provide useful practice.

17. 55.0 J/°C. See the definition of heat capacity of an object, Section 1-13.

$$\underline{?} \text{ cal/°C} = 0.220 \ \dfrac{\text{J}}{\text{g} \cdot °\text{C}} \times 250 \text{ g} = 55.0 \text{ J/°C}$$

18. $N \times M$ degrees Celsius. Exercises 61 through 66 in the text give useful practice in calculations involving heat transfer, specific heat, and heat capacity.

19. (b), (c), (d), (f), and (g). These properties are characteristic of the *substance* under study. The other properties listed describe *how much* of the substance is present, and can be different without changing *what* the substance is.

20. (c) and (g) are chemical properties; the substance would be changed into a different substance. The others are physical properties.

21. Definite Proportions. This is also called the Law of Constant Composition (Section 1-6).

22. compound. A compound is a pure substance consisting of two or more different elements in a fixed ratio. See Section 1-6 for an introduction to compounds.

23. oxygen and methane. From the boiling points, we can see that only methane and oxygen are above their boiling points (and hence are gases) at 21°C. Oxygen is an element and methane is a compound composed of carbon and hydrogen.

24.

Element Name	Element Symbol	Element Name	Element Symbol
hydrogen	H	phosphorus	P
carbon	C	potassium	K
calcium	Ca	sulfur	S
fluorine	F	zinc	Zn
oxygen	O	mercury	Hg

Learn all of the element names and symbols in Table 1-3. Make flash cards to help you study and then help you determine any names or symbols that need more review. Always capitalize single-letter element symbols; if the symbol consists of two or three letters, the first is capitalized and subsequent letters are always lower case.

Multiple-Choice

1. (a). A scientific (natural) law is a general statement about the observed behavior of matter to which *no* exceptions are known.

2. (c). Nitrogen is an element. (a) Table salt is composed of the elements sodium and chlorine. (b) Baking soda is composed of the elements sodium, hydrogen, carbon, and oxygen. (d) Carbon monoxide is composed of the elements carbon and oxygen.

3. (b). kilo-. (a) milli- = 0.001, (c) micro- = 0.000 001, (d) centi- = 0.01, (e) nano- = 0.000 000 001.
   You should know the names and meanings of all the prefixes in Table 1-6.

4. (b). liter. (a) A milliliter is 0.001 L. (c) A centiliter is 0.01 L. (d) A deciliter is 0.1 L. (e) A liter is 1.057 quart (a little larger than a quart). The value of each prefix is given in Table 1-6. See also Table 1-8.

5. (d). The first unit factor converts quarts to liters, and the second one converts liters to milliliters. The unit factors in (a) or (b) would not cancel the unit *quart*. The unit factors in (c) would not cancel the unit *liter*.

6. (b). Be careful of significant figures. $(1.609 \text{ km})^2 = 2.588881 \text{ km}^2$, which should be rounded to 4 significant figures, 2.589 km^2, The unit factor in (d) would not cancel the unit mi^2.

7. (c). 1/100th. (b) 1/1000 = milli-, (d) 100 = hecto-, (e) 1000 = kilo-

8. (a). Any two samples of a pure substance must have the same melting point. (b) No more information is needed. (c) If the substances had identical chemical formulas, they would be the same substance. (d) Different elements generally have different melting points. (e) Two different substances that have different melting points could have about the same density.

9. (a). Fermentation of wine converts sugar to alcohol and carbon dioxide, different substances. (b) Salt is already salt when it is dissolved in seawater. (c) The identity of a solid does not change when it is filtered. (d) The liquid water that condenses consists of the same molecules that make up the gaseous water in the air. (e) The graphite of the pencil is the same substance on the paper as it is in the pencil.

10. (e). (a) Since centi- (c) means 0.01, 100.0 cm = 100.0 × 0.01 m = 1.000 m. (b) Since kilo- (k) means 1000, 0.0010 km = 0.0010 × 1000 m = 1.000 m. (d) Since milli- (m) means 0.001, 1000 mm = 1000 × 0.001 m = 1.000 m.

11. (c). −31°F. °F = $(-35°C \times \dfrac{1.8°F}{1.0°C})$ + 32°F = (−63°F) + 32°F = −31°F

12. (d). The proportions of the elements in a compound do not vary. (a) Compounds can be heated. (b) Compounds can be compressed. (c) Some compounds are colorless.

13. (c). The volume of a substance depends upon the amount of the substance. (a) Since the density of a substance is a ratio of the mass and the volume, it does not vary with the amount of the substance. (d) The color of a substance does not depend upon the amount of the substance. (e) The chemical properties of a substance do not vary with the amount of the substance. (See Section 1-4.)

2 Chemical Formulas and Composition Stoichiometry

Chapter Summary

In this chapter you learn about the language that chemists use to describe substances. This language developed from the idea that all matter is composed of atoms. These atoms combine in various ways to form substances.

A chemical formula tells us (1) which elements are present in the compound and (2) the relative numbers of atoms of the elements involved. However, we cannot handle or count individual atoms. How can we interpret chemical formulas in terms of some quantity we *can* measure, such as the masses of the elements present in the compound? What is the relationship between the mass of an element or compound and the number of atoms or molecules it contains? How can we use the chemical formula of a substance to tell us the masses of elements contained in that substance? How are chemical formulas determined from measurements of the amounts of the elements present in compounds? How do we describe samples that are not pure?

These are some of the questions addressed in this chapter. Many of the ideas mentioned here are probably new to you and might sound somewhat complicated. Do not worry about that—in your study of this chapter, try to progress steadily through the material. New material often uses the skills you have learned earlier. Be sure to master the ideas and methods associated with one concept or one type of problem before you try to proceed. Scientists in many diverse areas of study use calculations and principles of the types you will learn in this chapter. Nearly every chapter in this text, or in any course in chemistry, will use the ideas and skills you will develop in your study of this chapter.

Do not be alarmed by the obviously quantitative or mathematical nature of the chapter material. The math skills needed here are minimal—but you must develop your problem-solving ability. As you study, keep in mind the approaches in Chapter 1 and the study suggestions in "To the Student" at the beginning of this guide. Strive to keep aware of the reasons for approaching the problems as we do. Do not just try to learn to work "that kind of problem" from memory, but be aware of the concepts used. In addition, *think* about your answers.

We indicate the chemical composition of a substance by its **chemical formula** (Section 2-1). In this formula, we use subscripts to tell the number of atoms of each element in a molecule. The **structural formula** shows the order in which these atoms link to each other. This section also introduces two other representations of molecules, the **ball-and-stick model** and the **space-filling model**. The concept of chemical formulas also helps us understand such important chemical laws as the **Law of Definite Proportions (Constant Composition)**.

Many compounds consist of collections of **ions**—atoms or groups of atoms that possess an electrical charge (Section 2-2). **Cations** have a positive charge, and **anions** have a negative charge. The formula for an **ionic compound** indicates the ratio of atoms (ions) that are present in the compound. Ionic compounds do not contain molecules. They consist of extended arrays of the cations and anions in stable arrangements with each type of ion closest to ions of opposite charge.

Later you will learn systematic ways to name compounds. In this chapter, you begin that study (Section 2-3). Because there are so many compounds to deal with, you cannot learn all the names and formulas you will need. It is a good idea to memorize those shown in Table 2-1. You will gradually learn the names and formulas for many other compounds as you encounter them. We write the name or formula of an ionic compound by combining the names or the formulas of the individual ions. You should memorize the names, formulas, and charges of the common ions given in Table 2-2. Be sure you include the charge when you write the formula of any ion. We use these formulas and charges to write the formulas and names of many ionic substances. Any compound is electrically neutral, so determine the formula of an ionic compound by finding the smallest ratio of ions that would balance each other's charge. Though the formula for an ionic compound usually does not include the charges on the ions, we must always keep them in mind.

Further study (Chapters 5 and 7) will enable you to predict which substances consist of molecules and which contain ions. Likewise, you will gradually acquire some knowledge as to whether a substance is likely to be a solid, a liquid, or a gas. Do not let this worry you at this stage. The main goal of this chapter is to learn how to obtain quantitative information from chemical formulas.

Section 2-4 introduces a topic that we use throughout chemistry. Experiments can measure relative atomic masses (usually called **atomic weights**). Then we put these on a convenient scale by convention. The numbers that we look up (for example, in the tables inside the front cover of the text) give us the relative masses (weights) of atoms of the elements.

One of the most useful ideas in chemistry, the **mole concept**, is introduced in Section 2-5. Just as we describe 12 of anything as a dozen, we use the word "mole" to describe 6.022×10^{23} of anything. This large number, called **Avogadro's number**, is the number of atoms in a mole of atoms, the number of molecules in a mole of molecules, or the number of formula units in a mole of formula units (or even the number of eggs in a mole of eggs). A mole of atoms, molecules, or formula units is a convenient amount of a substance to work with in the laboratory. The importance of the mole concept is that it lets us relate something we *cannot* measure directly (the number of atoms, molecules, or formula units in a sample) to something we *can* measure (the mass of the sample). Section 2-5 establishes the relationship between the atomic weight and the number of atoms in a given mass of an element.

Section 2-6 then extends this relationship to apply it to the masses of molecules (or more generally, **formula units**) and the number of molecules or formula units in a given mass of a substance. We add the atomic weights of the atoms indicated by the formula to determine the **formula weight** of any substance, also called the **molecular weight** for substances consisting of molecules. Then, one mole of the substance is an amount of that substance whose mass, measured in grams, is numerically equal to its formula weight.

Putting the mole idea together with the concept of relative atomic weights, we develop a way of measuring equal numbers of different atoms or molecules (or a desired ratio). As an analogy, suppose that we know that all apples are identical, and that an apple weighs one and one-half times as much as an orange, all of which are also identical. If we wanted to have a very large number of oranges and the same very large number of apples, we could avoid counting them out individually. We could just weigh out apples and oranges in a $\frac{3}{2}$-to-1 ratio by weight. For instance, 300 pounds of apples would contain the same number of individual fruits as would 200 pounds of oranges, or 90 tons of apples the same number as 60 tons of oranges. Thus, we have replaced the tedious (or on the atomic and molecular scale, impossible) task of counting individual things by the much more convenient approach of weighing the amounts needed. The rest of this chapter uses the application of this mole concept. Be sure you understand the ideas of Section 2-5 and 2-6 and can apply them with ease before you proceed. The time you spend learning the mole concept, understanding and using the ideas of the mole, the formula unit, the formula weight, and the molecular weight, and practicing the associated calculations will benefit you later. Exercises 32 through 46 at the end of the textbook chapter will provide you with much practice in applying the ideas of these two important sections.

Compounds are composed of more than one element. We sometimes need to know what fraction of the total mass of a sample is due to each type of atom present—questions such as "In 50 g of water, how much of the mass is due to hydrogen and how much is due to oxygen?" In Section 2-7, you learn how to interpret chemical formulas in this way. We commonly represent this information as the **percent composition** of the compound. (Remember that in chemistry, we always understood *percent* to mean *percent by mass*, unless we specify some other basis.) What we are actually doing is applying a generalization that is based on the analysis of thousands of compounds—the Law of Definite Proportions (Constant Composition).

Of course, sometimes we do not know the identity of the compound with which we are dealing. We can approach this problem by carrying out some experiments to find out which elements are present (**qualitative analysis**), and then other experiments to find out the amounts of the elements present (**quantitative analysis**). Using this information, we can determine the **simplest**, or **empirical**, **formula** of the compound. This approach, discussed in Section 2-8, is the reverse of the calculation of percent composition (Section 2-7). Such analysis and deduction of the formula is important in characterizing any new or unidentified compound.

Section 2-9 discusses how we determine the **molecular formula** of a compound. Often chemists determine the amount of each element by converting it to a compound and then measuring the mass of the new compound formed. The molecular formula is always an integer multiple of the simplest formula. The **Law of Multiple Proportions** summarizes the relationship between compositions of different compounds composed of the same elements.

In Section 2-10, you learn about some other important interpretations of chemical formulas. As you can see, each of these depends upon the concepts you learned in Section 2-5 through 2-9.

The final section of the chapter tells how we can describe samples that are not pure. When dealing with **percent purity**, we always mean percent *by mass*, unless we specify otherwise.

IMPORTANT. The concepts and types of problems in this chapter are basic to your further study of chemistry. Work many problems, and always think carefully about the concepts you are using, so you will build a sound foundation for what lies ahead.

Study Goals

Use these goals to help you organize your study. As in other chapters, some textbook Exercises related to each study goal are indicated.

Sections 2-1 (Review Section 1-2)

1. Be able to interpret a chemical formula in terms of:
   (a) the type and number of atoms present and
   (b) the relative masses of elements present.
   Work Exercises 4-10, 12, 13, 61, and 62.

2. Know the names and formulas of the molecular compounds in Table 2-1. *Work Exercises 12 and 13.*

Sections 2-2 and 2-3

3. Describe ionic compounds. Know the names, formulas, and charges of the ions listed in Table 2-2. Be able to combine these to write names and formulas for some ionic compounds. *Work Exercises 14 through 23.*

Sections 2-4 through 2-6

4. Use the concepts of mole, atomic weight, formula weight, and molecular weight to relate masses of substances to numbers of atoms, molecules, or ions present. *Work Exercises 11, 24 through 46.*

Section 2-7

5. Given the formula of a compound, calculate its percentage composition. *Work Exercises 47, 48, 61, and 62.*

Section 2-8

6. Given the elemental composition of a compound, determine the simplest formula of the compound. *Work Exercises 53 through 57, and 59.*

Section 2-9

7. Distinguish between the simplest, or empirical, formula and the molecular formula. Given information about molecular weight or the number of atoms in a molecule, determine its molecular formula. *Work Exercises 49 through 52, 58, and 60.*

8. Use data from combustion and similar analysis of a compound to find the amounts of elements present in the compound. *Work Exercises 65 through 70.*

Review Sections 2-1 and 2-7 through 2-9

9. Summarize the laws of chemical combination: the Law of Conservation of Matter, the Law of Definite Proportions (Constant Composition), and the Law of Multiple Proportions. Be sure you understand how observation of chemical reactions has been the basis each of these. Give examples and be able to use these laws in calculations. *Work Exercises 71 through 74.*

Section 2-10

10. Apply the concepts of Sections 2-4 through 2-9 to carry out a variety of calculations based on chemical formulas. *Work Exercises 75 through 82.*

Section 2-11

11. Describe and use information about the purity of a sample. *Work Exercises 83 through 88.*

General

12. Use the concepts of the chapter to recognize and solve a variety of types of questions. *Work **Mixed Exercises** 89 through 97.*

13. Answer conceptual questions based on this chapter. *Work **Conceptual Exercises** 98 through 110.*

14. Apply concepts and skills from earlier chapters to the ideas of this chapter. *Work **Building Your Knowledge** exercises 111 through 116.*

23

15. Exercises at the end of chapter direct you to sources outside the textbook for information to use in solving them. Work **Beyond the Textbook** *exercises 117 through 120.*

Some Important Terms in This Chapter

Some important new terms from Chapter 2 are listed below. Do not overlook the other new terms you find in the Key Terms list or in your reading of the chapter. Fill the blanks in the following paragraphs with terms from the list. Use each term only once.

allotropes	formula	molecular formula
anion	formula unit	molecular weight
atomic mass unit	formula weight	percent composition
atomic weight	ion	percent purity
Avogadro's number	ionic compound	polyatomic ions
cation	mole	simplest formula
composition stoichiometry		

If chemistry is like Greek to you, it may be because many chemical terms have been derived from Greek words. The Greek word *stoicheion* means "first principle or element," and *metron*, means "measure." Thus, when you measure the quantitative relationships among elements in compounds, you are using [1]_____.

The composition of a substance—the kind and number of atoms in each molecule or formula unit—is represented by its chemical [2]_____. Even elements may consist of molecules that have two or more atoms of the element. If the number of atoms of the element or the arrangement of the atoms in a molecule can vary, then these different molecules, called [3]_____, have different properties.

Atoms that have equal numbers of protons and electrons are neutral, but if the numbers are not the same, the atom, called an [4]_____, will have an electrical charge. Since there are two kinds of charges, positive and negative, there are two kinds of ions: a [5]_____ (that's pronounced kăt'-ī-ən, not kā'-shən), and an [6]_____ (pronounced ăn'-ī-ən, not ān'-yən), respectively.

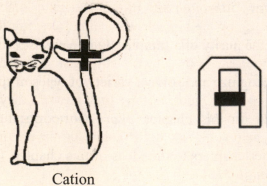

Cation

Compounds that are composed of ions do not contain molecules. The group of atoms represented by the chemical formula is called the [7]_____. Such a compound is called an [8]_____. Ions that consist of a single atom with a charge are monatomic ions; ions that consist of two or more joined atoms that together have a charge are called [9]_____.

The value of the relative atomic mass of an element, traditionally referred to as its [10]_____, does not depend upon the attraction of gravity. The [11]_____ or amu (sometimes simply u) is the unit for measuring the mass of an atom.

Atoms are very small, so they are hard to see or work with individually. Put a "bunch" of them together, about 6.022×10^{23} of them, and you have a chemist's "dozen," called a [12]_____. In fact, that number—602,200,000,000,000,000,000,000 in common notation—is so special that it's got a name, [13]_____, named for Amedeo Avogadro.

Just like the "weight" of an atom is called its *atomic weight*, the "weight" of a formula unit is called its [14]_____, and the "weight" of a molecule is called its [15]_____. These "weights," derived from the chemical formulas of the compounds and the atomic weights of the elements in the compound, express the proportions of each element in the compound and can be converted into the form of a percent by mass, specifically the [16]_____. On the other hand, by experimentally measuring the masses of the elements in a certain mass of the compound, the chemical formula of the compound may be calculated. The simplest ratio of the number of atoms of each element in the compound is the [17]_____. For molecular compounds, this formula, combined with a second experiment to measure the approximate molecular weight produces a formula, called the [18]_____, that provides more information about the molecule.

Just as the percent composition expresses the proportions of the elements in a compound, the proportion of an impurity in an impure substance can be expressed by the [19]_____. However, this value can vary, unlike the percent composition, which is constant. It describes a mixture, not a pure substance

Quotefall

Each puzzle below contains an important quote from Chapter 2. A black square indicates the end of a word. Words starting at the end of a line may continue in the next line. Punctuation in the statement is included in the boxes. Above the boxes in each vertical column are the letters that belong in the boxes, in a randomized order. Place the letters into the boxes directly below them to form words across. The letters do not necessarily go into the squares in the order listed. Use each letter only once. When all the letters are in their correct boxes, you will be able to read the complete quotation across the diagram from left to right, line by line. **Hint**: It may be helpful at times to fill in small words, like "the," to use up some letters to help in determining other longer words. Some letters have been "seeded" to help you get started. The solutions are on page 28.

Puzzle #1:

```
T E N T E B E T H A E H U O E A D N O L P F O H H L T G A N O T H D S S C E E F
C O E E E O E W E T S T M N B M N C M E O W R O T E E E M N P E B S A S F N S A T W
O E E E X P N N O S T I A L A N A P H B E O F F T R M R M O O R V S N E R A S N O E N O L
B H M L U R D S F A D C E A T I B E L O I T H O I M S U W O N L
M E L L M A M N T A E T Y B S L A P W H P I E N
                                    R   P               N : T
      L           ,           ,       R       R
  O       U     ,           O                       S
M         C           N                     N     A
  L     T               C                 P       D
      X       S           L         O   N       S .
```

Preliminary Test

As in other chapters, this test will check your understanding of basic concepts and types of calculations. Be sure to practice *many* of the additional textbook exercises, including those indicated with the Study Goals.

True-False

Mark each statement as true (T) or false (F).

_____ 1. I have seen a mole of water, but I have never seen a molecule of water.

_____ 2. When we say that the atomic weight of helium is 4.0, we mean that one atom of helium has a mass of 4.0 grams.

_____ 3. The number of oxygen atoms in 16 grams of oxygen is 6.022×10^{23}.

_____ 4. The number of oxygen molecules in 16.0 grams of oxygen is 6.022×10^{23}.

_____ 5. One mole of boron has a mass of $(10.8)(6.022 \times 10^{23})$ grams.

_____ 6. If we know only the simplest formula of an ionic compound and the required atomic weights, we can calculate the percent by mass of any element in the compound.

_____ 7. If we know only the molecular formula of a molecular compound and the required atomic weights, we can calculate the percent by mass of any element in the compound.

_____ 8. If we know only the percent composition by mass of a compound and the required atomic weights, we can calculate the simplest formula of the compound.

_____ 9. If we know only the percent composition by mass of a compound and the required atomic weights, we can calculate the molecular formula of the compound.

Short Answer

Answer with a word, a phrase, a formula, or a number with units as necessary.

1. The number of hydrogen atoms in a molecule of tetraethyl lead, $Pb(C_2H_5)_4$, is _____.

2. Carbon and oxygen combine to form two compounds. One is carbon dioxide, whose formula is _____, and the other is carbon monoxide, whose formula is _____.

3. The name of the compound HNO_3 is _____;
    the name of the compound CH_4 is _____.

4. The name of the compound HCl is _____ as a pure compound;
    it is _____ when dissolved in water,.

5. The formula for the sodium ion is _____, the formula for the magnesium ion is _____, and the formula for the copper(II) ion is _____.

6. The name of Ag^+ is _____, the name of Al^{3+} is _____, and the name of Fe^{2+} is _____.

7. The formula for the chloride ion is _____, the formula for the hydroxide ion is _____, and the formula for the sulfate ion is _____.

8. The name of CH_3COO^- is _____, the name of O^{2-} is _____, and the name of PO_4^{3-} is _____.

9. Write the chemical formula for each of the following compounds.
    (a) the compound formed from potassium ions, K^+, and sulfide ions, S^{2-}:
    _____
    (b) the ionic compound formed by calcium and bromine: _____
    (c) the compound named magnesium chloride: _____
    (d) the compound named iron(II) sulfide: _____

10. Write the name for the compound represented by each of the following chemical formulas.
    (a) Ag_2O _____
    (b) AlP _____
    (c) CuS _____
    (d) Fe_2S_3 _____
    (e) $Mg_3(PO_4)_2$ _____
    (f) Na_2CO_3 _____
    (g) $Fe_2(SO_4)_3$ _____

11. The number of nitrogen molecules in one mole of N_2 is _____;
    this amount of nitrogen contains _____ N atoms.

12. One mole of hydrogen peroxide, H_2O_2, contains the same number of hydrogen atoms as _____ mole(s) of water, H_2O.

13. One mole of hydrogen peroxide, H_2O_2, contains the same number of oxygen atoms as _____ mole(s) of water, H_2O.

14. Suppose you had a mole of gold.
    (a) Check the periodic table. How much does one mole of gold weigh, in grams?
    _____
    (b) How many atoms would it contain? _____

(c) According to the U.S. Census Bureau, the estimated population of the world on June 24, 2012 was 7,022,029,026. (http://www.census.gov/ipc/www/popclockworld.html) Divide the number of atoms of gold in part b equally among all the people of the earth. How many atoms would each one get? _____

15. Express the number of oxygen atoms in n moles of a substance whose formula is $C_aH_bO_c$, mathematically: _____.

16. To calculate the formula weight of a given compound from its known formula, _____ ___.

17. According to one theory of the origin of the universe, the age of the universe is about 13.7 billion years. (http://map.gsfc.nasa.gov/universe/uni_age.html) The number of seconds that have elapsed since the origin of the universe would be _____ times Avogadro's number.

18. Suppose we calculate the masses of carbon that would combine with 1.0 g of oxygen in the two compounds carbon monoxide and carbon dioxide. We could then use these two results as an illustration of the Law of _____.

19. The chemical formula we derive only from information about percent composition of the elements in the compound is the _____ or the _____ of the compound.

20. In order to tell whether the simplest formula, as derived from percent composition data, corresponds to the molecular formula, we need to make some independent measurement or determination of the _____ of the compound.

21. The molecular formula of methyl methacrylate, used in the manufacture of Plexiglas®, is $C_4H_6O_2$. What is its empirical formula? _____

22. A copper ore is 18.3% $CuFeS_2$. The mass of $CuFeS_2$ in 12.7 kg of the ore is _____ grams.

23. The mass of aluminum in 12.7 kg of ore that is 8.52% Al_2O_3 (and contains no other source of aluminum) is _____ grams.

Multiple Choice

_____ 1. What is the number of oxygen atoms in one molecule of sulfur trioxide?
  (a) 1   (b) 2   (c) 3   (d) 4   (e) 5

_____ 2. Which of the following compounds has the highest formula weight?
  (a) water   (b) propane   (c) ammonia   (d) sulfuric acid (e) acetic acid

_____ 3. A sample of matter that contains three kinds of atoms could be
  (a) a solution.
  (b) a heterogeneous mixture.
  (c) a compound.
  (d) a homogeneous mixture.
  (e) Any of the other answers could be correct.

_____ 4. How many grams of hydrogen would combine with 12.0 grams of carbon to form the compound ethylene, C_2H_4?
  (a) 1.0 g   (b) 2.0 g   (c) 4.0 g   (d) 8.0 g   (e) 12.0 g

_____ 5. Which of the following is *not* a correct description of 16.0 g of methane, CH_4?

    (a) one mole of methane

    (b) the amount of methane that contains 12.0 g C

    (c) $(16.0)(6.022 \times 10^{23})$ molecules of methane

    (d) the amount of methane that contains $(4)(6.022 \times 10^{23})$ hydrogen atoms

_____ 6. Suppose we have 100 g of each of the following substances. Which sample contains the largest number of moles?

    (a) H_2O, formula weight 18.0

    (b) HCl, formula weight 36.5

    (c) $AlCl_3$, formula weight 133.3

    (d) $MgCO_3$, formula weight 84.3

    (e) It is impossible to tell, unless we know what reaction will take place with the substance.

_____ 7. An unknown element, (call it X), combines with oxygen to form a compound known to have the formula XO_2.

We observe that 24.0 g of element X combines with 16.0 g of O to form this compound. What is the atomic weight of element X?

    (a) 48.0   (b) 32.0   (c) 24.0   (d) 16.0   (e) 12.0

_____ 8. What mass ratio of carbon to oxygen is contained in carbon dioxide, CO_2?

    (a) 0.375 g C/g O   (b) 1.33 g C/g O   (c) 2.67 g C/g O

    (d) 0.75 g C/g O   (e) 0.67 g C/g O

_____ 9. For a molecular substance, which of the following statements is true?

    (a) The molecular formula must be the same as the simplest formula.

    (b) The molecular formula must be proportional to the empirical formula, but with all integer subscripts multiplied by an integer greater than 1.

    (c) The molecular formula must be proportional to the empirical formula, but with all integer subscripts divided by an integer greater than 1.

    (d) The molecular formula may be the same as the simplest formula.

    (e) The molecular formula and the simplest formula are not necessarily related.

_____ 10. The simplest (empirical) formula of a compound is CH_2O.

By another experiment, the molecular weight of this compound is 180.

How many hydrogen atoms are in one molecule of the compound?

    (a) 2   (b) 6   (c) 8   (d) 12   (e) 18

_____ 11. Combustion of a 1.50-gram sample of a compound that contains only C, H, and O produced 3.00 g of CO_2 and 1.23 g of H_2O.

What is the simplest formula for the compound?

    (a) C_2H_4O (b) CHO   (c) C_2HO_4 (d) C_6HO_4 (e) C_3HO_2

Answers to Some Important Terms in This Chapter

1. composition stoichiometry
2. formula
3. allotropes
4. ion
5. cation
6. anion
7. formula unit
8. ionic compound
9. polyatomic ions
10. atomic weight
11. atomic mass unit
12. mole
13. Avogadro's number
14. formula weight
15. molecular weight
16. percent composition
17. simplest formula
18. molecular formula
19. percent purity

Answers to Quotefall

Puzzle #1:
The Law of Multiple Proportions: When two elements, A and B, form more than one compound, the ratio of the masses of element B that combine with a given mass of element A in each of the compounds can be expressed by small whole numbers.

Answers to Preliminary Test

Do not just look up answers. *Think about the reasons for the answers.*

True-False

1. True. Individual molecules are too small to see, but one mole of water is 18 g, which is about 1.2 tablespoons of water. (Can you carry out the conversions to show that one mole of water is 1.2 tablespoons? The density of water is 1.0 g/mL, there are 4 cups in a quart, there are 16 tablespoons in a cup, and there are more conversion factors in Table 1-8.)

2. False. What we mean is that it has a mass of 4.0 arbitrary units on a scale relative to which a carbon atom has a mass of 12.0 units. This is a situation where thinking about the size of the number helps to avoid a wrong answer. A newly minted U.S. nickel has a mass of about 5 g—do you think a single helium atom has a mass of 4 g?

3. True. Sixteen grams of oxygen is one mole of oxygen atoms. One mole of atoms always contains Avogadro's number of atoms.

4. False. Oxygen is a diatomic element (O_2), so 16.0 grams of oxygen, which contains 6.022×10^{23} oxygen atoms, would contain only half that many oxygen molecules, or 3.011×10^{23} O_2 molecules.

5. False. One mole of an element is the amount of the element whose mass in grams is numerically equal to the atomic weight. Thus, one mole of boron has a mass of 10.8 g. Think about the size of the number. You should remember that a mole of atoms of any substance is a convenient amount of material to handle in the laboratory. The number stated in the question, $(10.8)(6.022 \times 10^{23})$ g, is larger than the mass of the earth!

6. True. See Section 2-7.

7. True. See Section 2-7.

8. True. See Section 2-8.
9. False. Review the distinction between the molecular formula and the simplest formula, Section 2-9.

Short Answer

1. 20. $5 \times 4 = 20$
2. CO_2; CO. Know the names and formulas listed in Table 2-1.
3. nitric acid; methane. See Table 2-1.
4. hydrogen chloride; hydrochloric acid. See Table 2-1.
5. Na^+; Mg^{2+}; Cu^{2+}. You should know the names, formulas, and charges of the ions listed in Table 2-2. An ion formula written without the charge is incorrect.
6. silver ion; aluminum ion; iron(II) ion or ferrous ion. For some ion formulas, we indicate the charge by a Roman numeral in parentheses as part of the name, while for others this is not included. You will study naming ions more systematically in later chapters. For now, learn the list in Table 2-2.
7. Cl^-; OH^-; SO_4^{2-}. See Table 2-2.
8. acetate ion; oxide ion; phosphate ion.
9. (a) K_2S
   (b) $CaBr_2$
   (c) $MgCl_2$
   (d) FeS
10. (a) silver oxide
    (b) aluminum phosphide
    (c) copper(II) sulfide
    (d) iron(III) sulfide
    (e) magnesium phosphate
    (f) sodium carbonate
    (g) iron(III) sulfate
11. 6.022×10^{23} N_2 molecules; $(2)(6.022 \times 10^{23})$ or 1.204×10^{24} N atoms.
12. one. Each molecule of H_2O_2 has two hydrogen atoms, and each molecule of H_2O also has two hydrogen atoms, so one mole of H_2O_2 (6.022×10^{23} molecules) and one mole of H_2O each have $(2)(6.022 \times 10^{23})$ hydrogen atoms.
13. two. Each molecule of H_2O_2 has two oxygen atoms, but each molecule of H_2O has only one oxygen atom, so one mole of H_2O_2 (6.022×10^{23} molecules) has twice as many oxygen atoms as one mole of H_2O. Therefore, two moles of H_2O has the same number of oxygen atoms as one mole of H_2O_2.
14. (a) 197.0 g
    (b) 6.022×10^{23} atoms
    (c) 8.576×10^{13} atoms/person (0.000 000 028 05 g)
15. $(n)(c)(6.022 \times 10^{23})$ O atoms. There are 6.022×10^{23} molecules in each mole. Each molecule contains c O atoms.

16. Multiply the atomic weight of each element (given in the formula by the symbol) by the number of atoms of the element (given by its subscript) and then add these products together.

17. 7.18×10^{-7}, or 0.000 000 718. 13.7 billion = 1.37×10^{10};

$$1.37 \times 10^{10} \text{ yr} \times \frac{365.25 \text{ da}}{1 \text{ yr}} \times \frac{24 \text{ hr}}{1 \text{ da}} \times \frac{60 \text{ min}}{1 \text{ hr}} \times \frac{60 \text{ s}}{1 \text{ min}} = 4.32 \times 10^{17} \text{ s}$$

$$\frac{4.32 \times 10^{17}}{6.022 \times 10^{23}} = 7.18 \times 10^{-7}$$

Put another way, when the universe is *1.4 million times* as old as it is now, Avogadro's number of seconds will have elapsed. This gives some notion of what a large number this is. Recall, however, that Avogadro's number of carbon atoms has a mass of 12 g. That number of water molecules is about 1.2 teaspoon. This gives you an idea how small atoms and molecules are!

18. Multiple Proportions. See Section 2-9.

19. simplest formula; empirical formula

20. molecular weight or formula weight

21. C_2H_3O. Divide each subscript in $C_4H_6O_2$ by their largest common factor, which is 2.

22. 2320 g. Remember that percent can be interpreted as "parts per hundred."

The answer can be calculated as: $12.7 \text{ kg ore} \times \dfrac{1000 \text{ g ore}}{1 \text{ kg ore}} \times \dfrac{18.3 \text{ g CuFeS}_2}{100 \text{ g ore}}$

23. 573 g Al. The calculation is set up as follows:

12.7 kg

$$\text{ore} \times \frac{1000 \text{ g sample}}{1 \text{ kg ore}} \times \frac{8.52 \text{ g Al}_2\text{O}_3}{100 \text{ g sample}} \times \frac{1 \text{ mol Al}_2\text{O}_3}{102.0 \text{ g Al}_2\text{O}_3} \times \frac{2 \text{ mol Al atoms}}{1 \text{ mol Al}_2\text{O}_3} \times \frac{27.0 \text{ g Al}}{1 \text{ mol Al atoms}}$$

Multiple Choice

1. (c). Sulfur trioxide is SO_3. You should know the names and formulas in Table 2-1.

2. (d). You need to know the names and formulas in Table 2-1. Then calculate formula weight for each compound. Formula weights are: $H_2O = 18.0$; $C_3H_8 = 44.0$; $NH_3 = 17.0$; $H_2SO_4 = 98.1$; $CH_3COOH = 60.0$.

3. (e)

4. (b). From the formula, there are (4 × 1.0 g H)/(2 × 12.0 g C) or (1 g H)/(6 g C). Thus, 12 g of carbon would require 2 g of hydrogen. Notice that we do not need to know the reaction by which the ethylene is formed in order to answer the question. It is immaterial that hydrogen gas is diatomic (H_2), because we are dealing with the mass of hydrogen.

5. (c). Because 16.0 grams is one mole of methane, it contains 6.022×10^{23} molecules of CH_4.

6. (a). We do not need to calculate the actual number of moles. For each compound, the number of moles is equal to the number of grams divided by the formula weight. Thus, since all of the samples have the same mass, the compound with the lowest formula weight is present in the highest number of moles.

7. (a). According to the formula, one mole of X atoms would combine with 2 moles of O atoms, or 32.0 g of O. We observe that 24.0 g of X combines with 16.0 g of O, so it would

take 48.0 g of X to combine with 32.0 g of O at the same ratio. Hence, the atomic weight of X is 48.0.

8. (a). If you got answer (c), you probably calculated the inverse of the requested ratio. If you got (d), you forgot that there are two O atoms for every C atom. If you got (b), you made the mistake in (d), and also calculated the inverse ratio.

9. (d). See Section 2-9. The molecular formula is often equal to the empirical formula, but it may be a whole number multiple of the empirical formula.

10. (d). The formula weight of the simplest formula is 30 (12.0 + 2 % 1.0 + 16.0), and the molecular weight, the weight of the molecular formula, is six times that. Thus, the molecular formula must be $C_6H_{12}O_6$.

11. (a). Apply the methods of Section 2-8, as illustrated in Examples 2-16 and 2-17.

Step 1: $\underline{?}\,g\,C = 3.00\,g\,CO_2 \times \dfrac{12.01\,g\,C}{44.01\,g\,CO_2} = 0.818\,g\,C$

Step 2: $\underline{?}\,g\,H = 1.23\,g\,H_2O \times \dfrac{2.016\,g\,H}{18.02\,g\,H_2O} = 0.137\,g\,H$

Step 3: $\underline{?}\,g\,O = 1.50\,g\,sample = [0.818\,g\,C + 0.137\,g\,H] = 0.545\,g\,O$

Step 4:

Element	Mass of Element	Moles of Element (divide mass by AW)	Divide by Smallest	Simplest Formula
C	0.818 g	$\dfrac{0.818\,g}{12.01\,g/mol} = 0.0682\,mol$	$\dfrac{0.0681\,mol}{0.0341\,mol} = 2.00\,C$	
H	0.137 g	$\dfrac{0.137\,g}{1.008\,g/mol} = 0.136\,mol$	$\dfrac{0.136\,mol}{0.0341\,mol} = 3.99\,H$	C_2H_4O
O	0.545 g	$\dfrac{0.545\,g}{16.00\,g/mol} = 0.0341\,mol$	$\dfrac{0.0341\,mol}{0.0341\,mol} = 1.00\,O$	

3 Chemical Equations and Reaction Stoichiometry

Chapter Summary

In Chapter 2, we were concerned with one compound at a time. In Chapter 3, we answer some very important questions about chemical reactions, including the following: How do we use symbols to represent a chemical reaction? How can we interpret chemical equations so that they tell us about the amounts of substances that can react with one another or that can be formed? How do we tell when a chemical reaction reaches completion? Many important reactions occur in solution, so how can we conveniently express the amount of a substance present in a solution? How can we tell how much of a solution is required for a reaction to completely consume all reactants ("go to completion")?

We first represent a reaction in a quantitative form that we call the **chemical equation** (Section 3-1). To be consistent with the Law of Conservation of Matter, we must write each equation in **balanced** form, that is, with equal numbers of atoms of each element on the two sides of the equation. (We will see later that when the equation involves ions, the total charges on the two sides of the equation must also balance. For the applications in this chapter, all substances are written as neutral elements or compounds.) We then interpret the balanced chemical equation to find out the relative amounts of **products** and **reactants** involved in the reaction, initially in moles.

All calculations about amounts of reactants and products involved in reactions are based on the following idea (Section 3-2): *The coefficients in the balanced chemical equation tell the relative numbers of moles of those substances that react or are formed in the reaction.* From that starting point, we can "translate" the mole terminology into any other unit desired—numbers of molecules, masses of substances, or even gas volumes (as in Chapter 12). In this very important section, you will learn how to construct **unit factors** to represent the chemical equivalence of two substances. This central idea can then be expanded to solve many questions about chemical reactions. One of the most important of these is the question of finding which reactant limits the progress of a reaction, the **limiting reactant concept** (Section 3-3). In Section 3-4, we learn to use the idea of **percent yield** to describe reaction conditions in which less than the theoretically possible amount of a product is obtained. You can apply the same ideas to impure reactant samples. Section 3-5 deals with **sequential reactions**, which take place in a series of steps or individual reactions. As you study these topics, keep in mind the central idea stated above, and try to see how all these apparently different kinds of problems are just various applications of that central idea. In this way, you will avoid the trap of just memorizing how to work certain kinds of problems.

There are many reasons why mixtures are involved in chemical reactions; when the mixtures are homogeneous, we call them **solutions**. We usually do not add pure substances to one another to carry out a reaction. Sometimes the reaction forms products that are mixed with (dissolved in) other substances. Many substances are dangerous or difficult to handle when they are pure, or may not even exist as pure substances, so we supply them in the form of mixtures. The use of solutions makes it very easy to measure the amount of a reactant to add to a reaction mixture. We can use a solution to measure a very small amount of a reactant more precisely than we could with the pure substance. The remainder of this chapter deals with solutions and their use in chemical reactions.

First, we learn some methods for expressing concentrations of solutions (Section 3-6). A simple method of telling how much **solute** is contained in a given solution is in terms of **percent by mass**. (Remember that when we describe anything in chemistry in terms of percent, we mean "percent by mass" unless we specify otherwise.) As we have seen in this chapter, the relative amounts of substances involved in chemical reactions are most simply expressed in terms of moles. Further, it is more convenient to measure the volume than the mass of a liquid solution. **Molarity** relates the number of moles (which we need to know or calculate) to the volume of the solution (which we can easily measure, often from the mass). The calculations we need to do when we **dilute** a solution are discussed in Section 3-7.

We use molarity as the most convenient method for expressing the amount of a solution that contains the required amount of a substance for a chemical reaction. Calculations of this type, discussed in Section 3-8, are quite useful both in the laboratory and in practical applications. They will also be important in our discussion of the properties of solutions (Chapter 14), the rates of chemical reactions (Chapter 16), the extent to which reactions proceed before reaching equilibrium (Chapter 17 through 20), electrochemistry (Chapter 21), and many other topics.

Study Goals

Section 3-1
1. Balance simple chemical equations. Know what information is contained in a balanced chemical equation. Express the relative amounts of reactants used and products produced in terms of molecules, moles or masses. *Work Exercises 1 through 11.*

Section 3-2
2. Use the concepts of this chapter to do calculations based on balanced chemical equations. Convert moles, mass or number of particles of a given substance in a chemical equation to moles, mass or number of particles of another substance in the chemical equation. *Work Exercises 12 through 29.*

Section 3-3
3. Given the amounts of two or more reactants, determine which is the limiting reactant and which is (are) present in excess. Determine the extent of the reaction. *Work Exercises 30 through 39.*

Section 3-4
4. Given the amount of a limiting reactant present in a chemical reaction and the amount of a product formed, determine the percent yield. Given the percent yield and one of the other two variables, calculate the missing quantity. *Work Exercises 40 through 49.*

Section 3-5
5. Carry out calculations for a series of reactions or reaction steps. *Work Exercises 50 through 56.*

Section 3-6

6. Relate a given amount of solute in a given amount of solvent or solution to (a) percent by mass of the solute and (b) molarity. *Work Exercises 57 through 71.*

Section 3-7

7. Relate the concentration and volume of a solution prepared by dilution to the concentration and volume of the concentrated solution from which it was prepared. *Work Exercises 72 through 77.*

Section 3-8

8. Relate the amount of substance required or produced in a reaction to the concentration of a solution containing that substance. *Work Exercises 78 through 82.*

General

9. Recognize and work various kinds of problems from this chapter. *Work* **Mixed Exercises** *84 through 87.*

10. Answer conceptual questions based on this chapter. *Work* **Conceptual Exercises** *88 through 103.*

11. Apply concepts and skills from earlier chapters to the ideas of this chapter. *Work* **Building Your Knowledge** *exercises, 104 through 113.*

12. Exercises at the end of chapter direct you to sources outside the textbook for information to use in solving them. Work **Beyond the Textbook** *exercises 114 through 118.*

Some Important Terms in This Chapter

Some important new terms from Chapter 3 are listed below. Do not overlook the other new terms you find in the Key Terms list or in your reading of the chapter. Fill the blanks in the following paragraphs with terms from the list. Each term is only used once.

actual yield	percent by mass	sequential reactions
balanced	percent yield	solute
chemical equation	products	solution
dilution	reactants	solvent
limiting reactant	reaction stoichiometry	theoretical yield
molarity		

A chemist represents what happens in a chemical reaction by a kind of shorthand, a set of symbols that quickly tells the chemist a lot about the reaction, known as a [1]_____ _____. Since matter cannot be created or destroyed during a chemical reaction, the equation is not complete until it is [2]_____. The equation includes the substances that react, called the [3]_____, and the substances produced, called the [4]_____, separated by an arrow.

The use of measured amounts of some reactants and/or products and the balanced chemical equation to calculate the amounts of other reactants and/or products not measured is called [5]_____. Sometimes in these calculations the amounts of two or more reactants are given. It's not likely that these reactants will react completely, down to the last molecule. One of them will be used up first, and then the reaction will have to stop. Because no more product can be formed, this reactant limits the amount of product that can be made. It is the [6]_____. Reaction stoichiometry allows us to calculate the amount of product that could be produced from the amount(s) of reactant(s) that react. In theory, this amount of product could be produced, so this amount is called the [7]_____. In reality, there are various reasons why not all of the product calculated is produced. The measured amount of product actually obtained is the [8]_____. These two amounts, calculated (theoretical) and measured (actual), can be compared by dividing the actual by the theoretical and multiplying by 100, a percentage known as the [9]_____. In real situations, such as industrial processes in which chemicals are manufactured, the initial reactant (or reactants) does not directly produce the desired product. A series of reactions may produce that product through a number of intermediate substances. These reactions are [10]_____, because they must occur in a specific sequence.

A homogeneous mixture, or [11]_____, consists of two (or more) substances in which one, the [12]_____, dissolves in the other, the [13]_____. Unlike a compound, in which the proportions of the elements are always the same, the relative amounts of substances mixed together in a solution can vary greatly. Therefore, it is important to specify the concentration of the solution, the relative amounts of solute and solvent. The ratio of the masses of the solute and solution, multiplied by 100, produces the [14]_____. The ratio of the moles of solution to the volume (in liters) of the solution produces the [15]_____. If the concentration is too strong, you should consider [16]_____, adding more solvent to the solution.

Preliminary Test

As in other chapters, this test will check your understanding of basic concepts and types of calculations. Be sure to practice *many* of the additional textbook Exercises, including those indicated with the Study Goals.

True-False

Mark each statement as true (T) or false (F).

_____ 1. Before we can use a chemical equation to determine the amounts of different substances involved in the reaction, we must balance the equation.

_____ 2. In a balanced chemical equation, the total number of moles of reactants must equal the total number of moles of products.

_____ 3. The total number of grams of the reactants that react must equal the total the number of grams of products formed.

_____ 4. The total number of atoms in the reactants that react must equal the total number of atoms in the products.

_____ 5. The equation for the reaction involving the hypothetical substances A, B, C, and D:

$$A + 2B \rightarrow 3C + D$$

_____ implies that the products C and D are always produced in a three-to-one mole ratio.

_____ 6. The equation shown in question 5 implies that in any reaction involving A and B as reactants, A must be the limiting reactant.

_____ 7. The equation: $2Na + H_2O \rightarrow 2NaOH + H_2$ is balanced.

_____ 8. To balance the chemical equation: $KClO_3 \rightarrow KCl + O_2$ we change O_2 to O_3.

_____ 9. The balanced equation: $Ca + 2H_2O \rightarrow Ca(OH)_2 + H_2$
tells us that 1 mole of calcium atoms would react to give 1 mole of H_2 molecules.

_____ 10. According to the balanced equation in Question 9, 18 grams of water would be required to react with 40. grams of calcium.

_____ 11. When we are told the amounts of two different reactants that are put into a reaction mixture, we should assume that all of both reactants would be used up.

_____ 12. To tell how to prepare a 20.0% solution of a compound in water, we would have to know the formula or formula weight of the compound.

_____ 13. To tell how to prepare a 0.25 M solution of a compound in water, we would have to know the formula or formula weight of the compound.

_____ 14. If we wish to make a 1.00 M solution of sodium chloride (NaCl, formula weight 58.4), we would add 58.4 g of NaCl to 1.00 L of water.

Short Answer

Answer with a word, a phrase, a formula, or a number with units as necessary.

1. Balancing a chemical equation is based on the Law of
_____.

2. When the equation: $Ca(OH)_2 + HNO_3 \rightarrow Ca(NO_3)_2 + H_2O$
is balanced, the coefficient of $Ca(OH)_2$ is _____.

3. When the equation: $C_3H_8 + O_2 \rightarrow CO_2 + H_2O$
is balanced, the coefficient of O_2 is _____.

4. When the equation: $Al + CuSO_4 \rightarrow Al_2(SO_4)_3 + Cu$
is balanced, the coefficient of Cu is _____.

In questions 5 through 7, balance the equations with the smallest whole-number coefficients. Then find the sum of the coefficients in the balanced equation. Don't forget coefficients of 1.

5. $CS_2 + Cl_2 \rightarrow CCl_4 + S_2Cl_2$ _____

6. $C_7H_{16} + O_2 \rightarrow CO_2 + H_2O$ _____

7. $Ag + H_2S + O_2 \rightarrow Ag_2S + H_2O$ _____

8. Balance the following chemical equation:

$$Ca(OH)_2 + H_3PO_4 \rightarrow Ca_3(PO_4)_2 + H_2O$$

The unit factor that relates the number of moles of $Ca_3(PO_4)_2$ formed to the number of moles of $Ca(OH)_2$ reacting is _____.

9. Using the balanced chemical equation in Question 8, the unit factor that relates the mass of $Ca_3(PO_4)_2$ formed to the mass of $Ca(OH)_2$ reacting is _____.

10. Any chemical reaction must stop when one of the reactants is entirely consumed, even if some of the other reactants remain. The reactant that is consumed first is called the _____.

We place 2.0 moles of CS_2 and 7.5 moles of O_2 in a closed container and burn them, according to the balanced equation: $CS_2 + 3O_2 \rightarrow CO_2 + 2SO_2$

The next four questions, 11 through 14, refer to this reaction.

11. The limiting reactant is _____.
12. When the reaction stops, the reactant left over is _____; there would be _____ moles of this reactant left.
13. The number of moles of CO_2 formed would be _____.
14. The number of moles of SO_2 formed would be _____.

---

15. In the reaction in Question 8, the maximum mass of $Ca_3(PO_4)_2$ that can be prepared from 19.6 g of $Ca(OH)_2$ and 18.6 g of H_3PO_4 is _____ grams.
16. The maximum amount of a particular product that can be obtained from a reaction is called the _____.
17. The amount of a particular product that is actually isolated from a reaction is called the _____.
18. The percent yield of a chemical reaction is defined as _____.
19. The maximum percent yield that a reaction can give is _____.
20. If the reaction: $P_4 + 6I_2 \rightarrow 4PI_3$

gives a 76.3% yield, the mass of PI_3 that could be obtained from 250. g of I_2 and excess phosphorus is _____ grams.
21. In order to dilute a solution, we add more _____.
22. To make a 3.00 M HNO_3 solution, we would have to dilute 500. mL of a 12.0 M HNO_3 solution to a final volume of _____ mL.
23. The molarity of a solution that contains 0.025 70 mol of solute in 518 mL of solution is _____.
24. Suppose we know the molarity of an aqueous solution of sulfuric acid (formula weight 98.1) and we wish to express the concentration as mass percent sulfuric acid. The one additional value that would allow us to calculate the conversion is the _____ of the solution.

Multiple Choice

_____ 1. Ethylene, C_2H_4, burns in oxygen to produce carbon dioxide and water. The correct form of the chemical equation that describes this reaction is
  (a) $C_2H_4 + O \rightarrow CO_2 + H_2O$ (b) $2C_2H_4 + O_2 \rightarrow 2CO_2 + H_2O$
  (c) $C_2H_4 + 2O_2 \rightarrow 2CO + 2H_2O$ (d) $C_2H_4 + 3O_2 \rightarrow 2CO_2 + 2H_2O$

_____ 2. In the reaction: $CaCO_3 \rightarrow CaO + CO_2$

100. grams of calcium carbonate, $CaCO_3$, is observed to decompose upon heating into 56 grams of calcium oxide, CaO, and 44 grams of carbon dioxide, CO_2. This illustrates
  (a) the Law of Conservation of Matter.
  (b) the Law of Conservation of Energy.
  (c) the Law of Gravity.
  (d) the Law of Multiple Proportions.
  (e) the conversion of matter into energy.

_____ 3. If 5.0 grams of H_2 is reacted with 10.0 grams of O_2 in the reaction

$$2H_2 + O_2 \rightarrow 2H_2O$$

which would be the limiting reactant?
  (a) H_2 (b) O_2 (c) H_2O

_____ 4. We place 8.0 g of S in a closed container with 14.0 g of O_2. These elements then react as completely as possible, by the equation

$$2S + 3O_2 \rightarrow 2SO_3$$

At the end of the reaction, how many grams of S and O_2, respectively, would be left over unreacted? (Use these atomic weight: S = 32, O = 16.)
  (a) 2.0, 2.0   (b) 0.0, 6.0   (c) 2.0, 0.0   (d) 0.0, 2.0   (e) 0.0, 0.0

_____ 5. Two reactions that occur in the production of iron (Fe) are

$$2C + O_2 \rightarrow 2CO \quad \text{followed by} \quad Fe_2O_3 + 3CO \rightarrow 2Fe + 3CO_2.$$

Note that the carbon monoxide, CO, formed in the first reaction is then used in the second reaction. According to these equations, how many moles of oxygen, O_2, are required to produce 800 moles of Fe?
  (a) 300   (b) 600   (c) 900   (d) 1200
  (e) None of the preceding answers is correct.

_____ 6. What is the total mass of products formed when 3.2 grams of CH_4 is burned in air?

$$CH_4 + 2O_2 \rightarrow CO_2 + 2H_2O$$

  (a) 16 g   (b) 36 g   (c) 44 g   (d) 80. g   (e) 160. g

_____ 7. What is the percent yield of elemental sulfur if 7.54 grams of sulfur are obtained from the reaction of 6.00 grams of SO_2 with an excess of H_2S?

$$2H_2S + SO_2 \rightarrow 2H_2O + 3S$$

  (a) 76.1%   (b) 79.4%   (c) 83.7%   (d) 88.4%   (e) 91.4%

_____ 8. A 1.00% solution of NaCl (formula weight 58.4) weighing 200 grams contains how much NaCl?
(a) 1.00 moles   (b) 1.00 grams   (c) 0.584 moles   (d) 0.584 grams   (e)   2.00 grams

_____ 9. What is the molarity of a solution containing 3.50 grams of NaCl (formula weight 58.4) in 500. mL of water?
(a) 0.0598 M (b) 0.120 M (c) 0.0769 M (d) 0.0154 M (e) 0.600 M

_____ 10. We add 0.010 moles of $BaCl_2$ to enough water to make a total of 100 mL of solution. What is the molarity of the solution?
(a) 0.010 M (b) 0.10 M (c) 99.99 M (d) 0.0010 M
(e) Cannot be determined from the information given.

_____ 11. We wish to dilute some 18.4 M H_2SO_4 solution to make 600. mL of 0.100 M H_2SO_4 solution. How much of the 18.4 M solution should we start with?
(a) 1.84 mL   (b) 3.07 mL   (c) 3.26 mL   (d) 3.68 mL   (e) 4.00 mL

_____ 12. What mass of a solution that is 10.0% by mass H_2SO_4 will be required to react with 24.0 grams of NaOH according to the following equation?

$$H_2SO_4 + 2NaOH \rightarrow Na_2SO_4 + 2H_2O$$

(a) 29.4 g   (b) 294 g   (c) 586 g   (d) 58.6 g   (e) 1980 g

_____ 13. What volume of 3.00 molar sulfuric acid, H_2SO_4, is required to react with 250 grams of calcium carbonate, $CaCO_3$?

$$CaCO_3 + H_2SO_4 \rightarrow CaSO_4 + CO_2 + H_2O$$

(a) 208 mL   (b) 462 mL   (c) 767 mL   (d) 833 mL   (e) 946 mL

_____ 14. If 20.0 mL of 0.010 M H_3PO_4 solution is just neutralized by 60.0 mL of a $Ca(OH)_2$ solution, what is the molarity of the $Ca(OH)_2$ solution?

$$Ca(OH)_2 + H_3PO_4 \rightarrow Ca_3(PO_4)_2 + H_2O$$

(a) 0.010 M (b) 0.200 M (c) 0.030 M (d) 0.040 M (e) 0.0050 M

Answers to Some Important Terms in This Chapter

1. chemical equation
2. balanced
3. reactants
4. products
5. reaction stoichiometry
6. limiting reactant

7. theoretical yield
8. actual yield
9. percent yield
10. sequential reactions
11. solution

12. solvent
13. solute
14. percent by mass
15. molarity
16. dilution

Answers to Preliminary Test

Do not just look up answers. *Think about the reasons for the answers.*

True-False

1. True. "A balanced equation is a happy equation."
2. False. The total number of atoms of each element among the reactants must equal the total number of atoms of the same elements among the products.
3. True. This is one way to state the Law of Conservation of Matter.
4. True. This also follows from the Law of Conservation of Matter.
5. True. In fact, the equation could be read as, "One mole of A reacts with two moles of B to produce three moles of C and 1 mole of D."
6. False. The identification of the limiting reactant is determined by both the number of moles of each reactant that combine and the physical amounts of each reactant that are used.
7. False. Count the number of oxygen atoms on the two sides of this equation. Count the number of hydrogen atoms. You should be able to balance this equation.
8. False. We must not change the formula of any substance in the equation; this would mean changing the substances described. O_2 is ordinary oxygen, which we all breathe. O_3 is ozone, a very reactive, poisonous gas; it is not produced in the reaction of $KClO_3$.

   The balanced equation is: $2KClO_3 \rightarrow 2KCl + 3O_2$

9. True. The coefficients for both Ca and H_2 are assumed to be ones.
10. False. 36 g of water would be required. ($2 \text{ mol} \times \dfrac{18.0\,\text{g}}{1\,\text{mol}}$)
11. False. Remember that the reactants react only in the mole ratio indicated by the balanced equation, and therefore in a weight ratio that we can calculate. Study the limiting reactant type of problems in Section 3-3.
12. False. Remember that 20.0% is understood to mean "by mass." We add enough of the compound to the solvent that 20.0%, or 1/5, of the mass of any amount of solution would be due to the compound.
13. True. This method of expressing solution concentration deals with the number of moles. But we cannot directly measure how many moles of solute we are adding. Knowing the formula weight of the solute, we can convert number of moles needed to number of grams.
14. False. 1.00 *M* means that the solution contains 1.00 mole of solute (NaCl) per liter of *solution*, not per liter of solvent. Study the description of Figure 3-2. (The amount of water needed will be close to 1.00 liter, but not exactly equal.)

Short Answer

1. Conservation of Matter
2. 1. The balanced equation is: $Ca(OH)_2 + 2HNO_3 \rightarrow Ca(NO_3)_2 + 2H_2O$.
3. 5. The balanced equation is: $C_3H_8 + 5O_2 \rightarrow 3CO_2 + 4H_2O$.

4. 3. The balanced equation is: $2Al + 3CuSO_4 \rightarrow Al_2(SO_4)_3 + 3Cu$. It may be convenient to treat SO_4 (actually the sulfate ion, SO_4^{2-}) as a single group in balancing this equation.

5. 6. The balanced equation is: $CS_2 + 3Cl_2 \rightarrow CCl_4 + S_2Cl_2$

6. 27. The balanced equation is: $C_7H_{16} + 11O_2 \rightarrow 7CO_2 + 8H_2O$

7. 11. The balanced equation is: $4Ag + 2H_2S + O_2 \rightarrow 2Ag_2S + 2H_2O$

8. The balanced equation is: $3Ca(OH)_2 + 2H_3PO_4 \rightarrow Ca_3(PO_4)_2 + 6H_2O$

$$\frac{1 \text{ mole } Ca_3(PO_4)_2}{3 \text{ mole } Ca(OH)_2}$$

9. $$\frac{310.3 \text{ g } Ca_3(PO_4)_2}{3 \times 74.1 \text{ g } Ca(OH)_2}$$

10. limiting reactant. This concept is discussed in Section 3-3.

11. CS_2. Each mole of CS_2 requires 3 moles of O_2, so the 2.0 moles of CS_2 would require 6.0 moles of O_2. Extra O_2 is supplied, so CS_2 is the limiting reactant. Study Section 3-3.

12. O_2; 1.5 (7.5 mol at start – 6.0 mol used)

13. 2.0 (1 mol $CS_2 \rightarrow$ 1 mol CO_2, so 2.0 mol $CS_2 \rightarrow$ 2.0 mol CO_2)

14. 4.0 (1 mol $CS_2 \rightarrow$ 2 mol SO_2, so 2.0 mol $CS_2 \rightarrow$ 4.0 mol SO_2)

15. 27.4. $Ca(OH)_2$ is the limiting reactant. (Can you show this?)

$? \text{ g } Ca_3(PO_4)_2 = 19.6 \text{ g } Ca(OH)_2$

$$\times \frac{1 \text{ mol } Ca(OH)_2}{74.1 \text{ g } Ca(OH)_2} \times \frac{1 \text{ mol } Ca_3(PO_4)_2}{3 \text{ mol } Ca(OH)_2} \times \frac{310.3 \text{ g } Ca_3(PO_4)_2}{1 \text{ mol } Ca_3(PO_4)_2}$$

$$= 27.4 \text{ g } Ca_3(PO_4)_2$$

16. theoretical yield

17. actual yield

18. $$\frac{\text{actual yield of product}}{\text{theoretical yield of product}} \times 100\%.$$

Study Section 3-4, and be sure you know how to carry out related calculations.

19. 100%. This would mean that the reaction went to completion as written, stopping only when one of the reactants was completely used up, that there were no side-reactions to consume reactants, and that we were able to recover *all* of the product. You can imagine that such laboratory procedures are few.

20. 206 g.

$$\text{theoretical yield } PI_3 = 250. \text{ g } I_2 \times \frac{1 \text{ mol } I_2}{254 \text{ g } I_2} \times \frac{4 \text{ mol } PI_3 \text{ formed}}{6 \text{ mol } I_2 \text{ that react}} \times \frac{412 \text{ g } PI_3 \text{ formed}}{1 \text{ mol } PI_3}$$

$$= 270. \text{ g } PI_3 \text{ formed}$$

$$\% \text{ yield} = \frac{\text{actual yield}}{\text{theoretical yield}} \times 100\%$$

$$\text{actual yield} = \frac{\% \text{ yield} \times \text{theoretical yield}}{100\%} = \frac{76.3\% \times 270. \text{ g}}{100\%} = 206 \text{ g}$$

21. solvent.
22. 2000 mL. Calculate this as shown in Section 3-7, Example 3-21. It is not necessary to know the identity of the solvent to work this problem.

$$V_2 = \frac{V_1 M_1}{M_2} = \frac{500 \cdot \text{mL} \times 12.0\,M}{3.00\,M} = 2000 \text{ mL } (2.00 \times 10^{23})$$

23. 0.0496 M. You do not need to know what the solute is. See Section 3-6.

$$\frac{0.02570\,\text{mol}}{0.518\,\text{L}} = 0.0496\,M$$

24. density. See Example 3-20, which is a similar problem stated in a different way. Be sure you can work such problems. Work Exercises 70 and 71 at the end of the text chapter.

Multiple Choice

1. (d). Equation (a) shows oxygen as atoms instead of as diatomic molecules, and is not balanced. Equation (b) is not balanced. Equation (c) is balanced, but shows carbon monoxide (CO) rather than carbon dioxide as a product. The reaction in equation (c) could occur if there were a shortage of oxygen in the combustion process. Reaction (c) requires less O_2 per mole of C_2H_4 than does the "complete" combustion reaction to CO_2 given in (d). See Example 3-7 and the marginal note that starts with "Sometimes it is not possible..." on page 82 for a brief discussion of burning in the presence of excess O_2.

2. (a). 100. g $CaCO_3$ → 56 g CaO + 44 g CO_2

3. (b). Find this out by the method shown in Section 3-3.

$$? \text{ mol H}_2 = 5.0 \text{ g H}_2 \times \frac{1\,\text{mol H}_2}{2.0\,\text{g H}_2} = 2.5 \text{ mol H}_2$$

$$? \text{ mol O}_2 = 10.0 \text{ g O}_2 \times \frac{1\,\text{mol O}_2}{32.0\,\text{g O}_2} = 0.313 \text{ mol O}_2$$

To react completely 5.0 g H_2 requires

$$? \text{ mol O}_2 = 2.5 \text{ mol H}_2 \times \frac{1\,\text{mol O}_2}{2\,\text{mol H}_2} = 1.25 \text{ mol O}_2, \text{ less than the 0.313 mol O}_2 \text{ available}$$

4. (d). Sulfur is the limiting reactant, so it would be consumed first. Consuming 8.0 grams of sulfur by this reaction requires 12.0 grams of O_2, so there would be 2.0 grams O_2 left over.

$$? \text{ g O}_2 = 8.0 \text{ g S} \times \frac{1\,\text{mol S}}{32\,\text{g S}} \times \frac{3\,\text{mol O}_2}{2\,\text{mol S}} \times \frac{32\,\text{g O}_2}{1\,\text{mol O}_2} = 12.0 \text{ g O}_2; \; 14.0 \text{ g} - 12.0 \text{ g} = 2.0 \text{ g}$$

Study Section 3-3.

5. (b). Using the unit factor approach, you can show that 1200 moles of CO must go into the second reaction to produce 800 moles of Fe. For the first reaction to produce those 1200 moles of CO, 600 moles of O_2 would be required.

$$? \text{ mol O}_2 = 800 \text{ mol Fe} \times \frac{3\,\text{mol CO}}{2\,\text{mol Fe}} \times \frac{1\,\text{mol O}_2}{2\,\text{mol CO}} = 600 \text{ mol O}_2$$

6. (a). $\underline{?}\ g\ O_2 = 3.2\ g\ CH_4 \times \dfrac{1\ mol\ CH_4}{16.0\ g\ CH_4} \times \dfrac{2\ mol\ O_2}{1\ mol\ CH_4} \times \dfrac{32.0\ g\ O_2}{1\ mol\ O_2} = 12.8\ g\ O_2$

   $3.2\ g\ CH_4 + 12.8\ g\ O_2 = 16\ g$ of reactants $= 16\ g$ of products (Conservation of Matter)

7. (c). $\underline{?}\ g\ S = 6.00\ g\ SO_2 \times \dfrac{1\ mol\ SO_2}{64.1\ g\ SO_2} \times \dfrac{3\ mol\ S}{1\ mol\ SO_2} \times \dfrac{32.1\ g\ S}{1\ mol\ S} = 9.01\ g\ S$ (theoretical yield)

   % yield $= \dfrac{7.54\ g\ S}{9.01\ g\ S} \times 100 = 83.7\%$

8. (e). $\underline{?}\ g\ NaCl = 200.\ g\ soln \times \dfrac{1.00\ g\ NaCl}{100.\ g\ soln} = 2.00\ g\ NaCl$

9. (b). $\dfrac{\underline{?}\ mol\ HCl}{L\ soln} = \dfrac{3.50\ g\ NaCl}{0.500\ L\ soln} \times \dfrac{1\ mol\ NaCl}{58.4\ g\ NaCl} = 0.120\ mol\ NaCl/L\ soln = 0.120\ M$

   (Assuming that the volume of the solution = the volume of the water ±0.5 mL.)

10. (b). $\dfrac{\underline{?}\ mol\ BaCl_2}{L\ soln} = \dfrac{0.010\ mol\ BaCl_2}{0.100\ L\ soln} = 0.10\ mol\ BaCl_2/L\ soln = 0.10\ M$

11. (c). $V_1 = \dfrac{V_2 M_2}{M_1} = \dfrac{600.\ mL \times 0.100\ M}{18.4\ M} = 3.26\ mL$

12. (b).

   $\underline{?}\ g\ soln = 24.0\ g\ NaOH \times \dfrac{1\ mol\ NaOH}{40.0\ g\ NaOH} \times \dfrac{1\ mol\ H_2SO_4}{2\ mol\ NaOH} \times \dfrac{98.1\ g\ H_2SO_4}{1\ mol\ H_2SO_4} \times \dfrac{100\ g\ soln}{10.0\ g\ H_2SO_4}$

   $= 294\ g\ soln$

13. (d).

   $\underline{?}\ mL\ soln = 250.\ g\ CaCO_3 \times \dfrac{1\ mol\ CaCO_3}{100.1\ g\ CaCO_3} \times \dfrac{1\ mol\ H_2SO_4}{1\ mol\ CaCO_3} \times \dfrac{1\ L\ H_2SO_4}{3.00\ mol\ H_2SO_4}$

   $= 0.833\ L = 833\ mL$

14. (e).

   $\underline{?}\ M\ soln = 0.0200\ L\ H_3PO_4 \times \dfrac{0.010\ mol\ H_3PO_4}{1\ L\ H_3PO_4} \times \dfrac{3\ mol\ Ca(OH)_2}{2\ mol\ H_3PO_4} \times \dfrac{1}{0.0600\ L\ Ca(OH)_2}$

   $= 0.0050\ M$

4 The Structure of Atoms

Chapter Summary

In this chapter, we learn about the structure of atoms by studying questions such as the following. What are the particles that compose atoms? What is the arrangement of these particles in atoms? Are all atoms of a given element identical? If not, how do they differ and how are they alike? What evidence do we have for these descriptions? For a given element, how do we describe how many electrons the atom has, how these electrons are arranged within the atom, and how strongly they are held? (This last question is an especially important one, because electrons are the particles in atoms that are involved in chemical reactions.)

First, let us see how the ideas presented in this chapter were developed. It is not our intent to learn the history of chemistry for its own sake. But an appreciation of how the ideas about atoms and their architecture have been developed can help us to develop our own understanding in a gradual and logical way. If the text or your instructor were to begin by telling you all the information currently known or believed about atoms and subatomic particles, you would soon be overwhelmed, and would learn little. A presentation along historical lines can help us appreciate that chemistry is a living, developing science, with new evidence revealed and old evidence continually reassessed or reinterpreted. This presentation helps us to see that we describe the atom as we do, not because many of these properties of atoms or their particles can be observed directly, but because these descriptions are the best available explanations of our observations.

In modern chemistry an early significant concept about atoms was Dalton's theory, discussed in Section 1-2. Some features of Dalton's Atomic Theory are still useful to us in chemistry; other features have required modification in light of more recent observations. As you study this chapter, watch for this pattern of development of ideas.

Section 4-1 introduces us to the three fundamental particles that we will work with in our understanding of chemistry—the basic particle of positive charge (the **proton**), the fundamental particle of negative charge (the **electron**), and the neutral particle (the **neutron**). Much of our study of chemistry will be concerned with understanding the organization of these particles in atoms and the effect of this order on the physical and chemical properties of atoms. We first encounter the properties of the two charged subatomic particles (Sections 4-2 and 4-3) through a discussion of the experiments that led to the discovery and study of the electron and the proton. Section 4-4 discusses the classic experiment that led to the idea of the nuclear atom, the Rutherford "gold foil" experiment. In this description of the atom, the protons and neutrons (Section 4-6) are in the **nucleus**, an extremely small volume in the center of the atom. Thus, the nucleus of the atom contains all of the positive charge and nearly all of the mass of the atom. The electrons account for all of the negative charge but for only a tiny fraction of the mass of the atom; they are distributed throughout the otherwise empty space comprising the rest of the atom. The **atomic number** (Section 4-5) tells the number of protons in the nucleus of an atom of the element. It is important to remember that the atomic number characterizes each element uniquely—no two elements have the same atomic number, and two atoms with different numbers of protons in their nuclei must be different elements.

In Section 4-7, we learn how to describe different "kinds" of atoms according to the numbers of protons, neutrons, and electrons they contain. All atoms except the most common form of hydrogen contain both protons and neutrons in their nuclei. The neutrons contribute to the total mass of the atom but do not alter its electrical properties, so they do not affect chemical properties. The existence of different forms of a given element that have different numbers of neutrons in the nucleus thus gives rise to atoms that are chemically identical but have different masses. These are called **isotopes**. The total number of particles in the nucleus of a particular atom, the **mass number**, tells the total number of protons and neutrons in the nucleus. The **mass spectrometer**, which aided in determining this aspect of atomic structure, is helpful in understanding these concepts (Section 4-8). Most elements occur naturally as mixtures of isotopes, which behave the same chemically. Any chemical determination of the "atomic weight" can therefore measure only the **average** atomic weight of all isotopes present, weighted for the relative amounts of the isotopes. This idea, discussed in Section 4-9, is the basis for our present atomic weight scale. On this scale, the mass of one atom of the most abundant isotope of carbon is assigned a mass of exactly 12 **atomic mass units (amu or u)**.

We organize properties and reactions with the aid of the classification scheme known as the **periodic table**. We can use this table: (1) to organize our understanding of many properties and (2) to predict properties or reactions of elements with which we may not be familiar or for which the property or reaction may not have been studied. The term "periodic" arises because elements with similar properties recur somewhat regularly as we go through the elements by increasing atomic number. As you study this chapter, you should become increasingly aware of the regularities that many properties of the elements exhibit and with the similarities within groups of elements. Later (Sections 4-18 and 4-19, Chapters 5 and 6) we shall relate the periodic table to the electronic structures of atoms.

The development of this simple yet powerful correlation, described in Section 4-10, is one of the major intellectual achievements in the history of science. This section also explains the terminology we use to discuss the periodic table. Each horizontal row is referred to as a **period** and each vertical column is called a **group**. Here you are also introduced to the common names of some of the groups of elements—the **alkali metals**, **alkaline earth metals**, **halogens**, and **noble gases**.

The elements are often broadly classified as **metals**, **nonmetals**, or **metalloids**. The general physical and chemical properties that distinguish among these classes of elements are also discussed in Section 4-10. Metalloids have properties that are intermediate between those of the other two classes.

The most important aspect of atomic structure as it relates to chemistry is the arrangement of the electrons around the nucleus. The remainder of Chapter 4 is devoted to this topic. Most of our knowledge about electrons in atoms comes from a study of the ways in which matter interacts with light. Section 4-11 deals with the nature of light, which can sometimes be considered as a **wave** phenomenon (periodically varying electrical and magnetic fields—hence the name **electromagnetic radiation**). We learn that different colors of light correspond to different wavelengths or different frequencies of alternation of these fields. The other, equally important interpretation of light, applicable to other kinds of experiments, is in terms of a **particle** model. In this view, light consists of streams of particles called **photons**. For a given color of light, each photon consists of a particular amount of energy, called a **quantum**. Einstein's explanation of the **photoelectric effect** (Section 4-12) was an important early application of this view of light.

When excited atoms give off light (atomic emission spectra, Section 4-13), they give off only particular colors of light, i.e., only certain energies. This important observation shows that the atoms can undergo only certain specific energy changes. The **Bohr theory** of the atom introduces two ideas that are of central importance in our understanding of atoms: (1) an electron in an atom can only be at certain specific distances from the nucleus and (2) when it is at one of these allowed distances, it possesses a definite amount of energy. We shall modify and extend these ideas later. The simple Bohr theory of the atom assumed that the electrons travel in well-defined orbits; each orbit is characterized by a positive integer, n, known as a **quantum number**. Actually, the Bohr theory works well only for simple atoms like hydrogen. The Enrichment feature on pages 141-143 emphasizes how the spectral wavelengths depend on *differences* between energies within the atom.

Very small particles such as the electron can also be described in terms of both wave and particle nature, Section 4-14. The modern **quantum-mechanical picture of the atom**, which we shall use throughout our study of chemistry, is introduced in Section 4-15. This theory extends the ideas of quantization of electronic energy and describes them in a highly mathematical approach, called **quantum theory**. In this theory, each electron in the atom is treated mathematically as though it were a wave. The electron is described by a wave equation called the Schrödinger equation, later made more general by Dirac. Even though we do not need to be concerned with the mathematical details, we must describe the solution to these wave equations—the wave function. We will use the solutions to these equations frequently in our study of chemistry.

In your study of the important Sections 4-15 through 4-18, keep in mind that this more refined and elegant theory, even though it looks quite complex, is really just concerned with two features of an electron in the atom: *where the electron is most likely to be* and *what its energy is*. Each electron in the atom corresponds to one of the solutions of the wave equation. Each solution is described by a particular set of four numbers, n, ℓ, m_ℓ and m_s known as the quantum numbers. You need to become familiar with these numbers, their interpretations, and the rules for assigning them (Section 4-16) in order to determine the electron configuration.

Each valid set of n, ℓ, and m_ℓ defines one **atomic orbital**, which may be interpreted as a region of space in which there is a high probability of finding the electron. In Section 4-17 we see that each atomic orbital is identified by a valid set of the quantum numbers n (distance from nucleus), ℓ (orbital shape), and m_ℓ (orbital orientation). In a given atom, orbitals that have the same values of n and ℓ but that differ in their m_ℓ values have the same energy; such orbitals are described as **equivalent** or **degenerate** orbitals.

You must learn to use rules based on the values of the four quantum numbers to tell the arrangement of the electrons in any specified atom, the **electron configuration** (Section 4-18). This is the *most important topic you have yet encountered in chemistry*. It serves as a basis for further understanding of chemical properties and bonding. The application of the rules and procedures that you learn here will later enable you to understand and predict properties of elements, chemical bonding, formulas, shapes of molecules, reactivities, and properties of compounds. Practice writing electron configurations of atoms, in several different notations.

We can figure out the electron configuration of an element by imagining that we add the electrons one by one, putting each into the available orbital of lowest energy. This imaginary process is called the **Aufbau procedure**, from the German word meaning "to build up." It is important to know the usual order in which the orbitals fill: $1s$, $2s$, $2p$, $3s$, $3p$, $4s$, $3d$, $4p$, $5s$, $4d$, $5p$, $6s$, $4f$, $5d$, $6p$, $7s$, $5f$, $6d$, $7p$, ... A memory aid to help you reconstruct this order is given in Figure 4-31. But remember that this order is only a guide; the electron configurations of many atoms are not entirely predicted by this order. Whatever the order of the filling of the orbitals, two additional rules that affect this order and the number of electrons in the orbitals are the **Pauli Exclusion Principle** and **Hund's Rule** (Section 4-18).

We see in Section 4-19 that the periodic table, which was first introduced in Section 4-10 based on physical and chemical similarities, corresponds to grouping elements with similar electron configurations. In later chapters, we will see that elements with similar outer electron configurations are very similar in their physical and chemical properties. The periodic table is therefore an invaluable aid in classifying elements, their electron configurations, and their properties.

Obviously, the electron configurations of the elements are determined from theoretical concepts. We cannot directly observe the electrons in the atom, even with a microscope. However, the conclusions are supported by experimental evidence. Either the atoms within a substance have one or more unpaired electrons or all of their electrons are paired. This aspect of the electron configuration determines whether the substance is **paramagnetic** or **diamagnetic**, respectively (Section 4-20).

Study Goals

As in other chapters, some text Exercises related to the Study Goals are indicated.

Sections 4-2, 4-4, 4-5, 4-11, 4-13 through 4-15
1. Describe the experiments or ideas of the following scientists and their contributions to our knowledge of the structure of atoms: (a) Thomson, (b) Millikan, (c) Rutherford, (d) Moseley, (e) Rydberg, (f) Bohr, (g) Planck, (h) de Broglie, (i) Davisson and Germer, (j) Schrödinger, and (k) Dirac.

Sections 4-1 through 4-3 and 4-6
2. Describe the three fundamental particles that make up atoms. Describe the experimental evidence that led to the discovery of each of these three fundamental particles. *Work Exercises 1 through 5.*

Sections 4-2, 4-3, and 4-8
3. Describe the behavior of charged and uncharged particles and rays in electrical and magnetic fields. *Work Exercises 2 through 6, 12, and 13.*

Section 4-4
4. Describe the nuclear atom. *Work Exercises 7 through 11.*

Section 4-7
5. Distinguish between isotopes and atoms. *Work Exercises 15 and 16.*
6. From its nuclide symbol, tell the number and type of fundamental particles present in any atom. *Work Exercises 17 through 20 and 22 through 25.*

Sections 4-5 through 4-8
7. Distinguish among atomic number, mass number, atomic weight, and isotopic weight. Carry out calculations based on these quantities. *Work Exercises 14 through 38.*

Sections 4-7 and 4-8

8. Describe and illustrate the basic ideas of mass spectrometry. *Work Exercises 12, 13, 35, and 36.*

Section 4-10

9. Describe and understand what is meant by periodicity. Know and use the terminology of the periodic table. *Work Exercises 39 through 42 and 49.*

10. Know the contrasting physical and chemical properties of metals and nonmetals. Know where the metals, nonmetals, and metalloids are located in the periodic table. *Work Exercise 52.*

11. Know which elements are examples of each of the following: (a) alkali metals, (b) alkaline earth metals, (c) halogens, and (d) noble gases. Relate each of these classes to position in the periodic table. *Work Exercises 49 through 52.*

12. Use the position of an element in the periodic table to predict its properties and the formulas of compounds it forms. *Work Exercises 43 through 48.*

Section 4-11

13. Be able to carry out calculations to relate the descriptions of light to its wavelength, frequency, and energy. *Work Exercises 53 through 60, 63, and 64.*

Sections 4-11 and 4-12

14. Understand the wave and particle descriptions of light and explain why both descriptions are useful. *Work Exercises 61 and 62.*

Section 4-13.

15. Understand the relationship between the observation of atomic line spectra and the idea of distinct energy levels in atoms. Describe how the wavelength of light emitted or absorbed by an atom depends on the separations between its electronic energy levels. *Work Exercises 65 through 74.*

Section 4-14

16. Understand the basis for the description of particles, such as the electron, either as a wave or as a particle. *Work Exercises 75 through 78.*

Section 4-15

17. State and understand the basic ideas of quantum mechanics. *Work Exercise 79.*

Sections 4-15 and 4-16

18. Learn the four quantum numbers: what they represent and what values they can have. *Work Exercises 80 and 81.*

Sections 4-16 and 4-17

19. Understand the relationship between atomic orbitals and the corresponding values of the quantum numbers. Be familiar with the notation for identifying orbitals. *Work Exercises 82 through 93.*

Section 4-17

20. Describe atomic orbitals. Sketch the shapes of *s*, *p*, and *d* orbitals. Describe how the same kinds of orbitals in different energy levels differ. *Work Exercises 81, 82, and 94.*

21. Learn how the energies of orbitals depend on the quantum numbers.

Section 4-18

22. Learn and apply the rules for the filling of electrons in atomic orbitals. Write out the electron configurations of specified atoms. Write out the electron configurations of (especially) the first 36 elements, both with orbital notation ($\underline{\uparrow\downarrow}$) and with simplified

notation (using *s*, *p*, *d*, and *f*). *Work Exercises 95 through 109, 111 through 116, and 118 through 123.*

Section 4-19

23. Relate the electron configuration of an atom to its position in the periodic table. Use Table 4-9 to write electron configurations of elements. *Work Exercises 117, 124, and 125.*

Section 4-20

24. Understand how the magnetic properties of an element depend on the number of unpaired electrons in an atom of the element. *Work Exercises 110 and 126 through 129.*

General

25. Answer conceptual questions based on this chapter. *Work* **Conceptual Exercises** *130 through 141.*

26. Apply concepts and skills from earlier chapters to the ideas of this chapter. *Work* **Building Your Knowledge** *exercises 142 through 150.*

27. Exercises at the end of chapter direct you to sources outside the textbook for information to use in solving them. Work **Beyond the Textbook** *exercises 151 through 155.*

Some Important Terms in This Chapter

Some important new terms from Chapter 4 are listed below. (Be sure to study other new terms and review terms from preceding chapters, if necessary.) Fill the blanks in the following paragraphs with terms from the list. Each term is only used once.

alkali metals	halogens	paired
alkaline earth metals	Hund's Rule	Pauli Exclusion Principle
angular momentum quantum number		period
atomic mass unit	isotopes	periodic table
atomic number	magnetic quantum number	periodicity
atomic orbital	mass number	photons
Aufbau Principle	metals	principal quantum number
d orbital	metalloids	proton
electromagnetic radiation	neutron	quantum
electrons	noble gases	quantum numbers
electron configuration	nonmetals	*s* orbital
f orbital	nucleus	spin quantum number
frequency	nuclide symbol	wavelength
group	*p* orbital	

Atoms are believed to consist of an extremely small, extremely dense center, called the [1]_____, surrounded by a cloud of [2]_____. Most of the mass of the atom is concentrated in the nucleus, but the size of the atom is determined by the size of the electron cloud. Two kinds of subatomic particles are found in the nucleus: the [3]_____, which has a positive charge, and the [4]_____, which is neutral. The identity of the atom—what element it is—is determined by the number of protons in the nucleus. This number is the

[5]_____ of the element. Atoms of the same element can have different numbers of neutrons. These atoms are the same element because they have the same number of protons, but they have different masses. These atoms are [6]_____ of each other. The total number of protons and neutrons is called the [7]_____ because only the protons and neutrons significantly affect the mass of the atom. Isotopes have the same atomic number but different mass numbers. Different isotopes of the same element have the same chemical symbol, of course, so they are distinguished in print by their [8]_____, which adds the mass number of the isotope and the atomic number of the element as a superscript and a subscript, respectively, in front of the chemical symbol. This symbol can be used to determine the number of nucleons—protons and neutrons—in the nucleus. The mass of an individual atom is very small, so the unit used to measure it is the [9]_____, which is defined as $1/12$th of the mass of an atom of carbon-12, the isotope of carbon that has 6 protons and 6 neutrons. This unit is abbreviated amu or simply u.

Using and understanding chemistry efficiently requires an awareness of patterns among the physical and chemical properties of substances. One of these patterns is a regular periodic repetition of the properties of the elements when listed by increasing atomic number (originally, atomic weight). This repetition, this [10]_____, allowed Mendeleev and Meyer to design the first [11]_____, in which they grouped together elements with similar properties. Some of these groups are very distinctive; the elements in these "families" are very similar or their properties change in a very regular, predictable way. The first column on the periodic table consists of metals that are very reactive. These [12]_____ have to be stored under oil because they react so quickly with oxygen in the air. Next to them on the periodic table are the [13]_____, which are not quite as reactive as the alkali metals. These two family names are often confused because *alkaline* means *alkali*. Elements from these two columns form the strong inorganic bases. In the column on the far right-hand side of the periodic table are the most unreactive elements. They are called [14]_____ because they generally do not react to form compounds, as if they were aloof nobility, and they are all gases. Next to them is the family of the most reactive nonmetals, the [15]_____, which are used to make halogen lamps. In an even broader grouping, all of the elements can be divided into three categories: (a) [16]_____, which readily conduct electricity and heat, are shiny, and can be bent and hammered without breaking; (b) [17]_____, which do not easily conduct electricity and heat, are not shiny, and are brittle; and (c) [18]_____, which have properties in-between metals and nonmetals.

Scientists have been able to explore the structure of the atom by examining how matter interacts with light and other forms of [19]_____. Light can be thought of as traveling in waves. These waves can be characterized by the distance between any two adjacent peaks or troughs of the wave, called the [20]_____, represented by the Greek letter λ (lambda), or by the number of waves that pass a fixed point per second, called the [21]_____, represented by the Greek letter υ (nu). Curiously, light can also be thought of as consisting of particles, particles of light, called [22]_____ (which we all know about from science fiction torpedoes). The photons of light of each type (or color) have a certain specific amount of energy, called a [23]_____.

The type of light (color) absorbed or emitted by an atom is determined by the energy of the electrons in the atom. The energy of an electron can be characterized by four numbers. Because these numbers determine the energy of the photon of light absorbed or emitted they are called [24]_____. The first one, the [25]_____ _____, identifies the main energy level of the electron, and, therefore, the volume of space in which it moves. The second one describes the shape of the space in which the electron travels, so it's called the [26]_____ _____. It's related to the type of sublevel for the electron. The third number indicates the specific orbital of the electron within the sublevel, and is referred to as the [27]_____. The last number is called the [28]_____ because the electron is thought of as spinning, clockwise or counterclockwise.

The space in which an electron or a pair of electrons "orbit" the atom is known as an [29]_____. There are several types of orbitals, each having a different characteristic shape. An [30]_____ is spherical. A [31]_____ is said to be "dumbbell" shaped because it consists of two "lobes," rounded regions of space on either side of the nucleus. Generally a [32]_____ consists of four lobes in a clover-leaf shape (except for the d_{z^2}). An [33]_____ is even more complicated, consisting of either eight lobes or two lobes with two donut-shaped rings. If a pair of electrons are "in" any of these orbitals simultaneously, they must have opposite "spins." They are said to be [34]_____.

The physical and chemical properties of an element are determined by the energies of all the electrons in its atoms, what orbitals they are "in." A description of the arrangement of the electrons is called the [35]_____ of the atom. To determine this configuration, we are guided by a principle that suggests how it is "built up," the [36]_____; another that suggests that no two electrons in the same atoms can have the same set of four quantum numbers, the [37]_____ _____; and a third rule describing the distribution of electrons among multiple orbitals of the same sublevel, [38]_____.

A comparison of the electron configurations of the elements with their locations on the periodic table reveals many relationships. One of these is that all the elements that have the same highest occupied energy level (1, 2, 3, 4, 5, 6, or 7) are in the same [39]_____ on the periodic table. Also, elements that have the same number of electrons in the same type of sublevel (ns^1, ns^2, np^1, np^2, np^3, np^4, np^5, np^6, etc.) for the last sublevel listed in the Aufbau order are in the same [40]_____ on the periodic table.

Preliminary Test

As in other chapters, this test will check your understanding of basic concepts and types of calculations. Be sure to practice *many* of the additional textbook Exercises, including those indicated with the Study Goals.

True-False

Mark each statement as true (T) or false (F).

_____ 1. Most of the mass of an atom is due to the presence of electrons.

_____ 2. The figure "56" in the symbol $^{56}_{26}$Fe represents the total number of electrons and protons in the nucleus of an iron atom.

_____ 3. If we know that the atomic number of an element is 81, then the element must be thallium (Tl).

_____ 4. If we know that an element is thallium, then its atomic number must be 81.

_____ 5. If we know that an atom has mass number 101, then the element must be ruthenium (Ru).

_____ 6. In an atom, the protons and neutrons are in the nucleus, and the electrons are clustered together at some distance from the nucleus.

_____ 7. The characterization of cathode rays showed them to consist of streams of protons.

_____ 8. All cathode ray particles have the same charge/mass (e/m) ratio.

_____ 9. All canal ray particles have the same charge/mass (e/m) ratio.

_____ 10. In the Rutherford experiment, the scattering of streams of alpha particles by the atoms in a metal foil was studied.

_____ 11. The feature that distinguishes the atoms of different elements from each other is their mass number.

_____ 12. All naturally occurring elements exist as mixtures of isotopes.

_____ 13. Atomic numbers are always integers.

_____ 14. Mass numbers are always integers.

_____ 15. Masses of individual atoms are always integers.

_____ 16. Atomic weights are always integers.

_____ 17. The experimentally observed atomic weight of carbon is listed as 12.011, but this is due to a slight error in the definition of the atomic weight scale, because it should be exactly 12.

_____ 18. There are more metallic elements than nonmetallic ones.

_____ 19. Any element is either a metal or a nonmetal.

_____ 20. The properties of all elements in a group on the periodic table are identical.

_____ 21. Calcium is more metallic than iron.

_____ 22. Many colors of light are not visible to the human eye.

_____ 23. The longer the wavelength of light, the higher is its frequency.

_____ 24. The higher the frequency of light, the higher is the energy of one photon of the light.

_____ 25. "A particle is a particle and a wave is a wave, and never the twain shall meet." (Apologies to Rudyard Kipling)

_____ 26. The number of lines in the emission spectrum of an element is equal to the number of electronic energy levels in the atom.

_____ 27. Each combination of n, ℓ, and m_ℓ corresponds to an orbital.

_____ 28. When two electrons occupy the same orbital, they must have opposite spins.

_____ 29. In a given atom, no more than two electrons can occupy a single orbital.

_____ 30. The second main energy level in an atom can have any number of electrons up to 10.

_____ 31. Electrons can exist only in atoms.

_____ 32. All neutral atoms of any given element are identical in all respects.

_____ 33. The higher an atomic orbital's value of n, the higher is its energy.

Short Answer

Answer with a word, a phrase, a formula, or a number with units as necessary.

1. Fill in the following chart to indicate the effect of positively and negatively charged electrical plates on protons, electrons, and neutrons. Use the notation **A** for "attracted," **R** for "repelled," or **U** for "unaffected."

	proton	electron	neutron
positive plate	(a) ___	(b) ___	(c) ___
negative plate	(d) ___	(e) ___	(f) ___

2. The charge on the nucleus of a carbon atom is _____.

3. The fundamental particle with positive charge is the _____.

4. The fundamental particle with no charge is the _____.

5. In an atom, the _____ and the _____ are in the nucleus.

6. The two fundamental particles that have nearly the same mass are the _____ and the _____.

7. The two fundamental particles that have equal but opposite charges are the _____ and the _____.

8. Think about the statement, "The nucleus is a combination of protons and neutrons." Carefully consider the table of atomic numbers and observed atomic weight. The element that must have one isotope that is an exception to this statement is _____.

9. The symbol for an alpha particle is _____; this particle is now known to be identical to the ion that is designated by _____.

10. If a beam containing two kinds of particles having the same mass but different charges were sent through a magnetic field, the particles with the _____ charge would be deflected most.

11. If a beam containing two kinds of particles having the same charge but different masses were sent through a magnetic field, the particles with the _____ mass would be deflected most.

12. Atoms of the same element having different masses are called _____.

13. The symbol $^{23}_{11}$Na is called a _____ symbol.

14. In the symbol $^{23}_{11}$Na , the number 11 is called the _____ and represents the number of _____; it also represents the number of _____ for the neutral atom.

15. In the symbol $^{23}_{11}$Na , the number 23 is called the _____ and represents the _____.

16. The element indium (In) consists of two isotopes. These two isotopes are designated by the symbols $^{113}_{49}$In and $^{115}_{49}$In. The observed atomic weight of naturally occurring indium is 114.8. As a very good approximation, use 1 amu for the mass of each proton and neutron. The natural abundance of the two isotopes is _____ percent $^{113}_{49}$In and _____ percent $^{115}_{49}$In.

17. The numbers of subatomic particles in a neutral atom of palladium (Pd) with mass number 106 are ____ protons, ____ neutrons, and ____ electrons.

18. A statement of the periodic law is _____ _____ .

19. The two scientists credited with the development of the ideas of periodicity of elemental properties were _____ and _____.

20. The elements in Group 1A (1) are referred to as the _____.

21. Calcium and barium are members of Group ____, known as the _____ group.

22. The members of the halogen group are _____ .

23. The quantum number that designates the main energy level an electron occupies is ____, called the _____ quantum number.

24. The quantum number that designates the different possible sublevels within an energy level is ____, called the _____ quantum number.

25. The value of ____, called the _____ designates the shape of the orbital.

26. The value of ____, called the _____ quantum number, designates the spatial orientation of an orbital.

27. The quantum number that designates the "spin" of the electron is ____.

28. An orbital that is spherically symmetrical has $\ell =$ ____.

29. An orbital that is spherically symmetrical is designated as a(an) ____ orbital.

30. An orbital with $\ell = 1$ is designated as a(an) ____ orbital.

31. A p orbital has _____ (how many?) regions of high electron density.

32. In a single atom, the maximum number of electrons possible in a fourth main energy level is _____.

33. If two electrons in the same atom have the same values of n, ℓ, and m_ℓ, they must have _____ .

34. For any value of the principal quantum number (except $n = 1$) there are three p orbitals. These three p orbitals correspond to regions of high electron density in three different directions from the atomic center. The three p orbitals of the same energy level make angles of _____ with each other.

35. In the normal order of occupancy of electronic energy levels (Aufbau order), the sublevel occupied just after $3p$ is ____.

36. In the normal order of occupancy of electronic energy levels (Aufbau order), the sublevel occupied just after $4s$ is _____.

37. The ground state electron configuration of bromine (Br) is _____.

38. The ground state electron configuration of arsenic (As) is _____.

39. The number of unpaired electrons in a neutral atom of sulfur (S) in the ground state is _____.

40. The number of unpaired electrons in a neutral atom of iron (Fe) in the ground state is _____.

41. The possible values of the magnetic quantum number for a $3d$ orbital are _____.

42. Substances that are strongly attracted into magnetic fields are termed _____ substances. The atomic feature that is responsible for this behavior is the presence of _____ in atoms.

43. In the periodic table, a vertical column of elements is called a _____ or _____.

44. In the periodic table, a horizontal row of elements is called a _____.

45. For the elements in Group 3A (13), the configuration of the outermost electrons is _____.

46. For the elements in Group 6A (16), the number of unpaired electrons in each atom is _____.

47. In the periodic table, the nth period begins with the element whose outer electron configuration is _____ and ends with the element whose outer electron configuration is _____.

Multiple Choice

Select the one best answer for each question.

____ 1. Two isotopes of the same element must differ in their
   (a) atomic number.   (b) mass number.   (c) total charge.
   (d) number of protons.   (e) number of electrons.

____ 2. Given the unidentified isotope $^{61}_{27}X$, which one of the following statements is true?
   (a) The atom contains 61 neutrons.
   (b) The neutral atom contains 27 electrons.
   (c) The element must be promethium, Pm.
   (d) The atom contains 34 protons.
   (e) The atom contains a number of protons plus electrons equal to 61.

____ 3. Which one of the following statements about magnesium atoms is *not correct*?
   (a) All magnesium atoms have an atomic number of 12.
   (b) All magnesium atoms have the same number of protons.
   (c) All neutral magnesium atoms have 12 protons and 12 electrons.
   (d) All magnesium atoms have the same number of neutrons.
   (e) All atoms with an atomic number of 12 are magnesium atoms.

____ 4. If a neutral atom has an atomic number of 28 and a mass number of 60, it contains

(a) 28 neutrons.   (b) 32 electrons.   (c) 60 electrons.
(d) 32 protons.   (e) 60 nuclear particles (number of protons + number of neutrons).

_____ 5. Cathode rays are _____. These particles have the _____ charge/mass ratio of the three fundamental particles found in the atom.
(a) neutrons, largest   (b) protons, smallest   (c) electrons, largest   (d)   protons, largest
(e) None of the preceding answers is correct.

_____ 6. The discovery and characterization of cathode rays was important in the development of atomic theory because
(a) it indicated that all matter contains protons.
(b) it indicated that all matter contains alpha particles.
(c) it indicated that all matter contained electrons.
(d) it provided a basis for weight calculations in chemical reactions.
(e) it led to the suggestion of the existence of the neutron.

_____ 7. Which one of the following observations provided evidence that most of the mass of the atom, as well as all of the positive charge, is concentrated in a very small core, the nucleus?
(a) the analysis of X-ray wavelengths
(b) the scattering of alpha particles by a metal foil
(c) the deflection of ions in a mass spectrometer
(d) the results of the Millikan oil-drop experiment
(e) the existence of elements with noninteger values for atomic weights.

_____ 8. What is the number of electrons in 1 mole of carbon atoms?
(a) 6   (b) 12   (c) $6 \times 6.022 \times 10^{23}$
(d) $12 \times 6.022 \times 10^{23}$ (e) depends on the isotope of carbon

_____ 9. Suppose that a hypothetical element consists of a mixture of two isotopes. One isotope, having a mass of 44, is present in 70.0% atomic abundance, while the other isotope, having a mass of 46, accounts for the other 30.0%. We expect the experimentally determined atomic weight for this hypothetical element to be
(a) 44.0   (b) 46.0   (c) 45.0   (d) 44.6   (e) 45.4

_____ 10. In an atom
(a) the electrons contain most of the mass of the atom.
(b) the nucleus occupies most of the volume of the atom.
(c) all of the electrons and protons are in the nucleus.
(d) the neutrons are in the nucleus.
(e) the neutrons carry all of the negative charge.

_____ 11. When a beam of charged particles is sent through an electrical or magnetic field, the path of the particles becomes curved as the particles are deflected by the field. Which one of the following statements correctly relates the direction and the amount of this curvature to the properties of the particles?
(a) The direction of curvature depends on the charge; the amount of curvature depends on the e/m ratio.
(b) The direction of the curvature depends on the e/m ratio; the amount of curvature depends on the charge.

(c) The direction of curvature is independent of the charge; the amount of curvature depends on the e/m ratio.

(d) The direction of curvature depends on the e/m ratio; the amount of curvature is does not depend on the e/m ratio (the sign of the charge does not affect these properties).

(e) None of the above statements is correct.

_____ 12. The number of neutrons in one atom of the iron isotope $^{56}_{26}$Fe is _____.

(a) 56   (b) 26   (c) $56 - 26$ (d) $56 + 26$

(e) indeterminate unless we know the observed atomic weight of naturally occurring iron

_____ 13. The following are all proposals of Dalton's atomic theory. Which part of Dalton's atomic theory was shown to be *wrong* by the discovery of natural radioactivity?

(a) Atoms of a given element all weigh the same.

(b) Two or more kinds of atoms may combine in different ways to form more than one kind of chemical compound.

(c) Atoms combine in simple numerical ratios to form compounds.

(d) Atoms are permanent, unchanging, indivisible bodies.

(e) Compounds consist of collections of molecules made up of atoms bonded together.

_____ 14. Which of the following must represent an atom that is a *different element* from the others?

(a) an atom with 12 protons, 12 neutrons, and 12 electrons

(b) an atom with 12 protons, 12 neutrons, and 10 electrons

(c) an atom with 12 protons, 14 neutrons, and 12 electrons

(d) an atom with 10 protons, 12 neutrons, and 10 electrons

(e) All of the preceding are isotopes of the same element.

_____ 15. The positive rays that are sometimes found in cathode ray tubes are characterized as particles having a smaller charge-to-mass ratio for heavier elements. The positive ray particles that are formed when the gas in the tube is hydrogen have a larger observed charge-to-mass ratio than those for any other gas. Which one of the following assumptions explains these observations?

(a) The particles are all alpha particles.

(b) The particles are all identical, regardless of the gas in the tube.

(c) The particles are all ions formed by the removal of one or more electrons from neutral atoms of gases in the tube.

(d) Each gas consists of a mixture of at least two kinds of atoms.

(e) The positive ray particles are neutrons.

_____ 16. If a beam of each of the following were sent through a mass spectrometer, which one would follow the most curved path (that is, be deflected most)?

(a) neutral Ti atoms, mass number 47   (b) neutral Ti atoms, mass number 48

(c) Ti^+ ions, mass number 47   (d) Ti^{2+} ions, mass number 47

(e) Ti^{2+} ions, mass number 48

_____ 17. Which one of the following pairs of elements would be expected to be most similar chemically?

(a) Sn and Sr   (b) S and Se   (c) Ca and Na   (d) Ne and Na   (e) Ga and Ge

_____ 18. The formulas of the oxides of some second period elements are Li_2O, BeO, B_2O_3, and CO_2. Which one of the following is likely to be the correct formula for aluminum sulfide?
(a) Al_2S (b) AlS   (c) Al_2S_3 (d) AlS_2
(e) Cannot predict without knowing the percent by mass of the compound.

_____ 19. Which one of the following observations provided evidence that the electrons in atoms are arranged in distinct energy levels?
(a) the line spectra produced by gas discharge tubes
(b) the scattering of alpha particles by a metal foil
(c) the deflection of ions in a mass spectrometer
(d) the results of the Millikan oil-drop experiment
(e) the existence of elements with noninteger values for atomic weights

_____ 20. Suppose that an atom had available to it *only* the six energy levels indicated below. What would be the total number of lines in the emission spectrum of this element?

$$\rule{3cm}{0.4pt} \quad n = 6$$
$$\rule{3cm}{0.4pt} \quad n = 5$$
$$\rule{3cm}{0.4pt} \quad n = 4$$
$$\rule{3cm}{0.4pt} \quad n = 3$$
$$\rule{3cm}{0.4pt} \quad n = 2$$
$$\rule{3cm}{0.4pt} \quad n = 1$$

(a) 5   (b) 6   (c) 15   (d) 21   (e) 36

_____ 21. Bohr's theory of the hydrogen atom assumed that
(a) electromagnetic radiation is given off when the electron moves in an orbit around the nucleus.
(b) the electron can exist in any one of a set of discrete states (energy levels) without emitting radiation.
(c) because a hydrogen atom has only one electron, the emission spectrum of hydrogen should consist of only one line.
(d) the electron in a hydrogen atom can jump from one energy level to another without gain or loss of energy.
(e) energy, in the form of electromagnetic radiation, must be continually supplied to keep the electron moving in its orbit.

_____ 22. An electron in a hydrogen atom could undergo any of the transitions listed below by changing its energy. Which transition would give off (emit) light of the longest wavelength? (You will not need to calculate wavelengths, but think about the Rydberg equation or the Planck relationship.)
(a) $n = 3$ to $n = 2$ (b) $n = 3$ to $n = 1$ (c) $n = 2$ to $n = 1$
(d) $n = 2$ to $n = 3$ (e) $n = 1$ to $n = 3$

_____ 23. An orbital that has $n = 4$ and $\ell = 2$ is identified as a _____ orbital.
(a) $2f$ (b) $3s$ (c) $4s$ (d) $4p$ (e) $4d$

_____ 24. Which of the following is *not* a valid combination of quantum numbers?
  (a) $n = 3$, $\ell = 4$, $m_\ell = 0$, $m_s = +\frac{1}{2}$ (b) $n = 5$, $\ell = 4$, $m_\ell = 0$, $m_s = -\frac{1}{2}$
  (c) $n = 3$, $\ell = 2$, $m_\ell = -2$, $m_s = -\frac{1}{2}$ (d) $n = 2$, $\ell = 1$, $m_\ell = 1$, $m_s = +\frac{1}{2}$
  (e) $n = 4$, $\ell = 0$, $m_\ell = 0$, $m_s = +\frac{1}{2}$

_____ 25. Can an electron in an atom be described by the following set of quantum numbers:
$$n = 5, \ell = 3, m_\ell = -2, m_s = -\frac{1}{2}$$
  (a) No, because ℓ must be equal to $(n - 1)$.
  (b) No, because m_ℓ cannot be negative.
  (c) No, because n cannot be as large as 5.
  (d) No, because m_s must be positive.
  (e) Yes.

_____ 26. How many orbitals are there in a set of equivalent p orbitals (any p sublevel)?
  (a) 1   (b) 2   (c) 3   (d) 5
  (e) The answer depends on the value of n ($2p$, $3p$, etc.)

_____ 27. Which of the following orbitals has a shape that we describe as "spherically symmetrical?"
  (a) an s orbital   (b) a p orbital   (c) a d orbital
  (d) an f orbital   (e) all of these

_____ 28. In a given atom, how many electrons can occupy the $3p$ set of orbitals?
  (a) 0   (b) 2   (c) 3   (d) 6   (e) 10

_____ 29. In a given atom, how many electrons can occupy the $4d$ set of orbitals?
  (a) 0   (b) 2   (c) 3   (d) 6   (e) 10

_____ 30. In a given atom, how many electrons can occupy the $2d$ set of orbitals?
  (a) 0   (b) 2   (c) 3   (d) 6   (e) 10

_____ 31. Which one of the following atoms would be most strongly paramagnetic?
  (a) S   (b) Co   (c) Cr   (d) Cl   (e) Ne

_____ 32. Which one of the following elements has three half-filled $2p$ orbitals in the ground state electron configuration of the neutral atom?
  (a) C   (b) N   (c) O   (d) F   (e) Ne

_____ 33. A neutral atom has a ground state electron configuration that we can designate $1s^2\, 2s^2$. Which of the following statements about this atom is (or are) correct?
  (a) The element has an atomic number of 4 and is beryllium.
  (b) The atom contains four protons.
  (c) The atom has no unpaired electrons.
  (d) The atom has a total of two orbitals occupied.
  (e) All of the preceding statements are correct.

Answers to Some Important Terms in This Chapter

1. nucleus
2. electrons
3. proton
4. neutron
5. atomic number
6. isotopes
7. mass number
8. nuclide symbol
9. atomic mass unit
10. periodicity
11. periodic table
12. alkali metals
13. alkaline earth metals
14. noble gases
15. halogens
16. metals
17. nonmetals
18. metalloids
19. electromagnetic radiation
20. wavelength
21. frequency
22. photons
23. quantum
24. quantum numbers
25. principal quantum number
26. angular momentum quantum number
27. magnetic quantum number
28. spin quantum number
29. atomic orbital
30. *s* orbital
31. *p* orbital
32. *d* orbital
33. *f* orbital
34. paired
35. electron configuration
36. Aufbau Principle
37. Pauli Exclusion Principle
38. Hund's Rule
39. period
40. group

Answers to Preliminary Test

True-False

1. False. The electrons are the lightest of the three subatomic particles and contribute little to the total mass.
2. False. There are two reasons. First, the electrons are not in the nucleus, and second, 56 is the mass number, which represents the sum of the number of protons and the number of neutrons in the nucleus.
3. True. The atomic number uniquely defines the element.
4. True. It works both ways.
5. False. It could be an isotope of Ru or of another element. Mass number does not uniquely define the element. There are many cases of more than one element having isotopes with the same mass number.
6. False. The electrons are distributed throughout a relatively large empty space surrounding the nucleus.
7. False. Cathode ray particles are electrons.
8. True. They're all electrons.
9. False. The canal ray particles are the positive ions that remain when an atom loses one or more electrons. These have different masses, depending on the identity of the original atom.
10. True. Most of the alpha particles passed through undeflected, but a few were scattered at large angles.
11. False. The atomic number (i.e., number of protons in the nucleus) determines the identity of the element. This is contrary to the earlier Daltonian idea of the atom.

12. False. Twenty-six elements consist of only one naturally occurring, non-radioactive isotope: Be, F, Na, Al, P, Sc, V, Mn, Co, As, Rb, Y, Nb, Rh, In, I, Cs, La, Pr, Eu, Tb, Ho, Tm, Lu, Re, and Au. All except Be have odd atomic numbers.

13. True. The atomic number is the number of protons, and you can't have a piece of a proton.

14. True. The mass number is the total of the number of protons and the number of neutrons, and you can't have just part of either one of these fundamental particles.

15. False. See Sections 4-6 through 4-8. Carbon-12 is one exception because of the definition of the atomic mass unit.

16. False. The observed atomic weight of an element is the weighted average of the masses of the constituent isotopes. See the sample calculations in Section 4-9.

17. False. The atomic weight is defined so that the mass of the most common isotope of carbon (carbon-12) is exactly 12 amu. Naturally occurring carbon also contains small amounts of a heavier isotope (carbon-13), so the average atomic weight is not exactly 12. See Section 4-9.

18. True. Look at Table 4-5. How many metals are there in the first six periods? How many nonmetals? How many metalloids?

19. False. The metalloids have properties that are intermediate between those of metals and nonmetals. However, there is no standard definition of *metalloid*, and no universal agreement upon which elements should be classified as metalloids, so that certain elements that are classified as metalloid by some may be classified as metal or nonmetal by others.

20. False. Properties of the elements in a group are similar, but not identical. In fact, some properties, such as the melting point, change in a regular way through the group.

21. True. Metallic properties decrease from left to right across a row of the periodic table.

22. True. You have probably already heard of ultraviolet and infrared light. The light visible to the human eye is only a very small part of the entire electromagnetic spectrum.

23. False. See Section 4-11. Wavelength and frequency are inversely proportional to one another. This point is illustrated by Example 4-4.

24. True. See Section 4-11. Energy is directly proportional to frequency. Be sure you understand the significance of Example 4-5.

25. False. Both light and particles such as the electron can exhibit either wave or particle behavior. See Sections 4-11 through 4-14.

26. False. Each line in the spectrum is the result of a transition between two electronic energy levels in the atom.

27. True. But remember that orbitals with the same values of n and ℓ but different m_ℓ values have the same energy in an isolated atom.

28. True. One with $m_s = +\frac{1}{2}$, and one with $m_s = -\frac{1}{2}$.

29. True. This is described by the Pauli Exclusion Principle and the fact that m_s has only two possible values.

30. False. The second main energy level can have a maximum of eight electrons.

31. False. Recall that cathode rays are streams of electrons (Section 4-2) and that atoms can lose electrons to form ions (Section 4-3). Diffraction and scattering experiments on beams of electrons are mentioned in Section 4-14.

32. False. It is true that all atoms of a given element have the same number of protons; therefore, to be neutral they must also have the same number of electrons. However, it is

possible for the atoms to have different numbers of neutrons; thus they would represent different isotopes of the same element.

33. False. You should know the normal order of electron energy levels, Figures 4-30 and 4-31. For example, less energy is required to place electrons into the $4s$ sublevel than into the $3d$ sublevel. Remember, though, that the order of energies is sometimes different, depending on the occupancy of the orbitals. See Section 4-18, especially the discussion concerning the configurations of Cr and Cu.

Short Answer

1. (a) R; (b) A; (c) U; (d) A; (e) R; (f) U. Like charges repel, unlike charges attract, and particles having no charge are not affected by an electrical (or a magnetic) field.
2. 6+. Remember that the atomic number gives the number of protons in the nucleus. The only other particles in the nucleus are the neutrons, which are uncharged.
3. proton
4. neutron
5. protons; neutrons
6. proton; neutron
7. proton; electron
8. hydrogen. The experimentally observed atomic weight is only slightly greater than 1, so hydrogen must consist at least partially (in fact mostly) of an isotope with mass number 1. This isotope would have only one proton and no neutrons in the nucleus. In fact, this isotope of hydrogen is the only exception to the statement.
9. α; He^{2+}
10. larger. The deflection depends on the charge/mass ratio.
11. smaller
12. isotopes
13. nuclide
14. atomic number; protons; electrons
15. mass number; sum of number of protons and number of protons (or total number of nuclear particles)
16. 10; 90. Let x = fraction of atoms that have mass number 113; then $(1 - x)$ equals the fraction that have a mass number of 115. So
$$x(113) + (1 - x)(115) = 114.8$$
$$113x + 115 - 115x = 114.8$$
$$2x = 0.2$$
$$x = 0.10 \text{ and } (1 - x) = 0.90$$

17. 46; 60; 46
18. Physical and chemical properties of the elements vary in a periodic fashion with the atomic numbers of the elements.
19. Mendeleev; Meyer
20. alkali metals. Be sure you remember the common group names (Section 4-10).
21. 2A (2); alkaline earth metals
22. F, Cl, Br, I, and At. Of these, only the first four are common. Notice the terminology by which a *group* is sometimes referred to as a *family* of elements.

23. n; principal (note the spelling—not principle!)

24. ℓ, angular momentum

25. ℓ, angular momentum

26. m_ℓ, magnetic

27. m_s

28. 0

29. s

30. p

31. two. Each of the three orbitals in a p sublevel is "dumbbell" shaped.

32. 32. For $n = 4$, the possible values of ℓ are 0 (one s orbital), 1 (three p orbitals), 2 (five d orbitals), and 3 (seven f orbitals) for a total of 16 orbitals. Each orbital can hold 2 electrons for a total of 32 electrons possible in this energy level.

33. opposite values of m_s. Study the Pauli Exclusion Principle (Section 4-18).

34. $90°$

35. $4s$

36. $3d$

37. $1s^2\ 2s^2\ 2p^6\ 3s^2\ 3p^6\ 4s^2\ 3d^{10}\ 4p^5$. Remember that it would also be correct to write the electron configuration with all orbitals having the same value of principal quantum number appearing together. Thus, we could have written: $1s^2\ 2s^2\ 2p^6\ 3s^2\ 3p^6\ 3d^{10}\ 4s^2\ 4p^5$. This shows $3s$, $3p$, and $3d$ together, even though in the normal order $4s$ fills before $3d$.

38. $1s^2\ 2s^2\ 2p^6\ 3s^2\ 3p^6\ 4s^2\ 3d^{10}\ 4p^3$. See the additional comments in the answer to Question 37.

39. 2. The electron configuration is $1s^2\ 2s^2\ 2p^6\ 3s^2\ 3p^4$. In the $3p$ orbitals, the electrons are arranged as ↑↓ ↑ ↑. All other electrons are paired.

40. 4. The electron configuration is $1s^2\ 2s^2\ 2p^6\ 3s^2\ 3p^6\ 4s^2\ 3d^6$. In orbital notation, $3d^6$ represents ↑↓ ↑ ↑ ↑ ↑. All other electrons are paired.

41. $-2, -1, 0, +1, +2$. This is a d orbital, so $\ell = 2$. Thus m_ℓ can range from $-\ell$ to $+\ell$ by integers.

42. paramagnetic; unpaired electrons

43. group; family

44. period

45. $ns^2\ np^1$

46. 2. Each of these elements has the electron configuration $ns^2\ np^4$. The two s electrons are paired in a single orbital. The four p electrons are arranged as ↑↓ ↑ ↑. All lower energy levels are filled, so all of their electrons are paired.

47. ns^1; $ns^2\ np^6$

Multiple Choice

1. (b). The atomic number (a) = the number of protons (d) = the number of electrons (e), assuming the atoms are neutral; that is, the total charge (c) for both isotopes = 0. Isotopes have the same atomic number but different mass numbers.

2. (b). The subscript is the atomic number, which is the number of protons. If the atom is neutral, then the number of electrons is the same. As for (a), 61 is the sum of the number of neutrons and the number of protons. (c) The element with atomic number 27 is cobalt, Co.

(d) 34 is the number of neutrons. (e) The number of protons plus *neutrons* (not electrons) is equal to 61.

3. (d). The observed atomic weight of magnesium is greater than 24, so there must be an isotope of magnesium that has a mass number greater than 24. This isotope would have more neutrons than the principal isotope, $^{24}_{12}Mg$.

4. (e). Since the atomic number is 28, the number of protons is 28, and the number of electrons in the neutral atom is 28. The mass number, the total number of protons and neutrons in the nucleus, is 60, so the number of neutrons is $60 - 28 = 32$.

5. (c). See Section 4-2.

6. (c). See Section 4-2.

7. (b). See Section 4-4.

8. (c). The atomic number of carbon is 6, which tells us that *each* atom of C contains 6 electrons (regardless of which isotope we are considering). But 1 mole of carbon contains 6.022×10^{23} carbon atoms, so the total number of electrons is

$$1 \text{ mol C atoms} \times \frac{6.022 \times 10^{23} \text{ C atoms}}{1 \text{ mol C atoms}} \times \frac{6 \text{ electrons}}{1 \text{ C atom}}$$

9. (d). Calculate as in Example 4-2 of the text. Study Example 4-3, which uses the same ideas to solve for the percentages instead. $0.700(44 \text{ amu}) + 0.300(46 \text{ amu}) = 44.6$

10. (d). As for … (a) The electrons contribute very little to the mass of the atom. (b) The nucleus occupies a tiny fraction of the volume of the atom. (c) The protons and the *neutrons* are in the nucleus. The electrons are in the electron cloud outside the nucleus. (e) Neutrons are *neutral*. The electrons carry all of the negative charge.

11. (a). See Section 4-8.

12. (c). The subscript, 26, is the atomic number, which is the number of protons. In a neutral iron atom, the number of electrons is the same. The superscript, 56, is the mass number, which is the sum of the number of protons and the number of neutrons. Therefore, the number of neutrons is the mass number, 56, minus the atomic number, 26. The procedure is the same for each isotope of iron, regardless of the atomic weight of naturally occurring iron, except that the value for the mass number is different for each isotope.

13. (d). Natural radioactivity, discovered almost 90 years after Dalton proposed his theory, results from atoms changing, dividing.

14. (d). The number of protons (atomic number) determines the element. Answer (d) has a different atomic number from the others, so it must be a different element. Answers (a) and (c) are isotopes of the same element. Answer (b) is an ion formed from (a) by loss of two electrons, but it's still the same isotope of the same element.

15. (c). See Section 4-3.

16. (d). The larger the charge-to-mass ratio (e/m), the greater the deflection. Neutral atoms would not be deflected, because they have no charge. Answer (d) has twice as great a charge as (c), but the same mass, so (d) would be deflected more. Answer (e) also has a 2+ positive charge, but it has a larger mass than answer (d), so it has a *smaller* e/m ratio.

17. (b). Remember that the most pronounced similarities are for elements in the same group (column) of the periodic table.

18. (c). Al is in the same column as B, and S is in the same column as O.

19. (a). See Section 4-13.

20. (c). Each transition from a higher to a lower energy level gives rise to one line in the emission spectrum. An electron could go from level $n = 6$ to any of five lower-energy levels (five possible transitions), from $n = 5$ to any of four lower-energy levels (four possible transitions), and so on. There are 15 such downward transitions.

21. (b). See Section 4-13.

22. (a). Answers (d) and (e) require the absorption of light, not emission. Of the other answers, (a) corresponds to the smallest energy difference (remember that the energy levels get closer together as n increases.). Wavelength is inversely proportional to energy of the light.

23. (e). For (a) $n = 2$ and $\ell = 3$. For (b) $n = 3$ and $\ell = 3$. For (c) $n = 4$ and $\ell = 0$. For (d) $n = 4$ and $\ell = 1$.

24. (a). ℓ can be no larger than $(n - 1)$.

25. (e). Review the rules for the allowed values of the quantum numbers, Section 4-16.

26. (c). An s sublevel is just 1 orbital, a p is 3, a d has 5 orbitals, and an f consists of 7.

27. (a). See Section 4-17 regarding orbital shapes. (b) A p orbital is dumbbell shaped. (c) Most of the d orbitals are shaped like a four-leaf clover. (d) f orbitals have either 8 lobes or 2 lobes and 2 donut-shaped rings.

28. (d). A p sublevel has 3 orbitals, each of which can hold up to 2 electrons. $3 \times 2 = 6$

29. (e). A d sublevel has 5 orbitals, each of which can hold up to 2 electrons. $5 \times 2 = 10$

30. (a). There is no such thing as a $2d$ orbital. This would mean that $n = 2$ and $\ell = 2$, which would violate the rules for the allowed values of quantum numbers.

31. (c). Cr has 6 unpaired electrons, the largest number of any of the atoms listed.

32. (b). According to Hund's rule, electrons in the $2p$ sublevel will be spread as evenly as possible among the three orbitals. C has two $2p$ electrons and thus two half-filled orbitals. O has four $2p$ electrons, two paired in one orbital and two half-filled orbitals. F has five $2p$ electrons, four in two pairs and one half-filled orbital. Ne ha six $2p$ electrons, which fill all three orbitals.

33. (e)

5 Chemical Periodicity

Chapter Summary

In Chapter 4, we introduced the periodic table as a way of representing similarities in reactions and compounds of groups of elements. As we saw in Section 4-19, the marvelous utility of the periodic table in chemistry arises from the relationship of the properties of the elements to their electron configurations. As you study this chapter, you will see many correlations between the patterns that the properties of elements exhibit and the parallels between their electron configurations and the locations of the elements on the periodic table. When you finish, you should be able to interpret the periodic table to remind you of electron configurations and to predict or correlate properties of substances.

Before you begin your study of this chapter, review the basic terminology of the periodic table, as presented in Section 4-10. Each horizontal row is referred to as a **period**, and each vertical column is called a **group**. You were also introduced to the common names of some of the groups of elements—the **alkali metals**, **alkaline earth metals**, **halogens**, and **noble gases**—and the classification of elements as **metal**, **nonmetal**, or **metalloid**. Another way of classifying elements for analysis is in terms of the "last" electron added to the electron configuration in the *Aufbau* process we learned in Chapter 4. (Review Section 4-19, especially Table 4-9. You should be able to tell quickly by looking at the location of an element in the periodic table into which kind of orbital the last electron went.) Thus, elements in which this last electron is in an *s* or a *p* orbital (excepting the noble gases) are termed **representative elements**, those in which the last electron is in a *d* orbital are **transition elements** or **transition metals** (Note: *not* transition<u>al</u>), and those with the last electron in an *f* orbital are **inner transition elements** (Section 5-1).

Many physical and chemical properties are related to the locations of the elements in the periodic table. Some of these are discussed in Sections 5-2 through 5-6. In turn, these properties will help us to understand bonding between elements and the properties of the resulting compounds. Keep two goals in mind as you study these sections. The first is to be able to explain how each of these observed trends results from changes in electron configurations across a period and down a group. The second is the reverse—using its location in the periodic table to predict the properties of an element.

One property that is easily correlated with location in the periodic table is the size of the atom, usually described in terms of the **atomic radii** of the elements (Section 5-2). The correlations that emerge are that the atomic sizes of the (representative) elements within a period *decrease* from left to right across a period ($\Rightarrow$), while within a group the sizes *increase* from top to bottom ($\Downarrow$). Be sure that you understand how these trends in atomic radii are related to regular changes in electron configuration.

Chemical reactions result from the interaction of electrons of the elements involved. It is important to understand how tightly the electrons of different atoms are held, and how easily each kind of atom gains or loses electrons. When an atom loses one or more electrons, it forms a **positive ion** (**cation**). When an atom gains one or more electrons, it forms a **negative ion** (**anion**). The ease with which these two processes can occur can be measured as the **ionization energy** and/or the **electron affinity** of the element. Sections 5-3 and 5-4 correlate trends in these two properties with location in the table, and hence with electron configuration. These correlations reveal that the ease with which electrons are lost or gained depends both on the sizes of the atoms and on how effectively the nucleus can attract the outermost electrons. The ions that are formed are characteristically smaller (for cations) or larger (for anions) than the corresponding neutral atoms. These trends in ionic radii can also be related to the types of orbitals that the outermost electrons occupy (Section 5-5).

Section 5-6 introduces a very important concept that we shall often use for predicting and understanding chemical bonding. This is the **electronegativity scale**, a relative scale that expresses the tendency of an atom to attract electrons. You must remember the correlations of electronegativity with location in the periodic table. Electronegativities generally increase going from left to right across a period ($\Longrightarrow$) and decrease going from top to bottom within a group ($\Downarrow$). We do not apply this terminology to the noble gases; because of their extremely stable configurations, these gases have practically no tendency to gain or lose electrons.

Many reactions involve electron transfer. A formal way of keeping track of electron transfer is by the tabulation of **oxidation state** or **oxidation number** (Section 5-7). The rules for assigning oxidation states to elements in compounds will be quite useful in later study. You will use oxidation states to recognize electron transfer reactions, to help balance chemical equations for such reactions (Chapter 11), to study the electron-transfer reactions that take place in batteries and in electrolysis cells (Chapter 21), and to systematize many kinds of reactions (Chapters 25 through 28).

The common oxidation states given in Table 5-4 are important. You need to learn them carefully so that you can recall them easily—practice until you can assign oxidation states to all elements in any compound or ion whose formula you are given. Remember that the two terms *oxidation state* and *oxidation number* mean the same thing and are used interchangeably. Also, be sure that you understand that the oxidation state is a quantity that is determined "per atom." It is represented in the text by the notation $+n$ or $-n$ in a circle directly above the elemental symbol. By contrast, the ionic charge is the actual net electrical charge (in units of electronic charge) on the total ion, even if the ion is polyatomic. Ionic charge is represented in the text by the notation $n+$ or $n-$ above and to the right of the formula for the ion.

Oxidation state/number $\nearrow$ $\overset{\textcircled{\pm n}\,\textcircled{\pm n}}{X_x Y_y}{}^{n\pm}$ $\leftarrow$ Ionic charge

Oxygen and hydrogen combine with more elements than any other elements do. For this reason, a study of the reactions and compounds of these two elements provides an illustration of the concepts and classifications embodied in the periodic table. Section 5-8 describes the properties of the element hydrogen and some methods of its preparation. In its reactions, hydrogen combines with metals or nonmetals to form binary (two element) compounds. These range in their properties from the ionic **hydrides** (containing an active metal) to molecular compounds (containing a nonmetal or metalloid), through some compounds of intermediate character. The periodic table is the key to keeping track of the formulas and properties of these compounds.

Section 5-9 takes a similar approach to the properties, preparation, reactions, and compounds of oxygen. Most of this free element exists in the familiar diatomic form, O_2, sometimes called **dioxygen,** but usually just referred to as **oxygen.** The other **allotropic** form of this element, O_3, called ozone, is a very reactive gas. Oxygen combines with more elements than any other element, forming binary compounds (**oxides**) by direct reaction with all elements except the noble gases and a few of the less reactive metals, such as silver and gold. Oxygen does form compounds indirectly with several of these. You will study more about the reactions of this important element when we discuss the descriptive chemistry of other elements. The binary compounds of oxygen range in character from ionic (with metals) to molecular (with other nonmetals). In binary compounds with metals, oxygen can exist as the **oxide ion (O^{2-}),** as the **peroxide ion (O_2^{2-}),** or as the **superoxide ion (O_2^{-}).** Though the first of these is the most common, the tendency to form the others correlates to trends in the periodic table (Section 5-9). The three types of oxide reactions discussed here are: (1) the reaction of metal oxides (basic anhydrides) with water to form bases, (2) the reaction of nonmetal oxides (acid anhydrides) with water to form **ternary acids** (containing three elements—H, O, and a nonmetal), and (3) the reaction of these two types of oxides with one another to form salts. Note that all three of these occur without change of oxidation state.

The last two parts of Section 5-9 discuss **combustion reactions**; these are redox reactions in which oxygen combines rapidly with oxidizable materials in highly exothermic reactions with a visible flame. (We call this "burning.") Some aspects of the atmospheric pollution that results from the combustion of fossil fuels are discussed.

Study Goals

Be sure to practice many of the indicated end-of-chapter Exercises, in addition to answering the Preliminary Test in this guide.

Section 5-1
1.   Relate the electron configuration of an atom to the location of the element on the periodic table (also Goal 23 of Chapter 4). *Work Exercises 2, 3, 5, 6, and 9 through 11.*

Section 5-1 (Review Section 4-10)
2.   Give examples of each of the following groups of elements: (a) alkali metals, (b) alkaline earth metals, (c) halogens, (d) noble gases, (e) representative elements, (f) *d*-transition elements, (g) inner transition elements or *f*-transition elements, (h) lanthanides, and (i) actinides. Relate each of these classes to electron configuration and to location in the periodic table. *Work Exercises 1, 4, 7, and 8.*

Sections 5-2 through 5-6
3.   Summarize horizontal and vertical trends in the periodic table for each of the following properties. Relate each of these properties and their trends to electron configurations:
   (a) atomic radius (Section 5-2) *Work Exercises 12 through 16 and 44.*
   (b) ionization energy (Section 5-3) *Work Exercises 18 through 29 and 44.*
   (c) electron affinity (Section 5-4) *Work Exercises 30 through 34.*

(d) ionic radius (Section 5-5) *Work Exercises 17 and 35 through 40.*

(e) electronegativity (Section 5-6) *Work Exercises 41 through 45.*

Section 5-7

4. Know the rules for assigning oxidation states and be able to assign oxidation states to elements, whether free or in molecules, ions, or compounds. *Work Exercises 46 through 49.*

Sections 5-8 and 5-9

5. Know some preparations and properties of hydrogen (H_2) and oxygen (O_2). *Work Exercises 52, 60, 61, and 63.*

6. Know some of the reactions of (a) hydrogen and the hydrides and (b) oxygen and the oxides. *Work Exercises 53, 54, 57, 59, 64, 65, 68, and 69.*

7. Recognize and give examples of the following classes of compounds: (a) metal hydrides, (b) nonmetal hydrogen compounds, (c) metal oxides (basic anhydrides), peroxides, and superoxides, (d) nonmetal oxides (acidic anhydrides). Give the preparations and reactions of these classes of compounds. *Work Exercises 55, 56, 58, 62, 66, 67, 70, and 71.*

Sections 5-9 (last two subsections)

8. Know the basic features of combustion reactions. *Work Exercises 72 through 78.*

General

9. Use the periodic table as a basis to recognize and answer a variety of types of questions about elements and their properties. *Work Exercises 50 and 51.*

10. Answer conceptual questions based on this chapter. *Work* **Conceptual Exercises** *79 through 87.*

11. Apply concepts and skills from earlier chapters to the ideas of this chapter. *Work* **Building Your Knowledge** *exercises 88 through 93.*

12. Exercises at the end of chapter direct you to sources outside the textbook for information to use in solving them. Work **Beyond the Textbook** *exercises 94 through 97.*

Some Important Terms in This Chapter

Some important new terms from Chapter 5 are listed below. (Be sure to study other new terms and review terms from preceding chapters, if necessary. Some of these terms have already been introduced in Chapter 4.) Fill the blanks in the following paragraphs with terms from the list. Each term is only used once.

acid anhydride	electron affinity	noble gases
alkali metals	electronegativity	nonmetals
alkaline earth metals	*f*-transition elements	oxidation state
amphoteric	halogens	oxide
anion	hydride	peroxide
atomic radius	ionic radius	representative elements
basic anhydride	ionization energy	screening effect
cation	isoelectronic	superoxide
d-transition elements	metals	ternary acid
effective nuclear charge	metalloid	

The elements can be classified several different ways, broadly dividing the periodic table into two, three, or four categories or grouping by families of elements with similar properties.

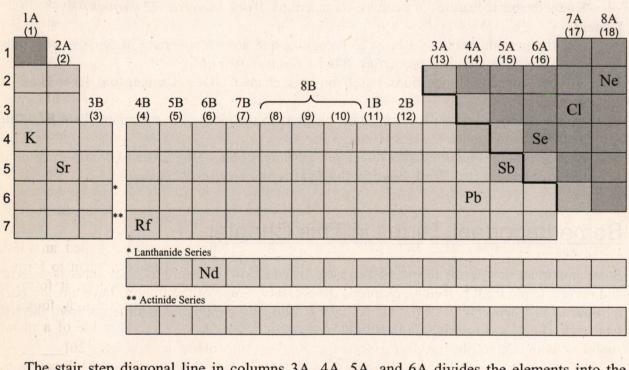

The stair step diagonal line in columns 3A, 4A, 5A, and 6A divides the elements into the [1]_____(▢), to the left and below the line; the [2]_____ (▨), to the right and above the line; and the [3]_____ (▢), generally on either side of the line. Another grouping divides elements by columns. The elements in Groups 1A, 2A, 3A, 4A, 5A, 6A, and 7A (the A group elements) are called [4]_____, the elements in Groups 3B, 4B, 5B, 6B, 7B, 8B, 1B, and 2B (the B group elements) are [5]_____, and the elements in the partial rows (the lanthanide series and the actinide series, actually between columns 3B and 4B) are labeled [6]_____. Then some individual columns of elements have family names. In Group 1A (1) are the [7]_____, Group 2A (2) has the [8]_____, Group 7A (17) has the [9]_____, and

Group 8A (18) has the [10]_____. Consequently, every element belongs to two or three categories.

For example,

1. Potassium (K) is a metal, a representative element, and an alkali metal,
2. Strontium (Sr) is a metal, a representative element, and an alkaline earth metal,
3. Rutherfordium (Rf) is a metal and a *d*-transition element,
4. Neodymium (Nd) is a metal and an *f*-transition element,
5. Lead (Pb) is a metal and a representative element,
6. Antimony (Sb) is a metalloid and a representative element,
7. Selenium (Se) is a nonmetal and a representative element,
8. Chlorine (Cl) is a nonmetal, a representative element, and a halogen, and
9. Neon (Ne) is a nonmetal and a noble gas.

The size of an atom, the [11]_____, we think of as the distance from its center to its "edge" (as if it had one). This size is affected by how much of the positive charge of the nucleus attracts the negative electrons of the outer shell, called the [12]_____, Z_{eff}. The nuclear charge "felt" by the outer-shell electrons is reduced by the [13]_____ or shielding effect of inner-shell electrons between the nucleus and the outer-shell electrons. The energy required to produce a positive ion by removing one of the outer-shell electrons from an atom is its [14]_____. Similarly, the attraction of an atom for adding an extra electron, which produces a negative ion, is its [15]_____. The positive ion produced is a [16]_____, and the negative ion is an [17]_____. The size of an ion is its [18]_____. A cation is always smaller than the neutral atom from which it is produced, and an anion is always larger than the neutral atom from which it is produced. But to compare the sizes of different ions, you must compare the members of an [19]_____ series, the group of atoms or ions that have the *same* number of *electrons*. The attraction of an atom for electrons, when it is chemically bonded to another atom, is its [20]_____. The apparent charge on an atom is its [21]_____, or *oxidation number*.

Hydrogen combines with an active metal to form a binary compound called an ionic [22]_____ (hydrogen + ide). Oxygen often combines with another element to form a binary compound called an [23]_____ (oxygen + ide). Sometimes, however, it forms a [24]_____, which contains the peroxide ion, O_2^{2-}. And sometimes it forms a [25]_____, which contains the superoxide ion, O_2^-. When an oxide of a metal combines with water, the product is often a base, so the oxide is called a [26]_____ ("without water") or basic oxide. When the oxide of a nonmetal dissolves in water, the product is often an acid, so the oxide is called an [27]_____ or acidic oxide. The acid formed is a [28]_____because it contains three elements—hydrogen, oxygen, and some other nonmetal. Oxides of some metals near the middle of the periodic table show some acidic and some basic properties, so they are [29]_____.

Preliminary Test

As in other chapters, this test will check your understanding of basic concepts and types of calculations. Be sure to practice *many* of the additional textbook Exercises, including those indicated with the Study Goals.

True-False

Mark each statement as true (T) or false (F).

_____ 1. Oxidation number and oxidation state are the same thing.
_____ 2. Each element has the same characteristic oxidation state in all its compounds.
_____ 3. Halogens have an oxidation state of –1 in all their compounds.

Short Answer

Answer with a word, a phrase, a formula, or a number with units as necessary.

1. The transition metals are elements that are characterized by electrons being added to the _____ in the Aufbau procedure.
2. The *f*-transition elements are characterized by electrons being added to the _____.
3. An element having the outer electron configuration ns^2np^3 would be found in Group _____.
4. The number of electrons in the outermost occupied shell of an atom of any Group 6A (16) element is _____.
5. The lightest three elements of Group 5A (15), arranged in order of increasing atomic size, are _____.
6. Because of the increasing nuclear charge going left to right across the representative elements of a period, the first ionization energies _____.
7. Because of the increasing nuclear charge going left to right across the representative elements of a period, the atomic radii _____.
8. The most stable ion formed by a member of the halogen group would have a charge of _____.
9. The number of electrons in the iodide ion, I^-, is _____.
10. The number of electrons is the calcium ion, Ca^{2+}, is _____.
11. Barium (Ba) would be _____ (more, less) likely to form a stable anion than a stable cation.
12. A cation is _____ (larger than, smaller than, the same size as) the neutral atom from which it is formed.
13. Atoms or ions that have the same number of electrons are called _____.
14. The neutral atom that is isoelectronic with Sr^{2+} is _____.
15. The sum of the oxidation states of all of the atoms in a neutral compound is equal to _____.
16. The sum of the oxidation states of all of the atoms in a polyatomic ion is equal to _____.

17. In most of its compounds, the oxidation state of hydrogen is ____, whereas oxygen usually exhibits the oxidation state ____.
18. The oxidation states of the elements in the compound K_2S are K ____ and S ____.
19. In the compound NaCl, the oxidation states of the elements are Na ____ and Cl ____.
20. In the compound $NaClO_3$, the oxidation states of the elements are Na ____, Cl ____, and O ____.
21. In the sulfate ion, SO_4^{2-}, the oxidation states of the elements are S ____ and O ____.
22. When the oxides of nonmetals dissolve in water, they form _____ solutions.
23. Basic anhydrides are the oxides of _____.
24. Strong soluble bases, such as the Group 1A (1) metal hydroxides, can often be prepared by the reaction of _____ with water.
25. Many ternary acids can be prepared by dissolving the appropriate _____ _____ in water.
26. An ionic hydride is a compound composed of _____ and _____.
27. A molecular hydrogen compound contains hydrogen and _____.
28. The general formula of the binary compounds of hydrogen with the halogens is _____ (denoting a halogen as X).
29. The general formula of the binary compounds of hydrogen with the Group 6A (16) nonmetals is _____ (denoting a Group 6A nonmetal as Y).
30. The general formula for the normal oxides of the alkali metals is _____ (denoting an alkali metal as M).
31. The two binary compounds that oxygen forms with hydrogen are:
_____ and _____.
(Give the name and formula of each.)
32. Of the two allotropic forms of oxygen, _____ is much more stable than _____. (Give name and formula of each.)
33. Metal peroxides contain the _____ ion, which has the formula _____; in this ion, the oxidation state of oxygen is _____.
34. Some nonmetallic oxides are referred to as _____ anhydrides, because they _____.
35. Many ionic metal oxides are called _____ anhydrides, because they _____.
36. Oxides that can dissolve in either acids or bases are termed _____.
37. The formula for sodium peroxide is _____.
38. The formula for hydrogen peroxide is _____.
39. The name of the ion O^{2-} is _____.
40. The name of the ion O_2^- is _____.
41. The name of the ion O_2^{2-} is _____.
42. The process in which oxygen combines rapidly with some materials in highly exothermic reactions, with a visible flame, is called _____.
43. The usual products of the complete combustion of hydrocarbons are
_____ and _____.
(Give the name and formula of each.)
44. Combustion reactions always give off _____.

75

45. Combustion of fuels containing sulfur produces _____ (name, formula), which is a very harmful atmospheric pollutant.

46. The substance produced in Question 45 is slowly oxidized in air to _____, which then combines with moisture in air to form _____.
(Give name and formula of each.)

Multiple Choice

_____ 1. For each of the noble gases except helium, the outermost occupied main energy level contains how many electrons?
(a) 1   (b) 2   (c) 6   (d) 8   (e) 18

_____ 2. Which one of the following statements concerning the element iodine is *not correct*?
(a) Iodine gives up electrons readily.
(b) Iodine is in the fifth period of the periodic table.
(c) Iodine has chemical properties similar to those of chlorine.
(d) Iodine is a halogen.
(e) One molecule of iodine contains two atoms.

_____ 3. Zirconium (Zr) is classified as
(a) a representative element.   (b) a *d*-transition element.
(c) a lanthanide.   (d) an actinide.   (e) a noble gas.

_____ 4. Which one of the following is a transition element?
(a) P   (b) Pd   (c) Ar   (d) U   (e) Na

_____ 5. Which one of the following is a representative element?
(a) P   (b) Pd   (c) Eu   (d) U   (e) V

_____ 6. An element with an outer electron configuration ns^2 (excepting helium) is classified as
(a) an alkali metal.   (b) an alkaline earth metal.   (c) a halogen.
(d) a transition element.   (e) an inner transition element.

_____ 7. When the properties of the elements with atomic numbers 119 and 120 are well characterized, they will probably fit into
(a) the actinide series of inner transition elements.
(b) the alkali metal group and the alkaline earth group.
(c) a new section of the periodic table, not yet represented.
(d) the noble gas group.
(e) the halogen group.

_____ 8. Atoms of the elements in the same A group in the periodic table have
(a) the same electronegativity.
(b) the same total number of electrons.
(c) the same size.
(d) the same number of electrons in the outer shell.
(e) the same number of protons.

_____ 9. Where in the periodic table are the least electronegative elements found? (Ignore the noble gases.)
(a) upper left   (b) upper right   (c) lower left
(d) lower right   (e) middle

76

_____ 10. Which one of the following elements would have the highest first ionization energy?
(a) Na   (b) Al   (c) Si   (d) S   (e) Cl

_____ 11. For which one of the following elements would the size of the neutral atom (atomic radius) be smallest)?
(a) K   (b) Na   (c) Mg   (d) Ca   (e) Sr

_____ 12. Cesium (Cs) would be expected to have a _____ first ionization energy and a _____ electronegativity.
(a) high, high   (b) high, low   (c) low, high   (d) low, low

_____ 13. The alkali metals have *low* first and *high* second ionization energies. The alkaline earth metals have relatively *low* values for both first and second ionization energies. What conclusion do these observations suggest?
(a) Both alkali metals and alkaline earth metals form very stable ions with charge 1+.
(b) Both alkali metals and alkaline earth metals form very stable ions with charge 2+.
(c) Alkali metals form stable 1+ ions, while alkaline earth metals form stable 2+ ions.
(d) Alkali metals form stable 1+ ions, while alkaline earth metals form stable 2– ions.
(e) Alkali metals form stable 1– ions, while alkaline earth metals form stable 2– ions.

_____ 14. Which one of the following isoelectronic ions would be expected to have the largest radius?
(a) S^{2-} (b) Cl^- (c) K^+ (d) Ca^{2+} (e) Sc^{3+}

_____ 15. Which one of the following pairs would correctly be described as *isoelectronic*?
(a) Na and Na^+ (b) Ca and Be   (c) Cu^+ and Zn^{2+} (d) P^- and P^{2-} (e) H and He

_____ 16. How many electrons are there in one Cr^{3+} ion?
(a) 24   (b) 52   (c) 27   (d) 21
(e) Cannot be answered without knowing the mass number.

_____ 17. The oxidation numbers (oxidation states) of the elements in $KClO_3$ are
(a) K = +3, Cl = +3, O = –2.   (b) K = –1, Cl = –5, O = +2.
(c) K = +1, Cl = +5, O = –2.   (d) K = +1, Cl = +3, O = –1.
(e) K = +1, Cl = +2, O = –3.

_____ 18. The mineral fluorite is an ionic substance consisting only of calcium and fluorine. What are the oxidation numbers of Ca and F in fluorite?
(a) Ca = –1, F = +1   (b) Ca = +1, F = +2   (c) Ca = +2, F = –2
(d) Ca = +2, F = –1   (e) Ca = +2, F = +9

_____ 19. Which one of the following compounds contains an element with an oxidation number of +5?
(a) HClO   (b) HNO_2 (c) $HClO_3$ (d) $HBrO_4$ (e) H_2SO_4

_____ 20. Which one of the following compounds contains an element with an oxidation number of +3?
(a) HClO   (b) HNO_2 (c) $HClO_3$ (d) $HBrO_4$ (e) H_2SO_4

_____ 21. Which one of the following compounds contains an element with an oxidation number of +6?
(a) HClO   (b) HNO_3 (c) $HClO_3$ (d) $HBrO_4$ (e) H_2SO_4

_____ 22. Each of the following is a known acid containing chlorine. In which one does chlorine have oxidation number +7?
(a) HCl   (b) HClO   (c) $HClO_2$ (d) $HClO_3$ (e) $HClO_4$

_____ 23. Manganese (Mn) exhibits a larger number of oxidation numbers in its compounds than any other of the Period 4 transition metals. The most important oxidation numbers of manganese are illustrated by the following oxides: MnO, MnO_2, and Mn_2O_7. What oxidation numbers of manganese are illustrated here?
(a) −2, −4, −7   (b) +2 and +7   (c) +2, +3, +9   (d) +2, +4, +7   (e) +2, +4, +14

_____ 24. Which one of the Group 6A (16) elements is *not* a solid at room temperature and pressure?
(a) oxygen   (b) sulfur   (c) selenium   (d) tellurium   (e) polonium

_____ 25. The element oxygen, in the form O_2, accounts for which of the following percentages, by volume, of dry air?
(a) about 20%   (b) about 40%   (c) about 50%   (d) about 60%   (e) about 80%

_____ 26. Which one of the following Group 6A (16) elements has the highest first ionization energy?
(a) oxygen   (b) sulfur   (c) selenium   (d) tellurium   (e) polonium

_____ 27. Some soluble bases can be formed by the reaction of a soluble metal oxide with water. For which one of the following metal oxides would this *not* be a likely way of preparing the corresponding hydroxide base?
(a) K_2O (b) CaO (c) Na_2O (d) BaO (e) FeO

_____ 28. Each of the following ions forms a stable hydroxide. For which one of these ions would the hydroxide *not* be amphoteric?
(a) Pb^{4+} (b) Sn^{4+} (c) Ga^{3+} (d) Be^{2+} (e) Ba^{2+}

_____ 29. Which one of the following is a molecular compound of hydrogen?
(a) RbH (b) SrH_2 (c) NaH (d) PH_3 (e) CaH_2

_____ 30. Arrange the following oxides in order of increasing ionic character: BaO, SiO_2, SO_2.
(a) $BaO < SiO_2 < SO_2$ (b) $SO_2 < BaO < SiO_2$
(c) $SiO_2 < SO_2 < BaO$ (d) $BaO < SO_2 < SiO_2$
(e) $SO_2 < SiO_2 < BaO$

_____ 31. Which one of the following compounds is a superoxide?
(a) Na_2O_2 (b) SrO (c) KO_2 (d) FeO (e) Cl_2O_7

_____ 32. For which one of the following elements is the normal oxide classified as most acidic?
(a) P   (b) Al   (c) Ca   (d) Na
(e) No other answer is correct, because an acid must contain hydrogen.

_____ 33. Reaction between which one of the following pairs is an example of a reaction between an acidic anhydride and a basic anhydride?
(a) CaO and SiO_2 (b) Mg and $CuBr_2$ (c) CaO and H_2O
(d) NH_3 and HNO_3 (e) KH and H_2O

Answers to Some Important Terms in This Chapter

1. metals
2. nonmetals
3. metalloids
4. representative elements
5. *d*-transition elements
6. *f*-transition elements
7. alkali metals
8. alkaline earth elements
9. halogens
10. noble gases
11. atomic radius

12. effective nuclear charge	18. ionic radius	24. peroxide
13. screening effect	19. isoelectronic	25. superoxide
14. ionization energy	20. electronegativity	26. basic anhydride
15. electron affinity	21. oxidation state	27. acid anhydride
16. cation	22. hydride	28. ternary acid
17. anion	23. oxide	29. amphoteric

Answers to Preliminary Test

Do not just look up answers. *Think about the reasons for the answers.*

True-False

1. True. See Section 5-7.
2. False. Most elements can exhibit various oxidation numbers in their compounds, and some elements can have many.
3. False. This is true only for their binary compounds with metals, H, or NH_4^+. See Table 5-4.

Short Answer

1. *d* orbitals
2. *f* orbitals
3. 5A. It might help you to remember that for the representative elements, the A group number is equal to the combined number of *s* and *p* electrons in the outer shell.
4. 6. Notice that the group number for representative elements also corresponds to the number of electrons in the outer shell.
5. N, P, As. Remember that within a group, atomic radii increase going down the group because of the filling of energy levels with higher *n*.
6. increase. Keep in mind that higher ionization energy means increased difficulty of removing an electron.
7. decrease
8. 1–. Each halogen has a configuration of ns^2np^5 when neutral, so with a gain of one electron, it would attain the noble gas configuration of ns^2np^6.
9. 54. The atomic number of iodine is 53. It gains one electron to form the I^- ion.
10. 18. A calcium atom has 20 electrons. It loses two electrons to form Ca^{2+} ion.
11. less. Ba has rather low first and second ionization energies but positive electron affinity. You could also look at its electronegativity, which is low. This means that is has a good tendency to lose electrons (form cations) and very little tendency to gain electrons (form anions).
12. smaller than. Understand the explanation in Section 5-5.
13. isoelectronic. The prefix "iso" means "same." Do you remember *iso*topes? Watch for further occurrences of this prefix in your study of chemistry.

14. Kr. To form an ion with a charge of 2+, a Sr atom must lose two of its original 38 electrons, to give 36 electrons in Sr^{2+}. A neutral Kr atom has 36 electrons.

15. 0. See the oxidation state rules in Section 5-7.

16. the charge on the ion. See Section 5-7.

17. +1; –2. The rare exceptions are pointed out in Table 5-4.

18. +1; –2. Remember that in a binary ionic compound, the oxidation states are equal to the charges on the ions. You should recall the rules in Section 5-7 and the common oxidation states in Table 5-4. The Group 1A (1) elements always have oxidation states of +1 in compounds; in binary compounds with less electronegative elements, the Group 6A (16) elements have oxidation state –2.

19. +1; –1

20. +1; +5; –2. First figure out Na (+1 in compounds) and O (–2). Then the sum of all oxidation states in a compound must be 0, so Cl must be +5.

21. +6; –2. As in the preceding question, recall that O has oxidation state –2. Then the sum of oxidation states in a polyatomic ion is equal to the charge on the ion, so S must have oxidation state +6.

22. acidic. See Section 5-9.

23. metals. See Section 5-9.

24. an alkali metal. Some such reactions occur at dangerously explosive rates.

25. nonmetal oxide. See Section 5-9.

26. hydrogen; an active metal

27. a nonmetal. Remember that the hydrides of the less active metals are intermediate in character between the ionic hydrides and the molecular hydrogen compounds. See the classification of the hydrides of the representative elements in Figure 5-5.

28. HX

29. H_2Y

30. M_2O. But you should be aware that the alkali metals sometimes form peroxides or superoxides in their direct reactions with O_2. See Table 5-5.

31. water, H_2O; hydrogen peroxide, H_2O_2

32. dioxygen, O_2; ozone, O_3

33. peroxide; O_2^{2-}; –1. Keep in mind that we present an ionic charge as $n-$ or $n+$, while the oxidation number or oxidation state is designated as $-n$ or $+n$. The peroxides are one of the few classes of compounds in which oxygen exhibits an oxidation number other than –2.

34. acid; react with water to give acid solution. How many examples of this behavior can you list?

35. basic; react with water to form bases. These are primarily the oxides of the alkali metals and those of the heavier alkaline earth metals.

36. amphoteric

37. Na_2O_2

38. H_2O_2

39. oxide

40. superoxide

41. peroxide

42. combustion

43. water, H_2O; carbon dioxide, CO_2

44. heat. These reactions are always exothermic. The usual reason for carrying out combustion reactions is to use the heat energy produced, rather than to make desirable products.
45. sulfur dioxide, SO_2
46. sulfur trioxide, SO_3; sulfuric acid, H_2SO_4

Multiple Choice

1. (d). All have the outer configuration ns^2np^6.
2. (a). Iodine is near the right side of the periodic table. Thus, it has a fairly high electronegativity, and does not give up electrons easily, but gains them readily.
3. (b). Be clear about these classifications, Section 5-1.
4. (b). The term "transition element" without further qualification refers to the d-transition elements.
5. (a). Phosphorus is in Group 6A (16).
6. (b). The elements in Group 2A (2) all have two electrons in the outer shell.
7. (b). To see this, you can look at the periodic table and to determine where elements 119 and 120 would fall. Alternatively, you can figure out their electron configurations. Element 119 would have outer shell $8s^1$, and would thus fall in Group 1A (1), with the other ns^1 elements. Element 120 would have outer shell $8s^2$, and would thus fall in Group 2A (2), with the other ns^2 elements.
8. (d). The same number as the group number, 1A-8A (ones digit of the group number, 13-18).
9. (c). Electronegativity increases left to right and decreases top to bottom.
10. (e). Higher ionization energy means greater difficulty of removing electrons. First ionization energy increases going left to right across a period. See Figure 5-2 and the accompanying discussion in terms of electron configurations.
11. (c). Atomic radii generally decrease left to right within a period and increase going down a group. But the changes in this property going down a group are much more pronounced than those going across a period. Thus, we know Mg would be *smaller* than Na because it is further to the right. We know that Mg would also be *smaller* than Ca, because it is higher. K is even *larger* than Na (because it is further down) and *larger* than Ca (because it is further left). Sr is further down than any other of these, so it is the largest. Study Section 5-2.
12. (d). Remember the meanings of these terms. A low ionization energy implies ease of removal of electrons, while a low electronegativity means a poor tendency to attract electrons to it. Both of these descriptions fit Cs, with an outer shell electron configuration $6s^1$.
13. (c). It's relatively easy to remove one electron from an alkali metal atom but much harder to remove the second one. It's relatively easy to remove two electrons from an alkaline earth metal atom.
14. (a). All of these ions have the same number of electrons (18). Sc has the highest nuclear charge (21+) to attract these electrons, and S has the lowest nuclear charge (16+). Thus, S^{2-} would have its electrons most loosely held, and S^{2-} would be the largest of these ions.
15. (c). Cu^+ has $29 - 1 = 28$ electrons, and Zn^{2+} has $30 - 2 = 28$ electrons.

16. (d). Neutral Cr has 24 electrons; Cr^{3+} has three fewer.
17. (c). For Cl: $(+1) + (x) + 3(-2) = 0$
18. (d). Ca is a alkaline earth metal and F is a halogen combined with a metal in a binary compound.
19. (c). Chlorine in this compound has an oxidation number of +5. $(+1) + (+5) + 3(-2) = 0$
20. (b). Nitrogen in this compound has an oxidation number of +3. $(+1) + (+3) + 2(-2) = 0$
21. (e). Sulfur in this compound has an oxidation number of +6. $2(+1) + (+6) + 4(-2) = 0$
22. (e). $(+1) + (+7) + 4(-2) = 0$
23. (d). MnO: $(x) + (-2) = 0$; MnO_2: $(x) + 2(-2) = 0$; Mn_2O_7: $2(x) + 7(-2) = 0$
24. (a). The melting and boiling points of the elements increase as you go down Group 6A (16).
25. (a). Oxygen gas makes up about $^1/_5$ of the atmosphere.
26. (a). The trend of decreasing first ionization energy going down the group prevails in all representative groups of the periodic table. Review Section 5-3.
27. (e). This method is applicable only to the oxides of the Group 1A (1) metals and the heavier Group 2A (2) metals. Study Section 5-9.
28. (e). Ba, from Group 2A (2), is the only active metal listed. See Figure 5-9.
29. (d). Hydrogen compounds of nonmetals are molecular. (See Section 5-8, Figure 5-5.)
30. (e). Increasing ionic character of oxides is correlated with increasing metallic character of the element with which oxygen is combined.
31. (c). Remember how to identify oxides, peroxides, and superoxides by determining the oxidation state of oxygen in the compound.
32. (a). Oxides of nonmetals are the most acidic.
33. (a). An acidic anhydride is an oxide of a nonmetal, like SiO_2, and a basic anhydride is an oxide of a metal, like CaO.

6 Some Types of Chemical Reactions

Chapter Summary

Any study of chemistry should include a discussion of the reactions that elements and compounds undergo. This part of the study is termed "descriptive chemistry." There are more than a hundred elements and at least sixty-nine million known compounds, each able to undergo many reactions. A study of this huge mass of information might seem hopeless. Fortunately, we can systematize this information in several ways to make it manageable. This chapter will help you start your study of descriptive chemistry by showing you several ways to organize the information and introducing you to some general terminology of descriptive chemistry.

Section 6-1 introduces you to the nature of some important substances, known as *acids*, *bases*, and *salts*. We shall study acids, bases and their reactions in more detail in Chapters 10, 11, 18, and 19. According to the view presented here, in aqueous solution an **acid** acts as a source of *hydrogen ions*, H^+, and a **base** is a source of *hydroxide ions*, OH^-. A **salt** is an ionic compound, consisting of a *cation* (other than H^+) and an *anion* (other than hydroxide ion, OH^-, or oxide ion, O^{2-}).

The fundamental reaction that acids undergo is **ionization**,

$$H_nX(aq) \rightarrow nH^+(aq) + X^{n-}(aq)$$

where n is usually 1, but sometimes 2 (e.g., sulfuric acid, H_2SO_4) or even 3 (e.g., phosphoric acid, H_3PO_4). The acid, a molecular substance, is separated into ions. Bases and salts already consist of ions in the pure state (solids). These ions are separated from each other when the compounds are dissolved in water (if they're soluble), a process known as **dissociation**,

$$CaCl_2(s) \rightarrow Ca^{2+}(aq) + 2Cl^-(aq)$$

Many substances dissolve in water to give solutions (*aqueous* solutions) that contain ions. Due to the mobility of these ions, such solutions are able to conduct electricity; these substances are called **electrolytes**. They range in this behavior from **strong** to **weak**. Aqueous solutions of other substances, **nonelectrolytes**, do not conduct electricity to any appreciable extent.

Some acids undergo ionization completely, or almost completely, in dilute aqueous solutions; these are referred to as **strong acids**. The seven common strong acids and their anions (H_nX and X^{n-} in the previous general representation) are listed in Table 6-1. *To continue this course with the best success, you must learn the names and formulas of these acids and their anions.* Some other acids undergo this reaction only slightly in dilute aqueous solution; these are referred to as **weak acids**. These are much more numerous, but a brief list of common weak acids and their anions appears in Table 6-2. In solutions of strong acids, very little of the molecular acid exists in solution, nearly all of it having been converted to the anion. In weak acid solutions, the acid exists almost entirely as nonionized acid molecules, with relatively little

83

of the anion formed. This is an example of a reaction that is strongly **reversible**, it occurs in both directions, forward and reverse, indicated by a double arrow ($\rightleftharpoons$). This chapter introduces a color convention that we use for representing acids and bases throughout the book, blue for acids and pink for their anions and bases.

Similarly, bases can be strong or weak. Many common bases are metal hydroxides. Most of these are not very soluble in water, so only a few common bases need to be considered. *You must learn the list of common strong bases in Table 6-3, so that you can recall them easily.* The fundamental reaction for the dissociation of strong bases can be written as

$$M(OH)_n(s) \rightarrow M^{n+}(aq) + nOH^-(aq),$$

where n is an integer, 1 for the alkali metals (Group 1A or 1) and 2 for the alkaline earth metals (Group 2A or 2). Some metal hydroxides dissolve so little that they release very few hydroxide ions; these are called **insoluble bases**. Ammonia, NH_3, is the only common **weak base**.

Do not confuse the terminology of "weak" and "strong" acids and bases with that of "dilute" and "concentrated" solutions. The terms "dilute" and "concentrated" refer to *how much* of a substance we put into the solution. We have already learned about two methods for describing the concentrations of solutions (Section 3-6). The terms "strong" and "weak" refer to the *fraction* of acid or base that ionizes to produce H^+ or OH^-. In Chapter 18 we will learn about more quantitative ways to describe the strengths of acids and bases.

The guidelines that summarize the solubilities of some common ionic compounds in water (Section 6-1 part 5 and Table 6-4) are very important. You need to learn these guidelines.

In Section 6-2, we learn several useful ways of writing chemical equations, depending on what feature of the reaction we wish to emphasize. The three ways are: (a) the **formula unit equation**, in which all compounds are represented by complete formulas; (b) the **total ionic equation**, in which the formulas are written to show the form in which each substance exists in solution; and (c) the **net ionic equation**, which includes only those species that actually react. In this last version, the **spectator ions** have been eliminated from the equation. This formalism for writing equations is useful for many types of reactions.

We must name compounds unambiguously so that we can communicate about chemistry. In Chapter 2, you saw an introduction to naming compounds. Now, you will expand on that by learning about a more systematic way to name compounds. Many of the rules for naming inorganic compounds require knowing the oxidation number of one or more of the elements in the compound (Section 5-7). You need to learn the names, charges, and formulas for the common ions appearing in Table 6-6. (You already know many of these from Chapter 2). Section 6-3 discusses the naming of **binary** compounds—those consisting of two elements. Most of the naming of compounds that you learn will be according to the systematic nomenclature of the IUPAC convention. But many nonsystematic traditional names are still widely used, and it will be necessary to learn to recognize some of these.

Ternary acids contain H, O, and one other element, usually a nonmetal. In Section 6-4, you will learn how to name ternary acids and their salts. You must learn the names of the "-ic" acids in Table 6-7. Then you can use these as a starting point to name many other ternary acids and their salts.

Oxidation-reduction reactions (Section 6-5) are those in which some substances undergo changes in oxidation number. For short, these are often called **redox** reactions. It is essential that you learn the terminology—**oxidation, reduction, oxidizing agent, reducing agent**—that is presented early in the section. An oxidizing agent: (1) gains (or appears to gain) electrons, (2) undergoes a decrease in oxidation number (that is, is reduced), and (3) oxidizes other substances. A reducing agent: (1) loses (or appears to lose) electrons, (2) undergoes an increase in oxidation number (that is, is oxidized), and (3) reduces other substances. Oxidation cannot occur without a corresponding reduction, and vice versa. You have already encountered many oxidation-reduction reactions earlier in this chapter. Throughout the text, we use the code colors blue for oxidation and red for reduction as a memory aid (remember *red* for *red*uction). A reaction in which one element is both oxidized and reduced is termed a **disproportionation reaction.** Acid-base neutralization reactions are *not* oxidation-reduction reactions, but many of the other types of reactions we have seen are frequently also redox reactions. The study of electron-transfer reactions (electrochemistry, electrolysis, voltaic cells, batteries, etc.), which is the subject of Chapter 21, also involves oxidation-reduction reactions. In addition to being able to recognize such reactions, you need to learn to identify which substance is oxidized and which is reduced. Be sure that you understand the terminology. We sometimes call the *compound* or the *polyatomic ion* that contains the element undergoing a *decrease* in oxidation number the oxidizing agent; the compound that contains the element undergoing an oxidation number *increase* is called the reducing agent.

Combination reactions, those in which two or more substances combine to form a single compound, are discussed in Section 6-6. This class of reactions can be further subdivided according to the nature of the reactants. Remember (Chapter 3) that the reactants are the substances that are consumed as the reaction proceeds (written to the left of the reaction arrow) and the products are those substances being formed (written to the right of the arrow). (1) In some combination reactions, the two reactants are both elements. Typical reactions are the formation of **hydrides** by the reaction of metals with hydrogen, formation of **oxides** by the reaction of metals or nonmetals with oxygen, and other reactions of metals with nonmetals. (2) Some other combination reactions have an element and a compound as reactants. (3) Two compounds can combine to form a single new compound.

Decomposition reactions (Section 6-7) are just the reverse of combination reactions; they contain a single reactant and more than one product. They can also be subclassified into three groups, this time in terms of the nature of the products. (1) In some decomposition reactions, both products are elements. (2) In other decomposition reactions, the products are an element and one or more compounds, although this type of reaction is not common. (3) A compound can decompose to form two or more compounds.

Displacement reactions (Section 6-8) are those in which one element displaces another from a compound. All displacement reactions are redox reactions. Examples of such reactions are the displacement of one metal from a compound by another, of hydrogen from a compound by a metal, and of one nonmetal by another. We call the metal (or nonmetal) most able to displace hydrogen or other metals (or other nonmetals) from a compound the *most active*. A ranking of the ability of metals and hydrogen to displace one another from compounds is known as the **activity series**, Table 6-9. In addition, the activities of the halogens decrease going down the group. You must be able to use these rankings to predict whether one particular element will displace another. In Chapter 21, we will see a quantitative approach to these rankings.

In **metathesis reactions** (Section 6-9), two compounds react to form two new compounds. The terminology of "changing partners" will help you to recognize metathesis reactions.

Acid-base reactions represent a very important kind of metathesis reaction. These are also called **neutralization reactions**, because the typical properties of acids and bases are neutralized. In such a reaction, *an acid reacts with a base to produce a salt and (usually) water*. The essence of a neutralization reaction is the combination of H^+ and OH^- ions to form water molecules. This is represented by the following net ionic equation for any acid-base neutralization reaction between a strong acid and a strong base to form a soluble salt and water:

$$H^+(aq) + OH^-(aq) \rightarrow H_2O(\ell)$$

The spectator ions (and hence the salt that is formed) are different depending on the identities of the acid and the base involved in the reaction. Similar net ionic equations for neutralizations involving weak acids, weak or insoluble bases, or insoluble salts are also discussed in this section. Some metathesis reactions result in the formation of a **precipitate** (an insoluble solid); these are called **precipitation reactions** (Section 6-9.2). The solubility guidelines help us recognize such reactions.

Finally, in a **gas-formation reaction** (Section 6-10), an insoluble or slightly soluble gas is formed from reactants that are *not* gases. Such gases include CO_2, SO_2, and H_2S.

One of your principal goals in this chapter should be to classify reactions according to the types presented here. The summary and examples in Table 6-11 in Section 6-11 will be very helpful. Remember that some reactions can be classified in more than one way. For instance, some acid-base neutralization reactions are also precipitation reactions; and displacement reactions, many combination reactions, and many decomposition reactions are also oxidation-reduction reactions.

This chapter brings together many of the concepts and skills of earlier chapters and applies them to some important types of reactions. In addition, it introduces you to ways of viewing, describing, and systematizing such reactions and to terminology commonly used in discussing the reactions. As you study the chapter, you must become familiar with the terminology by being aware of the precise meanings of each term. Try also to relate this material to what you already know from earlier chapters.

Study Goals

Use these goals to help you organize your study. As in other chapters, some textbook Exercises related to each study goal are indicated.

Section 6-1.1
1. Describe solutions of strong electrolytes, weak electrolytes, and nonelectrolytes, giving examples of each. *Work Exercises 1, 2, 11 through 15, 17, 19, and 20.*

Sections 6-1.1 through 6-1.4
2. Be familiar with the characteristics of acids, bases, salts, and their solutions. *Work Exercises 3, 4, 6, 9 through 11, and 24.*

Sections 6-1.2 through 6-1.4
3. Distinguish between strong acids and weak acids. Distinguish among strong bases, weak bases, and insoluble bases. *Work Exercises 5 through 11 and 14.*

4. Remember the names and formulas of the seven common strong acids and their anions listed in Table 6-1. Remember the names and formulas of the common weak acids and their

anions listed in Table 6-2. Remember the names and formulas of the eight strong bases in Table 6-3. *Work Exercises 5 through 12 and 18.*

Section 6-1.5

5. Know the solubility guidelines (Table 6-4) and apply them to identify relatively soluble and relatively insoluble compounds. *Work Exercises 13, 16, 21, 22, and 88 through 97.*

Section 6-2

6. Distinguish among and write formula unit, total ionic, and net ionic equations. *Work Exercises 55 and 56.*

Sections 6-3 and 6-4

7. Given their formulas, name binary and other common inorganic compounds by the IUPAC system. Given their IUPAC names, write the formulas of common inorganic compounds. Name common inorganic compounds by other commonly used traditional methods. *Work Exercises 27 through 48.*

Section 6-5

8. Know what is meant by each of the following terms: (a) oxidation, (b) reduction, (c) oxidizing agent, (d) reducing agent. Identify oxidizing agents and reducing agents in equations for redox reactions. *Work Exercises 49 through 54.*

Sections 6-6 and 6-7

9. Write balanced formula unit equations for combination and decomposition reactions. *Work Exercises 57 through 62.*

Section 6-8

10. Use the activity series (Table 6-9) to predict whether or not one element would displace another from a compound. Write equations for displacement reactions. *Work Exercises 63 through 74.*

Section 6-9

11. Predict the potential products of a metathesis reaction, predict whether or not the reaction will occur, and write the complete, balanced formula unit equation. *Work Exercises 77 through 84 and 95 through 98.*

Section 6-10

12. Identify gas-formation reactions. *Work Exercises 107 and 108.*

Sections 6-2 and 6-5 through 6-10

13. Write balanced formula unit equations, total ionic equations, and net ionic equations as appropriate for reaction types discussed in this chapter. *Work Exercises 53, 55, 56, 63, 64, 67, 75 through 84, 95, and 96.*

Sections 6-11

14. Classify chemical reactions as one or more of these types: (a) oxidation-reduction (redox), (b) combination, (c) decomposition, (d) displacement, (e) metathesis, (f) acid-base neutralization, (f) precipitation, (g) gas-formation. *Work Exercises 99 through 109.*

General

15. Answer conceptual questions based on this chapter. *Work **Conceptual Exercises** 112 through 126.*

16. Apply concepts and skills from earlier chapters to the ideas of this chapter. *Work **Building Your Knowledge** exercises 127 through 128.*

17. Use sources outside the textbook to work **Beyond the Textbook** *exercises 129 through 131.*

Some Important Terms in This Chapter

Some important new terms from Chapter 6 are listed at the top of the next page. (Be sure to study other new terms and review terms from preceding chapters, if necessary.) Fill the blanks in the following paragraphs with terms from the list. Each term is only used once.

acids	gas-formation reaction	oxidizing agent
activity series	metathesis reaction	precipitation reaction
bases	net ionic equation	reducing agent
binary compound	neutralization reaction	reduction
combination reaction	salts	strong electrolytes
decomposition reaction	nonelectrolytes	ternary compound
displacement reaction	oxidation	total ionic equation
electrolytes	oxidation-reduction reaction	weak electrolytes
formula unit equation		

The properties of compounds also have patterns. Many compounds dissolve in water. Some of these, called [1]_____, produce solutions that will conduct an electric current. Substances whose solutions do not conduct electricity are called [2]_____. Electrolytes can be further categorized by how well they conduct electricity. Good conductors of electricity are [3]_____; poor conductors of electricity are [4]_____. Most electrolytes are acids, bases, or salts. Electrolytes are believed to conduct electricity because they produce ions in solution. Substances that produce hydrogen ions, H^+, when dissolved are called [5]_____; and those that produce hydroxide ions, OH^-, in solution are [6]_____. Most other ionic compounds are called [7]_____, such as ordinary table salt.

The way a chemical equation is written depends upon the type of information we wish to emphasize. Writing chemical formulas for all of the compounds, as if they existed as neutral formula units, produces a [8]_____, also known as a molecular equation. However, when electrolytes are dissolved in water they separate into positive and negative ions. Representing all of the ions produced from the formula units produces the [9]_____, also known as the *complete ionic equation*. Generally, some of the ions are the same on both sides of the equation. These *spectator ions* do not need to be included in the equation, since they do not change. Leaving them out produces the [10]_____.

If an atom *loses* one or more electrons, it becomes more *positive*; that is, its oxidation number increases, a process known as [11]_____. Conversely, if an atom *gains* one or more electrons, it becomes more *negative*; that is, its oxidation number decreases, a process known as [12]_____. In chemical reactions, oxidation cannot occur by itself; neither can reduction occur by itself. Electrons are matter, and, therefore, can be neither created nor destroyed. Thus, oxidation and reduction always occur together in a type of chemical reaction known as an [13]_____, or simply, *redox reaction*. Just as a secret agent is a person who *does* something, substances that are oxidized or reduced *do* something. However, for every action, there's an equal and opposite reaction. A substance that *reduces* a substance is a [14]_____. In the process, this agent is oxidized. Similarly, a substance that *oxidizes* a substance is an [15]_____ _____. At the same time, this agent is reduced.

There are several other types of chemical reactions. Two or more substances are *combined* to form a compound in a [16]_____. Conversely, a compound *decomposes* to produce two or more simpler substances in a [17]_____ _____. One element *displaces* another element from a compound in a [18]_____. Two compounds in aqueous solution, exchange ionic partners to form new compounds in a double displacement or [19]_____ _____. An acid and a base *neutralize* each other's properties in a [20]_____. Two soluble compounds combine to produce an insoluble product that *precipitates* in a [21]_____. An insoluble or slightly soluble *gas forms* in a [22]_____. A pattern that helps us predict the occurrence of displacement reactions is a list of the metals and hydrogen in order of decreasing chemical *activity*, a list called the [23]_____.

Compounds can also be grouped by the number of elements in the compound, two or more than two. A compound composed of two elements is a [24]_____, and a compound composed of more than two elements is a [25]_____.

Preliminary Test

As in other chapters, this test will check your understanding of basic concepts and types of calculations. Be sure to practice *many* of the additional textbook exercises, including those indicated with the Study Goals.

True-False

Mark each statement as true (T) or false (F).

_____ 1. All aqueous solutions are good conductors of electricity.
_____ 2. There are more weak acids than strong acids.
_____ 3. All chemical reactions go to completion.
_____ 4. All of the common strong bases are metal hydroxides.
_____ 5. All metal hydroxides are strong bases.
_____ 6. A reaction in solution that forms an insoluble solid product is called a precipitation reaction.
_____ 7. Acid-base neutralization reactions can also be precipitation reactions.
_____ 8. Lighter metals can always displace heavier ones.

Short Answer

Answer with a word, a phrase, a formula, or a number with units as necessary.

1. An acid is a substance that produces _____ (name, formula) in aqueous solution.

2. An acid that does not very efficiently produce H^+ in aqueous solutions is called a (an) _____.

3. The name and formula of the anion of nitric acid is _____.

4. The name and formula of the anion of acetic acid is _____.

5. The name and formula of the acid of which cyanide ion is the anion is _____ _____.

6. All of the strong bases listed in the text chapter are hydroxides of the elements of Groups _____ and _____.

7. Organic acids contain the atom grouping _____.

8. A base, when dissolved in water, produces _____ (name, formula).

9. When a strong acid is dissolved in water, its ionization takes place _____.

10. When a weak acid is dissolved in water, its ionization takes place _____.

11. Classify each of the following substances as soluble or insoluble in water.

    (a) KNO_3 _____
    (b) $SrCl_3$ _____
    (c) AgI _____
    (d) $Al(OH)_3$ _____
    (e) $(NH_4)_2S$ _____
    (f) FeS _____
    (g) $CaCO_3$ _____
    (h) HBr _____
    (i) $PbCl_2$ _____
    (j) $NaSCN$ _____
    (k) $Ca(OH)_2$ _____
    (l) $AgNO_3$ _____
    (m) $Mg(CH_3COO)_2$ _____
    (n) $Zn(ClO_4)_2$ _____
    (o) H_3PO_4 _____
    (p) $Ba_3(PO_4)_2$ _____
    (q) MgS _____
    (r) $(NH_4)_2CO_3$ _____
    (s) KOH _____
    (t) Rb_2SO_4 _____

12. A reaction in which the elements or ions may be described as "changing partners" is called a _____ reaction.

13. A reaction in which one element reacts with a compound to give a new compound and an element is called a _____ reaction.

14. A reaction in which one compound breaks apart to produce one or more elements or simpler compounds is called a _____ reaction.

15. A reaction in which two or more elements or compounds come together to form a single new compound is called a _____ reaction.

16. An insoluble solid that is formed in a reaction is called a _____.

17. A reaction in which an insoluble solid is formed is called a _____ reaction.

18. When silver nitrate solution and sodium sulfate solution are mixed, silver sulfate precipitates, leaving a solution containing sodium ions and nitrate ions. The balanced formula unit equation for this reaction is _____ .
    This reaction may be classified as a _____ reaction.
    This reaction _____ (is, is not) an oxidation–reduction reaction.

90

19. Solid lead(IV) chloride, when heated to 100ºC, reacts to give solid lead(II) chloride and chlorine gas. The balanced formula unit equation for this reaction is:

    _____ .

    This reaction may be classified as a _____ reaction.
    This reaction _____ (is, is not) an oxidation-reduction reaction.

20. When the ammonia fumes above a solution are allowed to mix with the hydrogen chloride fumes above a hydrochloric acid solution, a "smoke" (actually a colloid of finely divided solid particles—see Chapter 14) of ammonium chloride forms. The balanced formula unit equation for this reaction is _____ .
    This reaction may be classified as a _____ reaction.
    This reaction _____ (is, is not) an oxidation-reduction reaction.

21. In the following equations, A, B, X, and Y represent hypothetical elements, while H and O are hydrogen and oxygen, respectively. Classify each of the following as one or more of the following: a combination reaction, a displacement reaction, a decomposition reaction, a metathesis reaction, an acid-base neutralization reaction, a gas-formation reaction.

    (a) $XO + H_2O \rightarrow H_2XO_2$ _____

    (b) $A + X_2 \rightarrow AX_2$ _____

    (c) $A + BX \rightarrow AX + B$ _____

    (d) $2AX + Y_2 \rightarrow 2AY + X_2$ _____

    (e) $HX + BOH \rightarrow H_2O + BX$ _____

    (f) $AX \rightarrow A + X$ _____

    (g) $A + B \rightarrow AB$ _____

    (h) $AX + BX \rightarrow ABX_2$ _____

    (i) $2A_2O \rightarrow 2A_2 + O_2$ _____

    (j) $2HX + B(OH)_2 \rightarrow 2H_2O + BX_2$ _____

    (k) $AX + BY \rightarrow AY + BX$ _____

    (l) $AXO_3 \rightarrow AO + XO_2$ _____

    (m) $2A + 2HX \rightarrow 2AX + H_2$ _____

    (n) $2A + X_2 \rightarrow 2AX$ _____

    (o) $2AX + X_2 \rightarrow 2AX_2$ _____

    (p) $A_2 + 2BX \rightarrow 2AX + B_2$ _____

    (q) $AO + BO \rightarrow ABO_2$ _____

    (r) $2AXO_3 \rightarrow 2AX + 3O_2$ _____

22. When an acid and a base act to neutralize one another, the products are _____ and _____ .

23. The name and formula of the salt that results from the neutralization of calcium hydroxide and nitric acid are _____ .

24. The formula for magnesium acetate is _____ .

25. The net ionic reaction for all neutralization reactions involving strong acids and strong bases to form soluble salts in water is _____ .

26. For the neutralization reaction of sulfuric acid and potassium hydroxide, the balanced formula unit equation is _____ , the total ionic equation is _____ , and the net ionic equation is _____ . The spectator ions are _____ .

27. A process in which electrons are lost is termed _____ .

28. A substance that loses electrons in a reaction is termed a (an) _____ .

29. A substance that undergoes an increase in oxidation number in a reaction is termed a (an) _____ .

30. The substance that is reduced in a reaction is termed a (an) _____ .

31. In any reaction in which reduction occurs, _____ must also take place.

32. Another name for an oxidation-reduction reaction is a _____ .

_____

The reaction in which nitrogen and hydrogen react to give ammonia is an oxidation-reduction reaction in which all substances are gases. Questions 33 through 36 refer to this reaction.

33. The unbalanced formula unit equation that describes this reaction is:

_____ .

34. In this reaction, nitrogen undergoes a change in oxidation number from _____ to _____; hydrogen undergoes a change in oxidation number from _____ to _____ .

35. The balanced formula unit equation that describes this reaction is:

_____ .

36. This reaction can also be classified as a _____ reaction.

_____

37. An oxidizing agent is a species that _____ electrons.

38. When an element in a substance undergoes an algebraic decrease in oxidation number, we say that the substance is being _____ . In such a reaction, it acts as the _____ agent.

39. The usual product formed when sulfuric acid acts as an oxidizing agent is SO_2; in this process, the oxidation number of sulfur changes from ____ to ____ .

40. When permanganate ion acts as an oxidizing agent in acidic solution, the usual product formed is manganese(II) or manganous ion, Mn^{2+}; in this process, the oxidation number of manganese changes from ____ to ____ .

41. When permanganate ion acts as an oxidizing agent in basic solution, the usual product formed is manganese dioxide or manganese(IV) oxide, MnO_2; in this process, the oxidation number of manganese changes from ____ to ____ .

42. When $Sn(OH)_3^-$ is converted to $Sn(OH)_6^{2-}$, the oxidation number of the tin changes from ____ to ____; it is being _____ .

43. A binary compound is one that consists of _____ elements.

44. A ternary compound is one that consists of _____ elements.

45. A ternary acid is one that contains H, O, and _____ .

_____

92

Write the formulas for the compounds or ions named in Questions 46 through 71.

46. cupric chloride _____

47. copper(II) chloride _____

48. sodium acetate _____

49. sulfate ion _____

50. ammonium ion _____

51. permanganate ion _____

52. sodium hydrogen sulfate _____

53. sodium sulfate _____

54. sodium sulfite _____

55. sodium nitrate _____

56. sodium nitrite _____

57. lithium cyanide _____

58. iron(III) sulfite _____

59. sodium dichromate _____

60. sodium chromate _____

61. sulfurous acid _____

62. sulfuric acid _____

63. hypobromous acid _____

64. bromous acid _____

65. bromic acid _____

66. perbromic acid _____

67. hydrobromic acid _____

68. phosphoric acid _____

69. carbonic acid _____

70. magnesium carbonate _____

71. potassium hydrogen carbonate _____

Name the compounds or ions whose formulas appear in Questions 72 through 97.

72. $CaSO_4$ _____

73. $FeSO_4$ _____

74. $Fe_2(SO_4)_3$ _____

75. $NaMnO_4$ _____

76. $Fe(CN)_3$ _____

77. $Ca(NO_2)_2$ _____

78. CdS _____

93

79. CrF_3 _____

80. AsO_4^{3-} _____

81. ClO_4^- _____

82. Zn^{2+} _____

83. $CoSO_3$ _____

84. $Ba(SCN)_2$ _____

85. CH_3COO^- _____

86. $NaHCO_3$ _____

87. $Fe_3(PO_4)_2$ _____

88. $(NH_4)_2CO_3$ _____

89. NiO _____

90. $AgCl$ _____

91. $Hg_2(CN)_2$ _____

92. $HF(aq)$ _____

93. $HI(aq)$ _____

94. $Hg(NO_2)_2$ _____

95. H_3PO_3 _____

96. HIO_3 _____

97. HNO_3 _____

Multiple Choice

_____ 1. Which one of the following is *not* a strong acid?

(a) HF (b) HCl (c) HBr (d) HI (e) $HClO_4$

_____ 2. Which one of the following is *not* a weak acid?

(a) HNO_3 (b) $(COOH)_2$ (c) H_2SO_3 (d) HSO_3^- (e) CH_3COOH

_____ 3. What salt would be formed by the complete neutralization of $Mg(OH)_2$ with HCl?

(a) $MgCl$ (b) $MgCl_2$ (c) Mg_2Cl (d) HMg (e) $ClOH$

_____ 4. What salt would result from the complete neutralization reaction between sulfuric acid and potassium hydroxide?

(a) K_2SO_4 (b) KSO_4 (c) $K_2(SO_4)_2$ (d) KS (e) K_2S

_____ 5. The reaction: $2HCl + Ca(OH)_2 \rightarrow 2H_2O + CaCl_2$ can be described as

(a) a neutralization reaction. (b) a displacement reaction.

(c) a disproportionation reaction. (d) a precipitation reaction.

(e) an oxidation–reduction reaction.

_____ 6. What salt would result from the neutralization reaction between phosphoric acid and calcium hydroxide?
(a) $CaPO_4$ (b) $Ca_3(PO_4)_2$ (c) $Ca_2(PO_4)_3$ (d) $Ca(PO_4)_2$ (e) Ca_3PO_4

_____ 7. The following statements describe properties of substances.
Which one is *not* a typical acid property?
(a) It has a sour taste.
(b) It reacts with metal oxides to form salts and water.
(c) It reacts with acids to form salts and water.
(d) Its aqueous solutions conduct an electric current.
(e) It reacts with active metals to liberate H_2.

_____ 8. When heated in the presence of oxygen, O_2, the gas carbon monoxide, CO, is converted to carbon dioxide gas, CO_2. This reaction is best described as
(a) a displacement reaction. (b) a neutralization reaction.
(c) a precipitation reaction. (d) a redox reaction.

_____ 9. In the reaction: $2Na(s) + 2H_2O(\ell) \rightarrow 2Na^+(aq) + H_2(g) + 2OH^-(aq)$
sodium metal
(a) is acting as an acid.
(b) is the oxidizing agent and is being reduced.
(c) is the oxidizing agent and is being oxidized.
(d) is the reducing agent and is being oxidized.
(e) is the reducing agent and is being reduced.

_____ 10. In the reaction: $Cr_2O_7^{2-}(aq) + 3H_2S(aq) + 8H^+(aq) \rightarrow 2Cr^{3+}(aq) + 2S(s) + 7H_2O(\ell)$
which element is being oxidized?
(a) Cr (b) H (c) O (d) S
(e) None, because this is not an oxidation-reduction reaction.

_____ 11. In the reaction of Question 10, what is the reducing agent?
(a) $Cr_2O_7^{2-}$ (b) H_2S (c) H^+ (d) H_2O
(e) There is no reducing agent, because this is not a redox reaction.

_____ 12. In the reaction: $Mg(s) + Cl_2(g) \rightarrow MgCl_2(s)$ chlorine gas acts as
(a) an anion. (b) a cation. (c) an oxidizing agent. (d) an acid. (e) a reducing agent.

_____ 13. The reaction of Question 12 could be classified as
(a) a precipitation reaction. (b) a neutralization reaction.
(c) a displacement reaction. (d) a redox reaction.

_____ 14. In the reaction: $Ca_3(PO_4)_2(s) + 3H_2SO_4(aq) \rightarrow 2H_3PO_4(aq) + 3CaSO_4(s)$
which element is being oxidized?
(a) Ca (b) P (c) S (d) O
(e) None, because this is not an oxidation–reduction reaction.

_____ 15. Consider the following equation that is involved in the rusting of iron:

$$4Fe + O_2 \rightarrow 2Fe_2O_3$$

Which one of the following statements about this equation is *not* correct?

(a) This is an example of an oxidation–reduction reaction.
(b) Metallic iron is a reducing agent.
(c) O_2 is an oxidizing agent.
(d) Metallic iron is reduced.
(e) The total gain in oxidation number of one element equals the total loss of oxidation number of another element.

_____ 16. In the net ionic equation:

$$5SO_3^{2-} + 6H^+ + 2MnO_4^- \rightarrow 5SO_4^{2-} + 2Mn^{2+} + 3H_2O$$

which element is being reduced?

(a) S   (b) O   (c) H   (d) Mn
(e) None, because this is not a redox reaction.

_____ 17. In the reaction: $2SO_2 + O_2 \rightarrow 2SO_3$

what is the oxidizing agent?

(a) SO_2 (b) O_2 (c) SO_3
(d) None, because this is not a redox reaction.

_____ 18. In the reaction: $Mg + H_2O \rightarrow MgO + H_2$

what is the reducing agent?

(a) Mg   (b) H_2O (c) MgO   (d) H_2
(e) None, because this is not a redox reaction.

_____ 19. One of the reactions that take place in a copper smelter to remove unwanted compounds from the ore is:

$$CaCO_3 + SiO_2 \rightarrow CaSiO_3 + CO_2$$

In this reaction, what is the oxidizing agent?

(a) $CaCO_3$ (b) SiO_2 (c) $CaSiO_3$ (d) CO_2
(e) None, because this is not a redox reaction.

_____ 20. Copper(I) oxide can be prepared by boiling copper(I) chloride with a metal hydroxide. In this reaction, what is the oxidizing agent? The net ionic equation for this reaction is

$$2CuCl + 2OH^- \rightarrow Cu_2O + 2Cl^- + H_2O$$

(a) CuCl   (b) OH^- (c) Cu_2O (d) Cl^- (e) None, because this is not a redox reaction.

_____ 21. The following reaction (not balanced) is carried out in acidic solution. What is the oxidizing agent?

$$SO_3^{2-} + MnO_4^- \rightarrow SO_4^{2-} + Mn^{2+}$$

(a) SO_3^{2-} (b) MnO_4^- (c) SO_4^{2-} (d) Mn^{2+} (e) None (not a redox reaction)

_____ 22. In the reaction of Question 21, which species is described as the reducing agent?

(a) SO_3^{2-} (b) MnO_4^- (c) SO_4^{2-} (d) Mn^{2+}
(e) None, because this is not a redox reaction.

_____ 23. The reaction: $Cr_2O_7^{2-} + H_2S \rightarrow Cr^{3+} + S$

is carried out in acidic solution. What element is being oxidized?
(a) Cr   (b) H   (c) O   (d) S   (e) None, because this is not a redox reaction.

_____ 24. In the reaction of Question 23, which substance is described as the oxidizing agent?
(a) $Cr_2O_7^{2-}$ (b) H_2S (c) Cr^{3+} (d) S   (e) None, because this is not a redox reaction.

_____ 25. Which of the following is the formula for cobalt(II) nitrite?
(a) Co_2N (b) Co_2NO_2 (c) $Co(NO_2)_2$ (d) $CoNO_2$ (e) $Co(NO_3)_2$

_____ 26. Which of the following is the formula for iron(II) sulfite?
(a) $FeSO_4$ (b) FeS   (c) Fe_2SO_3 (d) $FeSO_3$ (e) $Fe(SO_3)_2$

_____ 27. What is the name of the ion SO_4^{2-}?
(a) sulfide   (b) sulfite   (c) sulfate   (d) sulfur tetroxide
(e) None of the preceding answers is correct.

_____ 28. Which of the following is the formula for iron(II) oxide?
(a) FeO   (b) FeO_3 (c) Fe_2O (d) Fe_2O_3 (e) Fe_3O_2

_____ 29. What is the name of the ion SO_3^{2-}?
(a) sulfide   (b) sulfite   (c) sulfate   (d) sulfur trioxide
(e) None of the preceding answers is correct.

_____ 30. Which of the following is the formula for nickel(II) sulfide?
(a) NiS   (b) Ni_2S (c) NiS_2 (d) $NiSO_4$ (e) Ni_2SO_4

_____ 31. What is the name of the ion NO_3^-?
(a) nitride   (b) nitrite   (c) nitrate   (d) nitrogen trioxide
(e) None of the preceding answers is correct.

_____ 32. Which of the following is the formula for manganese(II) hydroxide?
(a) MnO   (b) MnOH   (c) $Mn(OH)_2$ (d) Mn_2OH (e) $Mn_3(OH)_2$

_____ 33. Which of the following is the formula for iron(III) sulfate?
(a) $Fe_2(SO_4)_3$ (b) Fe_3SO_4 (c) $Fe(SO_3)_2$ (d) Fe_3SO_3 (e) $Fe_3(SO_4)_2$

_____ 34. Which of the following is the formula for chromium(III) nitrate?
(a) $CrNO_3$ (b) Cr_3NO_3 (c) $Cr(NO_3)_3$ (d) $Cr_2(NO_2)_3$ (e) $Cr(NO_2)_2$

NOTE: It takes a great deal of practice to learn to deal comfortably with the kind of material in this chapter. This is especially true for recognizing reaction types and identifying oxidizing and reducing agents. You need to work many of the text exercises suggested with the Study Goals to be able to recall and use this material easily.

Answers to Some Important Terms in This Chapter

1. electrolytes
2. nonelectrolytes
3. strong electrolytes
4. weak electrolytes
5. acids
6. bases
7. salts
8. formula unit equation
9. total ionic equation
10. net ionic equation
11. oxidation
12. reduction
13. oxidation-reduction reaction
14. reducing agent

15.	oxidizing agent	18.	displacement reaction	23.	activity series
16.	combination reaction	19.	metathesis reaction	24.	binary compound
17.	decomposition reaction	20.	neutralization reaction	25.	ternary compound
		21.	precipitation reaction		
		22.	gas-formation reaction		

Answers to Preliminary Test

Do not just look up answers. *Think about the reasons for the answers.*

True-False

1. False. Only solutions that contain significant numbers of ions conduct electricity. Read about electrolytes and nonelectrolytes, Section 6-1.1.
2. True. There are seven common strong acids (Table 6-1) and many more weak acids, a *few* of which are listed in Table 6-2. You should know the seven common strong acids.
3. False. Many reactions can occur in both directions, so they do not proceed completely to convert all of the reactants to products. Such reactions are called reversible reactions (Section 6-1.3).
4. True. These are listed in Table 6-3. You should know this list of bases.
5. False. Many metal hydroxides are essentially insoluble in water, so they cannot produce significant amounts of OH^- in solution.
6. True. "A precipitate precipitates in a precipitation reaction."
7. True. If the salt produced is insoluble, it forms a precipitate. An example is the neutralization reaction between sulfuric acid, H_2SO_4, and barium hydroxide, $Ba(OH)_2$. This reaction forms $BaSO_4$, which is an insoluble salt (refer to the solubility guidelines, Section 6-1.5 and Table 6-4). Can you think of other examples? However, since the metals that form soluble hydroxides mostly form salts that are also soluble, it is rare for an acid-base neutralization reaction to be a precipitation reaction also.
8. False. See the activity series, Table 6-9 and study Section 6-8.2.

Short Answer

1. hydrogen ion, H^+
2. weak acid
3. nitrate, NO_3^-. Remember that the anion of an acid is what is left when the acid gives up H^+ to water. See Section 6-1.2.
4. acetate, CH_3COO^-. Formulas for ions written without the proper charge are incorrect.
5. hydrocyanic acid, HCN. You can find out the acids corresponding to an ion by adding one (or more) H^+. Remember to adjust the charge.
6. Group 1A (alkali metals); the heavier members of Group 2A (alkaline earth metals).
7. —COOH.
8. hydroxide ion, OH^-.

9. completely
10. very slightly
11. You should *learn* and use the solubility guidelines in Section 6-1.5, Table 6-4.

(a) soluble	(f) insoluble	(k) soluble	(p) insoluble
(b) soluble	(g) insoluble	(l) soluble	(q) soluble
(c) insoluble	(h) soluble	(m) soluble	(r) soluble
(d) insoluble	(i) insoluble	(n) soluble	(s) soluble
(e) soluble	(j) soluble	(o) soluble	(t) soluble

12. metathesis
13. displacement
14. decomposition
15. combination
16. precipitate
17. precipitation
18. $2AgNO_3(aq) + Na_2SO_4(aq) \rightarrow Ag_2SO_4(s) + 2NaNO_3(aq)$; precipitation; is not. You can always check to find out whether a reaction is a redox reaction by determining the oxidation numbers of all elements on both sides of the equation. In this equation, there are no changes in oxidation number.
19. $PbCl_4(s) \rightarrow PbCl_2(s) + Cl_2(g)$; decomposition; is. Lead and part of the chlorine change in oxidation number.
20. $NH_3(g) + HCl(g) \rightarrow NH_4Cl(s)$; combination; is not.
21. (a) combination   (j) metathesis (a neutralization reaction)
    (b) combination   (k) metathesis
    (c) displacement   (l) decomposition
    (d) displacement   (m) displacement and gas-formation
    (e) metathesis (a neutralization reaction)   (n) combination
    (f) decomposition   (o) combination
    (g) combination   (p) displacement
    (h) combination   (q) combination
    (i) decomposition and gas-formation   (r) decomposition and gas-formation

    It would be helpful for you to find one or more examples of each of these reactions in the textbook. You might need to look somewhere other than Chapter 6 for some of them.
22. water; a salt (or reverse the order). Some salts are insoluble and precipitate from solution; others are soluble, remaining in solution in ionized form.

    NaCl, a compound that we commonly call "salt," is only one member of an entire class of compounds that are collectively called "salts" in correct chemical usage.
23. calcium nitrate, $Ca(NO_3)_2$. Review the names of the anions of the common strong acids (Table 6-1) and the common weak acids (Table 6-2). You must be familiar with the names of some common ions that you studied in Table 2-2, Section 2-3.
24. $Mg(CH_3COO)_2$
25. $H^+(aq) + OH^-(aq) \rightarrow H_2O(\ell)$
26. formula unit: $H_2SO_4(aq) + 2KOH(aq) \rightarrow 2H_2O(\ell) + K_2SO_4(aq)$

    total ionic: $[2H^+(aq) + SO_4^{2-}(aq)] + [2K^+(aq) + OH^-(aq)] \rightarrow$

$$2H_2O(\ell) + [2K^+(aq) + SO_4^{2-}(aq)]$$

net ionic: $H^+(aq) + OH^-(aq) \rightarrow H_2O(\ell)$

spectator ions: K^+, SO_4^{2-}

27. oxidation
28. reducing agent
29. reducing agent. Remember that a substance that undergoes an increase in oxidation number is said to be oxidized, and that the substance that is oxidized is the reducing agent.
30. oxidizing agent
31. oxidation
32. redox reaction
33. $N_2(g) + H_2(g) \rightarrow NH_3(g)$. It is incorrect to represent nitrogen gas as N or hydrogen as H, because they do not exist in those forms. Each is diatomic as the free element.
34. 0, –3; 0, +1.
35. $N_2(g) + 3H_2(g) \rightarrow 2NH_3(g)$
36. combination
37. gains
38. reduced; oxidizing
39. +6; +4
40. +7; +2
41. +7; +4
42. +2; +4; oxidized
43. two
44. three
45. another element, usually a nonmetal

---

46. $CuCl_2$
47. $CuCl_2$
48. $NaCH_3COO$
49. SO_4^{2-}
50. NH_4^+
51. MnO_4^-
52. $NaHSO_4$
53. Na_2SO_4
54. Na_2SO_3
55. $NaNO_3$
56. $NaNO_2$
57. $LiCN$
58. $Fe_2(SO_3)_3$
59. $Na_2Cr_2O_7$
60. Na_2CrO_4
61. H_2SO_3
62. H_2SO_4
63. $HBrO$
64. $HBrO_2$
65. $HBrO_3$
66. $HBrO_4$
67. $HBr(aq)$
68. H_3PO_4
69. H_2CO_3
70. $MgCO_3$
71. $KHCO_3$

NOTE: Where two answers are given for Questions 72 through 97, the first is the IUPAC name, which is usually the preferred one.

72. calcium sulfate
73. iron(II) sulfate or ferrous sulfate
74. iron(III) sulfate or

ferric sulfate
75. sodium permanganate
76. iron(III) cyanide or ferric cyanide

77. calcium nitrite
78. cadmium sulfide
79. chromium(III) fluoride or chromic fluoride
80. arsenate ion
81. perchlorate ion
82. zinc ion
83. cobalt(II) sulfite or cobaltous sulfite
84. barium thiocyanate
85. acetate ion
86. sodium hydrogen carbonate or sodium bicarbonate
87. iron(II) phosphate or

ferrous phosphate
88. ammonium carbonate
89. nickel(II) oxide or nickelous oxide
90. silver chloride
91. mercury(I) cyanide or mercurous cyanide
92. hydrofluoric acid
93. hydroiodic acid
94. mercury(II) nitrite or mercuric nitrite
95. phosphorous acid
96. iodic acid
97. nitric acid

Multiple Choice

1. (a). Remember that HF is the only one of the hydrohalic acids (binary acids that involve halogens) that is not a strong acid.

2. (a). Notice that (b), (c), and (e) are listed as common weak acids. Sulfite ion, SO_3^{2-}, is listed as one of the anions derived from the weak acid H_2SO_3; this means that HSO_3^- must also be able to act as a weak acid (very weak).

3. (b). $Mg(OH)_2 + 2HCl \rightarrow MgCl_2 + 2H_2O$

4. (a). $H_2SO_4 + 2KOH \rightarrow K_2SO_4 + 2H_2O$

5. (a). An acid, HCl, reacts with a base, $Ca(OH)_2$, to form water and a salt, $CaCl_2$.

6. (b). Calcium ion is Ca^{2+} and phosphate ion is PO_4^{3-}. These combine in a 3:2 ratio to form the salt $Ca_3(PO_4)_2$. $2H_3PO_4 + 3Ca(OH)_2 \rightarrow Ca_3(PO_4)_2 + 6H_2O$

7. (c). An acid reacts with *bases* to form salts and water.

8. (d). The oxidation number of C changes from +2 to +4, while the oxidation number of O changes from 0 (in the element O_2) to –2.

9. (d). The oxidation number of Na goes up from 0 to +1.

10. (d). Determine the oxidation number of each element on both sides of the equation. Sulfur goes from –2 in H_2S to 0 in S (the free element).

11. (b). Remember that the compound containing the element that is oxidized is called the reducing agent.

12. (c). The oxidation number of Cl is reduced from 0 to –1.

13. (d). A solid product is formed, but the reaction does not take place in solution, so this is not a precipitation reaction.

14. (e). No element undergoes a change in oxidation number. You can always check oxidation numbers to find out whether a reaction is a redox reaction.

15. (d). Iron is *oxidized* from 0 to +3.

16. (d). Look for changes in oxidation numbers. Mn changes from +7 to +2.

17. (b). The oxidation number of O is reduced from 0 to –2.

18. (a). The oxidation number of Mg is increased from 0 to +2.

19. (e). Write down the oxidation numbers of all elements on both sides of the equation. No element changes oxidation number, so this is not a redox reaction.
20. (e). Oxidation numbers: $Cu = +1$, $Cl = -1$, $O = -2$, and $H = +1$, on both sides.
21. (b). Remember that the compound or polyatomic ion that contains the element that is reduced is the oxidizing agent. Mn is reduced from +7 to +2. Permanganate ion, MnO_4^-, is a common laboratory oxidizing agent.
22. (a). S is oxidized from +4 to +6.
23. (d). S is oxidized from –2 to 0.
24. (a). Dichromate ion, $Cr_2O_7^{2-}$, is another common laboratory oxidizing agent. Cr is reduced from +6 to +3.
25. (c). (a) would be cobalt$(\frac{3}{2})$ nitride, (b) would be cobalt$(\frac{1}{2})$ nitrite, (d) would be cobalt(I) nitrite, and (e) is cobalt(II) nitrate.
26. (d). (a) is iron(II) sulfate, (b) is iron(II) sulfide, (c) would be iron(I) sulfite, and (e) would be iron(IV) sulfite.
27. (c). (a) is S^{2-}, (b) is SO_3^{2-}, (d) would be SO_4 (no charge).
28. (a). (b) would be iron(VI) oxide, (c) would be iron(I) oxide, (d) is iron(III) oxide, and (e) would be iron$(\frac{4}{3})$ oxide.
29. (b). (a) is S^{2-}, (c) is SO_4^{2-}, (d) is SO_3 (a compound).
30. (a). (b) would be nickel(I) sulfide, (c) would be nickel(IV) sulfide, (d) is nickel(II) sulfate, and (e) would be nickel(I) sulfate.
31. (c). (a) is N^{3-}, (b) is NO_2^-, and (d) is NO_3 (no charge).
32. (c). (a) is manganese(II) oxide, (b) would be manganese(I) hydroxide, (d) would be manganese$(\frac{1}{2})$ hydroxide, and (e) would be manganese$(\frac{2}{3})$ oxide.
33. (a). (b) would be iron$(\frac{2}{3})$ sulfate, (c) would be iron(IV) sulfite, (d) would be iron$(\frac{2}{3})$ sulfite, and (e) would be iron$(\frac{4}{3})$ sulfate.
34. (c). (a) would be chromium(I) nitrate, (b) would be chromium$(\frac{1}{3})$ nitrate, (d) would be chromium$(\frac{3}{2})$ nitrite, and (e) is chromium(II) nitrite.

7 Chemical Bonding

Chapter Summary

Chapter 7 begins your study of chemical bonding with a discussion of how atoms are held together in compounds. There are two extremes in the possible modes of bonding: (1) **ionic bonding**, in which one atom or group of atoms completely gives up or **transfers** one or more electrons to another, resulting in charged atoms or groups of atoms (ions), which attract one another, and (2) **covalent bonding**, in which electrons are **shared** between two atoms. We can correlate many properties of compounds with the predominant type of bonding that holds them together (page 250). You should recognize, however, that these are the extremes in a continuous gradation of bonding, with nearly all bonds being a mixture of both ionic and covalent, transfer and sharing.

Chemical bonding usually involves only the outermost or incompletely filled shells of electrons. We often concentrate our attention on these **valence electrons**. A notation for representing and keeping track of the numbers of valence electrons in atoms, the **Lewis dot formula**, is correlated with the location of the element in the periodic table (Section 7-1). Because this representation is somewhat more cumbersome and less useful for *d*- or *f*-transition elements, it is generally used only for representative elements.

The important Section 7-2 contains a detailed description of **ionic bonding**. The main ideas are as follows:

(1) The elements with low electronegativities (and low ionization energies) readily lose electrons, giving them up to the elements with high electronegativities and quite negative electron affinities. The atoms form ions, which then attract one another because of their opposite charges.

(2) The number of electrons an atom gains or loses to form a stable ion depends on its electron configuration. The gain or loss results in a more stable configuration, usually a noble gas configuration for representative elements.

This lets us understand and predict the formulas of simple **binary** (two-element) compounds, as summarized in Table 7-2. Do not try to memorize this table. Instead, understand *why* the formulas of the compounds are as shown in this table, using the electron configurations of the elements involved. The electron configuration diagrams and Lewis dot formulas that are used in the discussion of the section will help you learn to predict chemical formulas.

The rest of this chapter presents an introduction to **covalent bonding**. This type of bonding occurs between atoms that have the same or similar electronegativities; they attain more stable configurations only by sharing electron pairs. When *one pair* of electrons is shared, we refer to the bond as a **single bond**; when *two pairs* are shared, it is a **double bond**; and when *three pairs* are shared, it is a **triple bond**.

The sharing of electron pairs to form covalent bonds results in a balance of attractive and repulsive forces. The drive to overlap orbitals to share electrons is the attractive force, as

103

described in the discussion of Section 7-3, while the repulsion is due to the like charges of the two nuclei. For molecules more complicated than H_2, there is additional repulsion from the unshared electrons on the two atoms. The potential energy of the pair of atoms at various distances can be represented by curves such as those in Figure 7-4. The distance at which the curve indicates a minimum energy is the distance at which the bonded combination of atoms is most stable. The depth of the curve below the zero of potential energy represents the strength of the bond. This is the amount of energy that must be supplied to separate the bonded pair of atoms into two isolated atoms.

The **bond length** of a covalent bond is that internuclear distance at which the attractive and repulsive forces balance and the bond is most stable. The **bond energy** is the difference between the energy of the bonded atoms and total energy of the separate atoms. Tables 7-3 and 7-4 allow you to compare the average bond energies and bond lengths of bonds among some key elements.

The **Lewis formula** is a convenient notation to help us count valence electrons and propose stable bonding arrangements. In this diagram, shared and unshared electrons are each represented by a dot; alternatively, each shared electron pair can be represented by a single line, a dash. You should also be familiar with the following points of terminology. The terms **monatomic**, **diatomic**, **triatomic**, and **polyatomic** refer to *how many atoms* are in the molecule; for instance, H_2 and HF are diatomic molecules, O_3 (ozone) and SO_2 are triatomic, and so on. The term **homonuclear** means that the molecule is composed entirely of one kind of atom; O_2, H_2, and O_3 are each homonuclear. The term **heteronuclear** means that the molecule contains more than one kind of atom; HF and SO_2 are heteronuclear. The same terms can be applied to ions.

Sections 7-5 through 7-9 contain very important ideas, designed to expand your understanding of Lewis formulas. Lewis formulas are helpful for keeping track of the number and kinds of bonds in molecules and polyatomic ions. Section 7-6 introduces the **octet rule**, a very useful aid to writing Lewis formulas. This rule can be expressed in the simple mathematical relationship $S = N - A$. It summarizes the observation that representative elements usually achieve a noble gas electron configuration in most of their compounds. This generalization is known as the octet rule because all of the noble gases except helium have eight electrons in their outer shells. An obvious exception among the representative elements is hydrogen. It attains a noble gas configuration (that of He) by having only two electrons in its valence shell. You should work through examples in the chapter to understand the application of this rule. Drawing Lewis formulas for molecules and ions is essential to understanding molecular structure and to predicting polarities of molecules. Development of the skills presented in Section 7-6 is worth a great deal of practice.

A formalism that is helpful in selecting the correct Lewis formula is the determination of formal charges. **Formal charge** is the *hypothetical* charge on an atom in a molecule or polyatomic ion. It is calculated by counting bonding electrons as though they were equally shared between the two bonded atoms. Then the most likely formula for a molecule, the one that is the most energetically favorable, is usually the one in which the formal charge on each atom is zero or as near zero as possible.

We must not extend the octet rule too far; it has limits. There are many stable compounds in which elements attain some number of electrons other than eight in their outer shell. Section 7-8 lists elements and compounds that are subject to such limitations, and then explains several such limitations. One type of limitation involves atoms that are so far from having eight electrons in their valence shell that they cannot easily attain an octet by electron sharing, yet they do not readily form ionic compounds. Elements at the tops of Groups 2A and 3A behave in this way. In a second type of limitation, a molecule or ion has an odd number of electrons. Elements that involve *d* or *f* electrons in their bonding produce a third class of limitation. The chemistry of transition elements, which involves mostly their *d* electrons, is discussed further in Chapters 25 and 27.

As we develop our understanding of chemical bonding in molecules, we find that some molecules or ions cannot be properly described by a single Lewis formula. For such a structure, we introduce the idea of **resonance** (Section 7-9), a concept that describes the true structure as the average of several Lewis formulas. This is made possible by the **delocalization** of the bonding electrons over several atoms, rather than electron pairs being shared explicitly between two atoms.

The atoms in a heteronuclear molecule may not attract the shared electrons equally (Section 7-10). This leads to an imbalance in the electron distribution in the bond. We refer to such a bond as a **polar** covalent bond, in contrast to **nonpolar** covalent bonds, in which the bonding electrons are shared equally. Such an arrangement produces a **dipole** (or **bond dipole**). Section 7-11 extends the idea of bond polarity to describe the **dipole moment** of the molecule. We shall see in Chapter 8 that the experimental measurement of this quantity gives valuable information about the three-dimensional arrangements of atoms in molecules. We should view nonpolar covalent bonding and ionic bonding as the two ideal extremes of chemical bonding, with all intermediate stages also possible (Section 7-12). Some compounds, K_2SO_4, for example, exhibit both ionic and strong covalent interactions.

Study Goals

Be sure you work many of the suggested Exercises. Those that involve writing Lewis formulas for ions and molecules (Study Goals 7, 11, 12, and 13) are especially important.

Chapter Introduction
1. Distinguish between ionic and covalent bonding. *Work Exercises 1 through 3.*
2. Know some of the major differences in physical properties between ionic and covalent compounds. *Work Exercises 12 and 13.*

Section 7-1
3. Know how to write Lewis dot formulas for atoms of representative elements. Understand how these are related to periodic table group numbers. *Work Exercises 4 and 5.*

Section 7-2
4. Predict whether bonding between a given pair of elements or in a given compound would be primarily covalent or ionic. *Work Exercises 6 through 10.*
5. Relate the stable ions formed by elements to their electron configurations and to their locations in the periodic table. *Work Exercises 11, 20, and 21.*
6. Predict the formulas of binary ionic compounds from the locations of their constituent representative elements in the periodic table. *Work Exercises 14 through 19.*

7. Write equations for the reactions of representative elements to form binary ionic compounds. Be able to represent reactants and products with (a) simple chemical symbols and formulas, (b) electron configuration diagrams, and (c) Lewis formulas.

Section 7-3

8. Understand what forces are responsible for the formation of covalent bonds. *Work Exercises 22, 23, and 25.*

9. Distinguish among single, double, and triple bonds. *Work Exercise 24.*

Section 7-4

10. Compare the bond lengths and bond energies of bonds between two different elements and between the same two elements joined by single, double, and triple bonds. *Work Exercises 26 and 27.*

Sections 7-5, 7-6, and 7-7

11. Draw Lewis formulas for molecules and polyatomic ions that contain only representative elements. Recognize which ones do not follow the octet rule. *Work Exercises 28 through 42 and 48 through 56.*

Section 7-7

12. Be able to write formal charges for atoms in covalently bonded molecules and ions. Use these as a guide to assess whether Lewis formulas may be correct. *Work Exercises 43 through 46.*

Section 7-9

13. Recognize molecules and ions for which resonance structures are possible; draw appropriate Lewis formulas. *Work Exercises 57 through 62.*

Sections 7-10 through 7-12

14. Distinguish between polar covalent bonds and nonpolar covalent bonds. *Work Exercises 67, 68, 70, and 72 through 74.*

15. Be able to tell whether the bonding between two elements is ionic, polar covalent, or nonpolar covalent. You should be able to do this either with numerical values of electronegativity or based on the locations of the elements in the periodic table. Understand the continuous range of bonding possibilities. *Work Exercises 67 through 81.*

General

16. Recognize and work various kinds of problems from this chapter. *Work **Mixed Exercises** 63 through 66.*

17. Answer conceptual questions based on this chapter. *Work **Conceptual Exercises** 82 through 90.*

18. Apply concepts and skills from earlier chapters to the ideas of this chapter. *Work **Building Your Knowledge** exercises 91 through 97.*

19. Exercises at the end of chapter direct you to sources outside the textbook for information to use in solving them. Work **Beyond the Textbook** exercises 98 through 101.

Some Important Terms in This Chapter

Some important new terms from Chapter 7 are listed below. Fill the blanks in the following paragraphs with terms from the list. Each term is only used once.

bond energy	formal charge	polar bond
bond length	ionic bonding	polyatomic ion
bonding electrons	Lewis dot formula	resonance
chemical bonding	Lewis formula	single bond
covalent bonding	lone pairs	triple bond
dipole moment	nonpolar bond	valence electrons
double bond	octet rule	

In compounds and other groups of atoms, atoms are held together by attractive forces referred to as [1]_____. There are two kinds of chemical bonding. When one or more electrons is *transferred* from one atom to another, the atoms are held together by [2]_____. When atoms *share* pairs of electrons, they are held together by [3]_____.

Chemical bonding is illustrated in very simplified form by representing each atom by a [4]_____, a technique developed by G. N. Lewis. In these representations, the dots represent [5]_____, electrons in the highest occupied energy level.

A monatomic ion is a single atom with a charge. A [6]_____ is a group of atoms that has a charge. A covalent bond may be formed by sharing one, two, or three pairs of electrons. If one pair of electrons is shared, the bond is a [7]_____. If two pairs of electrons are shared, the bond is a [8]_____. If three pairs of electrons are shared, the bond is a [9]_____.

The length of a line between two atoms that are bonded together is the [10]_____. The energy released when the bond is formed is its [11]_____.

The bonding in a molecule or polyatomic ion can be illustrated by a [12]_____ _____, based upon to the Lewis dot representations used for individual atoms. One of the fundamental principles for drawing Lewis formulas is the [13]_____, which refers to the very stable eight valence electron configurations of the noble gases.

The electrons represented by dots in Lewis formulas are either [14]_____ _____, shared between the bonded atoms, or [15]_____, so called because they are associated with only one of the atoms. Sometimes two or more Lewis formulas together are used to represent the structure of substance. These diagrams are known as [16]_____ structures. Choosing the most likely structure for a group of atoms can be aided by calculating the hypothetical charge on an atom in the molecule or polyatomic ion, the [17]_____.

Covalent bonds are of two types, depending upon the electronegativities of the atoms bonding. Unless the electronegativities are the same, or very close to the same, the two bonded atoms do not share the bonding electrons equally; the bond is a [18]_____, meaning that the distribution of electrons is not equal or uneven. If the electronegativities are about the same, the electrons are shared more or less equally; the bond is a [19]_____. The more complete name for polar bond is polar covalent bond—it is still a covalent bond. A polar bond has a *dipole*, resulting from the separation of positive and negative charges. The degree of polarity of a molecule can be expressed by the [20]_____, μ, which is directly related to the distance separating the charges and the magnitude of the charge.

Preliminary Test

These preliminary test questions are quite basic. They are intended to check your understanding of the main ideas and terminology of the chapter. You will need to work many additional Exercises at the end of Chapter 7.

Short Answer

Answer with a word, a phrase, a formula, or a number with units as necessary.

1. The Lewis dot formula for an atom of calcium is _____.
2. The Lewis dot formula for an atom of phosphorus is _____.
3. The Lewis dot formula for any halogen atom (represent it as X) is _____.
4. The formula for the ionic compound of magnesium and oxygen is _____.
5. The formula for the ionic compound of rubidium and sulfur is _____.
6. The formula for the ionic compound of calcium and nitrogen is _____.
7. In the formation of a covalent bond, the principal "attractive force" is _____ _____.
8. In the formation of a covalent bond, the principal "repulsive forces" are due to _____ _____ and to _____.
9. When two atoms combine to form a stable chemical bond, the _____ of the bonded pair of atoms is lower than that of the isolated atoms.
10. When two atoms are joined by a single bond, they share _____ electrons.
11. When two atoms are joined by a double bond, they share _____ electrons.
12. Which of the following molecules would be described as diatomic: He, H_2, Cl_2, HCl, CS_2, NH_3, H_2O, S_8, P_4? _____
13. Which of the molecules in Question 12 would be described as homonuclear? _____ Which as heteronuclear? _____
14. A single bond is _____ (longer/shorter) and its energy is _____ (greater/less) than a double bond, which is _____ (longer/shorter) and its energy is _____ (greater/less) than a triple bond.
15. When the bonding pair of electrons in a covalent bond is equally shared between the two atoms, the bond is called _____; when one atom attracts the shared pair of electrons more strongly than the other, the bond is called _____.
16. A covalent bond is polar if the two atoms involved differ significantly in their _____ values.
17. In which of these molecules are the covalent bonds more polar, CH_4 or CF_4? _____

18. Consider a series of bonds in which carbon is bonded to another second period element: C–O, C–F, C–N, and C–C. Place these bonds in order of decreasing polar (or ionic) character (most polar first): _____.

Each of the following compounds or ions obeys the octet rule. Apply the relationship $S = N - A$ to each of them and draw the Lewis formula; draw each Lewis formula, first using only dots and then using dashes to represent bonding electrons.

Compound or ion	N	A	S	Lewis formula
19. CH_4				
20. CCl_4				
21. NH_3				
22. NH_4^+				

Compound or ion	N	A	S	Lewis formula
23. CH_3Cl				
24. CS_2				
25. PF_3				
26. SO_4^{2-}				

27. HCN (C is the
    central atom)

28. C_2H_2

29. N_2H_4

Multiple Choice

Questions 1 through 3 refer to an atom that has the configuration $3s^23p^4$ for its valence electrons in the ground state.

_____ 1. To which group in the periodic table does the element belong?

       (a) 2A   (b) 3A   (c) 6A   (d) 7A   (e) noble gas

_____ 2. How many of the outer-shell orbitals of the neutral atom are half-filled?

       (a) 1   (b) 2   (c) 3   (d) 4   (e) 0

_____ 3. What is this element?

       (a) O   (b) S   (c) P   (d) Cr   (e) Se

_____ 4. How many valence electrons are there in an isolated carbon atom?

       (a) 0   (b) 1   (c) 2   (d) 4   (e) 6

_____ 5. The electron configuration $1s^22s^22p^63s^23p^6$ could represent

       (a) an atom of Ar.   (b) a Cl^- ion.   (c) a K^+ ion.

       (d) any of the ions or atoms mentioned.   (e) none of the ions or atoms mentioned.

_____ 6. What is the ground state electron configuration of the S^{2-} ion?

       (a) $1s^22s^22p^63s^23p^4$ (b) $1s^22s^22p^63s^23d^6$ (c) $1s^22s^22p^63s^23p^2$

       (d) The same as the ground state configuration of a Ne atom.

       (e) The same as the ground state configuration of an Ar atom.

_____ 7. A barium (Ba) atom would form its most stable ion by the

       (a) gain of one electron.   (b) gain of two electrons.   (c) loss of one electron.

       (d) loss of two electrons.   (e) loss of 56 electrons.

_____ 8. How many electrons are in the $4s$ orbital of the Ca^{2+} ion?

       (a) 3   (b) 2   (c) 1   (d) 0   (e) None of the preceding answers is correct.

_____ 9. All of the following represent stable ions. Four of the five have a noble gas configuration. Which one does *not* have a noble gas configuration?

       (a) Fe^{3+} (b) Br^- (c) S^{2-} (d) Ba^{2+} (e) K^+

_____ 10. The ion Cl^- can best be described as

       (a) a metal ion.   (b) a monatomic anion.   (c) a monatomic cation.

       (d) a neutral ion.   (e) a Lewis dot ion.

_____ 11. How many electrons are present in the valence shell of fluorine in the compound KF?

       (a) 6   (b) 7   (c) 8   (d) 9   (e) 10

_____ 12. Predict the formula of an ionic compound containing potassium (K) and selenium (Se).
     (a) KSe (b) K_2Se (c) KSe_2 (d) K_2Se_2 (e) K_3Se_2

_____ 13. Which one of the following is the best Lewis dot formula for Na_2O?

     (a) $2Na^+$, $[:\ddot{O}:]^-$ (b) $2Na\cdot^+$, $[:\ddot{O}\cdot]^+$ (c) $2Na^-$, $[:\ddot{O}:]^+$ (d) $2Na^+$, $[:\ddot{O}:]^{2-}$

     (e) None of the preceding answers is correct, because Na and O have similar electronegativities and probably do not form ions.

_____ 14. Which of these pairs of elements combine with the most covalent (least polar) bonding?
     (a) Ca and N   (b) Fe and Cl   (c) Be and I   (d) C and O   (e) Ca and S

_____ 15. Which one of the following might be expected to be a stable compound?
     (a) ArH (b) NaO (c) $BaSe$ (d) CCl_2 (e) Ca_2O

_____ 16. A metal (call it X) in Group 1A of the periodic table forms a compound with an element (call it Y) of Group 6A. The compound would most likely have the formula _____ and would be _____.
     (a) X_2Y, ionic   (b) X_2Y, covalent   (c) XY_2, ionic
     (d) XY_2, covalent   (e) XY_6, covalent

_____ 17. Which one of the following is the best description of covalent bonding?
     (a) Bonding that occurs between elements with very different electronegativities, and with transfer of electrons.
     (b) Bonding that occurs between elements with very different electronegativities, and with sharing of electrons.
     (c) Bonding that occurs between elements with similar electronegativities, and with transfer of electrons.
     (d) Bonding that occurs between elements with similar electronegativities, and with sharing of electrons.
     (e) Bonding that occurs between ions and elements, and with sharing of electrons.

_____ 18. Which one of the following substances would be expected to be the best conductor of electricity in the liquid state (i.e., when melted or condensed if not already a liquid)?
     (a) CH_4 (b) PCl_3 (c) SiO_2 (d) H_2O (e) BaF_2

_____ 19. Which one of the following substances would be expected to have the highest melting point?
     (a) NaF (b) CH_4 (c) H_2O (d) HF (e) BF_3

_____ 20. On the usual Pauling scale of electronegativities, the electronegativity of selenium (Se) is 2.4, while that of chlorine (Cl) is 3.0. Based on these values, we should expect the bonding between Se and Cl to be
     (a) ionic.   (b) covalent and nonpolar.
     (c) covalent and polar with the Se end of the bond a little negative.
     (d) covalent and polar with the Se end of the bond a little positive.
     (e) unstable in any circumstances.

_____ 21. Which one of the following diagrams correctly indicates the polarity of the bond?
     (a) $\overset{\longrightarrow}{C-H}$ (b) $\overset{\longrightarrow}{Cl-C}$ (c) $\overset{\longrightarrow}{F-H}$ (d) $\overset{\longrightarrow}{F-C}$
     (e) None of the preceding answers is correct.

_____ 22. Nitrogen can form single covalent bonds with boron, carbon, nitrogen, oxygen, and fluorine. The bond would be expected to be polar with the bonding electrons displaced most strongly away from the nitrogen in the case when it is bonded to

(a) boron.   (b) carbon.   (c) nitrogen.   (d) oxygen.   (e) fluorine.

_____ 23. Nitrogen and hydrogen each exist as two naturally occurring isotopes. This means that molecules of ammonia, NH_3, could exist having several different masses. How many different masses of ammonia molecules would be possible?

(a) 1   (b) 2   (c) 4   (d) 6   (e) more than 6

_____ 24. Which one of the following molecules contains both covalent and ionic bonding?

(a) KF   (b) CO_2 (c) CH_2 (d) NH_4F (e) C_2H_4

_____ 25. Which one of the following molecules contains *no* polar bonds?

(a) HF   (b) CCl_4 (c) O_2 (d) NH_3 (e) PCl_5

_____ 26. What is the total number of electrons shared in the NH_3 molecule?

(a) 2   (b) 3   (c) 4   (d) 6   (e) 8

_____ 27. What is the total number of valence electrons that appear in a Lewis formula for C_3H_5ClO? (It is not necessary to draw the Lewis formula in order to figure this out!)

(a) 10   (b) 20   (c) 30   (d) 40   (e) 50

_____ 28. What is the number of shared electrons that appear in the Lewis formula for C_3H_5ClO?

(It is not necessary to draw the Lewis formula in order to figure this out, either!)

(a) 10   (b) 20   (c) 30   (d) 40   (e) 50

_____ 29. Which one of the following covalent substances contains a triple bond?

(a) H_2S (b) NH_3 (c) C_2H_2

(d) H_2CO (Hint: C is the central atom)   (e) HOCl (Hint: Cl and H are both bonded to O)

_____ 30. The ion NH_4^+ can best be described as

(a) an anion.   (b) a cation.   (c) a Lewis dot ion.

(d) a monatomic ion.   (e) a homonuclear ion.

_____ 31. Which one of the following contains only one unshared pair of valence electrons?

(a) H_2O (b) NH_3 (c) CH_4 (d) NaCl   (e) BF_3

_____ 32. Which one of the following statements concerning resonance is *not* correct?

(a) Resonance involves more than one Lewis formula to represent the bonding.

(b) Resonance involves the sharing of electrons in delocalized orbitals.

(c) The true energy of a molecule in which resonance is important is lower than that of any of the hypothetical molecules described by the contributing Lewis formulas.

(d) Resonance is important in understanding the structures of some ions as well as the structures of neutral molecules.

(e) An example of the use of the concept of resonance in understanding bonding is the explanation why the three C–O bonds in CO_3^{2-} are observed to be nonequivalent.

Answers to Some Important Terms in This Chapter

1. chemical bonding
2. ionic bonding
3. covalent bonding
4. Lewis dot formula
5. valence electrons
6. polyatomic ion
7. single bond
8. double bond
9. triple bond
10. bond length
11. bond energy
12. Lewis formula
13. octet rule
14. bonding electrons
15. lone pairs
16. resonance
17. formal charge
18. polar bond
19. nonpolar bond
20. dipole moment

Answers to Preliminary Test

Short Answer

1. Ca: But remember that it is arbitrary on which side of the symbol we write the dots, so :Ca or Ċa or Ça would also be correct representations. We usually do indicate each pair of electrons as a pair of dots.

2. :Ṗ· Phosphorus is in Group 5A.

3. :Ẍ· The halogens are in Group 7A.

4. MgO. Each magnesium atom gives up two electrons ($3s^2$) to form a Mg^{2+} ion. Each oxygen atom accepts two electrons (changing $2s^22p^4$ to $2s^22p^6$) to form an O^{2-} ion. Thus, there are equal numbers of Mg^{2+} and O^{2-} formed and the chemical formula is MgO.

5. Rb_2S. Reason as in Question 4. Rb is a Group 1A element (alkali metal), and S is in Group 6A.

6. Ca_3N_2. Similar reasoning as for Question 4.

7. the tendency of atoms to attain stable electron configurations by overlapping their orbitals so as to share electrons.

8. repulsion of the nuclei of the two atoms; repulsion of the unshared pairs of electrons of the two atoms.

9. potential energy

10. two (or one pair of)

11. four (or two pairs of)

12. H_2, Cl_2, HCl. Remember that the term *diatomic* means that the molecule consists of two atoms, whether or not they are the same kind of atom.

13. homonuclear: He, H_2, Cl_2, S_8, P_4; heteronuclear: HCl, CS_2, NH_3, H_2O. The term *homonuclear* means "consisting of only one kind of atom," whereas *heteronuclear* means "consisting of more than one kind of atom."

14. longer; less; longer; less

15. nonpolar; polar

16. electronegativity

17. CF_4. It will be helpful to remember that C and H have almost the same electronegativity values. You must also remember the trends in electronegativity with respect to location on the periodic table. Fluorine is the most electronegative element.

18. C–F > C–O > C–N > C–C. Bond polarity depends on the difference in electronegativity between the two bonded atoms. Remember the trends in electronegativities (Section 5-6). Each of the other elements mentioned here is more electronegative than carbon, so C–C is the least polar bond shown, and C–F is most polar.

Compound or ion	N	A	S	Lewis formula
19. CH_4	16	8	8	
20. CCl_4	40	32	8	
21. NH_3	14	8	6	
22. NH_4^+	16	8	8	
	(Remember to account for the + charge by decreasing A by 1)			
23. CH_3Cl	22	14	8	
24. CS_2	24	16	8	
25. PF_3	32	26	6	
26. SO_4^{2-}	40	32	8	
	(Remember to account for the 2– charge by increasing A by 2)			
27. HCN	18	10	8	
28. C_2H_2	20	10	10	
29. N_2H_4	24	14	10	

Multiple Choice

1. (c). The total number of valence electrons is 6.
2. (b). The one orbital of the $3s$ sublevel has a pair. The three orbitals of the $3p$ sublevel have four electrons, 1 pair and 2 singles.
3. (b). The element in Group 6A and the third period is sulfur.

4. (d). Carbon is in Group 4A.
5. (d). The total number of electrons is 18. Argon is atomic number 18. Chloride ion has 17 + 1 = 18 electrons. Potassium ion has 19 − 1 = 18 electrons.
6. (e). The sulfide ion has two more electrons than the neutral atom, atomic number 16, so it has 16 + 2 = 18 electrons. That's the same as a neutral atom of argon. (a) is the electron configuration of the neutral sulfur atom. (b) uses the $3d$ sublevel, rather than the $3p$. (c) would be the electron configuration of the S^{2+} ion. Neon, in (d) has only 10 electrons.
7. (d). Alkaline earth metals have two valence electrons relatively far from the nucleus and rather loosely held. They would rather easily be lost.
8. (d). The neutral calcium atom has two electrons in the $4s$ orbital, but both of them are lost in the formation of the calcium ion.
9. (a). A neutral iron atom has 26 electrons. The iron(III) ion has lost three of them, leaving 23. Vanadium is *not* a noble gas.
10. (b). Remember the meanings of the terms. An anion is a negative ion and a cation is a positive ion. The term "monatomic" means "consisting of one atom." No ion could be neutral, by definition.
11. (c). In this compound, fluorine exists as the fluoride ion, F^-.
12. (b). Since potassium is in Group 1A, the potassium ion has a charge of 1+ (K^+), and since selenium is in Group 6A, the selenium ion has a charge of 2− (Se^{2-}). The simplest combination that produces a neutral compound is two potassium ions for each selenium ion (K_2Se).
13. (d). One check you can perform is that the total charge in an ionic compound must be equal to zero, which eliminates answer (a). Answer (b) makes no sense because a sodium atom with one valence electron would not be an ion, nor would an oxygen atom with six valence electrons. Likewise, (c) can be eliminated because Na does not form a stable negative ion and O would not form a positive ion—and these Lewis formulas are wrong anyway. Answer (e) is wrong in its statement of properties.
14. (d). This is the pair of elements with the smallest electronegativity difference of any listed. Remember how trends in electronegativity are associated with location in the periodic table.
15. (c). In none of the others do all of the atoms attain a stable configuration.
16. (a). Group 1A metals form ions with a charge of 1+, and nonmetals in Group 6A form ions with a charge of 2−. Therefore, two X ions will combine with each Y ion to form a neutral compound, which would be, most likely, ionic because the electronegativities of Group 1A metals are all much less than Group 6A nonmetals.
17. (d). Elements with very different electronegativities (a and b) tend to form ionic compounds. Covalent bonding does not involve ions.
18. (e). BaF_2 is the only ionic substance listed. Pure water, which is covalent, is a *poor* conductor of electricity—the commonly observed ability of water to conduct electricity is due to small amounts of dissolved ions.
19. (a). NaF is the only ionic substance listed.
20. (d). For selenium and chlorine, the difference in electronegativities is 0.6, a rather small difference. Thus, the bond will not be ionic but polar covalent. Since the selenium has the larger electronegativity, it will be the positive end of the dipole.

21. (e). The dipole arrow should point *from* the less electronegative atom *toward* the more electronegative atom. For each of the other answer choices (a-d), the arrow is pointing toward the *less* electronegative atom.

22. (e). Fluorine is the most electronegative element, so it has the greatest attraction for electrons. The bond with oxygen, (d), would also have the bonding electrons displaced away from the nitrogen, but not as strongly. The bonds with boron, (a), and carbon, (b), would have the bonding electrons displaced *toward* the nitrogen, since boron and carbon are less electronegative than nitrogen. The bond with another nitrogen atom, (c), would be nonpolar.

23. (e). There are eight possible combinations. Can you write all of them? Some of the molecular weights are very similar because the two isotopes of nitrogen and hydrogen both differ by about 1 amu.

24. (d). The bonding within the ammonium ion, between the nitrogen atom and each of the hydrogen atoms, is covalent; but the bond between the ammonium ion and the fluoride ion is ionic. The bond between potassium and fluoride, (a), is ionic, and the bonds between carbon and oxygen, (b), between carbon and hydrogen, (c) and (e), and between carbon and carbon (e) are all covalent.

    $$\left[\begin{array}{c} H \\ | \\ H-N-H \\ | \\ H \end{array}\right]^{+} F^{-}$$

25. (c). Each of the other molecules listed contains covalent bonds between elements of unequal electronegativities.

26. (d). Use the formula $S = N - A$.

27. (c). This is A in the formula $S = N - A$. [(3 carbon atoms) × (4 valence electrons each)] + [(5 hydrogen atoms) × (1 valence electron each)] + [(1 chlorine atom) × (7 valence electrons)] + [(1 oxygen atom) × (6 valence electrons)].

28. (b). This is S in the formula $S = N - A$. $N = 50$ and $A = 30$ (Question 27).

29. (c). $H-C\equiv C-H$. As for the rest, (a) $H-\overset{..}{\underset{..}{S}}:$, (b) $H-\overset{..}{\underset{H}{N}}-H$, (d) $H-\overset{..}{\underset{H}{C}}=\overset{..}{\underset{..}{O}}$, and (e) $H-\overset{..}{\underset{..}{O}}-\overset{..}{\underset{..}{Cl}}:$

30. (b). An cation is a positively charged ion.

31. (b). $H-\overset{..}{\underset{H}{N}}-H$. As for the rest, (a) $H-\overset{..}{\underset{..}{O}}:$, (c) $H-\overset{H}{\underset{H}{C}}-H$, (d) $Na^{+}[:\overset{..}{\underset{..}{Cl}}:]^{-}$, and (e) $:\overset{..}{F}-\overset{:\overset{..}{F}:}{\underset{:\overset{..}{F}:}{B}}-\overset{..}{F}:$

32. (e). The three C–O bonds in CO_3^{2-} are equivalent.

8 Molecular Structure and Covalent Bonding Theories

Chapter Summary

This chapter deals with shapes of molecules and ions and with the orbitals that atoms use to share electrons. Several important ideas form a unifying theme throughout the chapter. Section 8-1 gives an overview of the main themes of this chapter. Pay special attention to the boxed Important Note on page 288, and consult your instructor, if necessary, about the order in which you should study the chapter material. The numbered steps and the flow chart on page 289 summarize the procedures you will use in this chapter to analyze the structure and bonding in any polyatomic molecule or ion. Writing the Lewis formula for a molecule or ion is an essential starting point for the material in this chapter, so review that procedure from Chapter 7.

Each atom in a polyatomic molecule or ion arranges the electron pairs in its valence shell in a way that minimizes the repulsion of these electron pairs. This is the central idea of the **valence shell electron pair repulsion (VSEPR) theory**, Section 8-2. Section 8-3 discusses one important consequence of the shape of a molecule—its polarity, or dipole moment. The experimental measurement of molecular polarity can give us valuable information about the three-dimensional arrangement of atoms in molecules. In some molecules, the arrangement of the polar bonds causes the bond dipoles to cancel, leaving the entire molecule nonpolar (dipole moment equal to zero). In other molecules, the bond dipoles are arranged so they do not cancel, leaving the entire molecule polar (dipole moment not equal to zero). The conditions on page 293 and the flow chart in Figure 8-1 summarize the approach for predicting whether a polyatomic molecule is polar or nonpolar. This approach will be used throughout this chapter.

To reach minimum energy (maximum stability), atoms in molecules often modify their orbitals to an arrangement that is different from what they have in isolated atoms. In the terminology of the **valence bond (VB) theory**, Section 8-4, this mixing or **hybridization** of atomic orbitals to form **hybrid orbitals** allows us to correlate the bonding with the observed molecular shapes.

We can consider the geometry of molecules and ions to fall into one of several types, depending on the number of atoms to which a central atom is bonded and the number of unshared electron pairs that this central atom has (Table 8-1). For each of these categories, we can use the VSEPR theory to predict the three-dimensional arrangement of groups of electrons around the central atom. This, in turn, allows us to predict the arrangement of atoms around the central atom. The valence bond theory is used to discuss the number of electrons occupying each type of hybrid orbital and the overlap of these orbitals (Table 8-2). Sections 8-5 through 8-12 discuss the most common bonding arrangements in terms of these two theories.

As you study this material, look ahead to the summary of molecular and electronic geometries in Section 8-16 and in Table 8-4. You must be careful to distinguish clearly between

electronic geometry ("Where are the *valence-shell electron pairs* around the central atom?") and the **molecular geometry** ("Where are the other *atoms* around the central atom?"). As you will see, there are only five common answers to the first question; the electronic geometry around the central atom is usually either:

1. linear with sp hybrid orbitals (Section 8-5),
2. trigonal planar with sp^2 hybrid orbitals (Section 8-6),
3. tetrahedral with sp^3 hybrid orbitals (Sections 8-7 through 8-10),
4. trigonal bipyramidal with sp^3d hybrid orbitals (Section 8-11), or
5. octahedral with sp^3d^2 hybrid orbitals (Section 8-12).

After you have determined the *electronic geometry* for a molecule or ion, then you can determine its *molecular geometry* by considering which of the electronic locations attach the central atom to other atoms and which ones are merely unshared electron pairs on that central atom. For example (Sections 8-7 through 8-9), molecules that have four regions of high electron density (electron groups) all have tetrahedral electronic geometry, but their molecular geometries could be tetrahedral (e.g., CH_4), pyramidal (e.g., NH_3), or angular (e.g., H_2O). Table 8-3 (Section 8-13) summarizes how the presence of lone pairs on the central atom affects molecular geometry. Once you have found the overall geometries by using the two theories presented in this chapter, you can often understand deviations from the ideal theoretical values by considering the stronger repulsion of unshared pairs for other electron pairs, and by considering bond polarities.

With the understanding about orbitals we have developed in this chapter, we can describe how multiple bonding occurs. Double bonds are discussed in detail in Section 8-14 and triple bonds in Section 8-15. Be sure you understand the difference between **sigma** and **pi** bonds, and which combinations of orbitals can form each kind of bond.

Study Goals

Much practice is required to master the bonding descriptions in this chapter. Work *many* of the exercises at the end of the chapter in the text to accomplish this mastery.

Section 8-2
1. Outline the significance and main ideas of the valence shell electron pair repulsion (VSEPR) theory. *Work Exercises 1 and 2.*
2. Identify a central atom and count its electron groups. *Work Exercises 22 and 23.*

Section 8-4
3. Outline the significance and main ideas of the valence bond (VB) theory. *Work Exercises 7 through 9, 13 through 15, 17, and 46.*
4. Name and describe the hybrid orbitals that correspond to each possible number of electron groups. *Work Exercises 9 through 12, 16, 18, and 19.*

Sections 8-2 and 8-5 through 8-12
5. Distinguish between electronic geometry and molecular geometry. *Work Exercises 3, 4, 20, and 21.*

Parts B of Sections 8-5 through 8-12

6. Use the VSEPR theory to describe or predict the electronic geometry *and* the molecular (or ionic) geometry of a molecule or polyatomic ion. *Work Exercises 24 through 45.*

Sections 8-3, 8-5 through 8-12 (Review Sections 7-10 and 7-11)

7. Knowing the electronic structure and shape of a molecule, determine whether it is polar or nonpolar. *Work Exercises 37 through 45.*

Parts C of Sections 8-5 through 8-12 and Section 8-16

8. Use valence bond structures to describe the bonding in molecules and ions having one central atom. *Work Exercises 48 through 50, and 53.*

Sections 8-6 through 8-15, Problem-Solving Tip on pages 303-304

9. Use valence bond structures to describe the bonding in molecules and ions that have more than one central atom. *Work Exercises 51, 52, 54, and 55.*

Section 8-14 and 8-15

10. Use valence bond structures to describe how double and triple bonding occur. *Work Exercises 58 through 64.*

11. In the valence bond description, distinguish between sigma and pi bonds. *Work Exercises 56, 57, 60 through 64.*

General

12. Use your understanding of this chapter to recognize and solve a variety of types of questions about structures, bonding, and molecular polarities. *Work* **Mixed Exercises** *69 through 78.*

13. Answer conceptual questions based on this chapter. *Work* **Conceptual Exercises** *79 through 86.*

14. Apply concepts and skills from earlier chapters to the ideas of this chapter. *Work* **Building Your Knowledge** *exercises, 87 through 91.*

15. Exercises at the end of chapter direct you to sources outside the textbook for information to use in solving them. Work **Beyond the Textbook** *exercises 92 through 94.*

Some Important Terms in This Chapter

Some important new terms from Chapter 8 are listed below. Fill the blanks in the following paragraphs with terms from the list. Each term is only used once. Study and review other new terms.

angular	nonpolar molecule	trigonal bipyramidal
central atom	octahedral	trigonal planar
electronic geometry	pi (π) bonds	trigonal pyramidal
hybrid orbitals	polar molecule	valence bond theory
hybridization	regions of high electron density	valence shell
linear	sigma (σ) bonds	valence shell electron
molecular geometry	tetrahedral	pair repulsion theory

Lewis formulas are useful for simplified descriptions of chemical bonding because they depict the valence electrons, which are generally the only electrons involving in bonding. They are found in the [1]_____, the highest occupied energy level or "outer shell." Many simple molecules consist of one atom that is bonded to several (two to six) other atoms. The atom in the center is called the [2]_____. The arrangement of the other atoms around this central atom can often be predicted by the [3]_____, a theory that proposes that pairs of valence electrons, since they are all negatively charged, repel each other, so they try to get as far apart from each other as possible. The resulting geometries are largely determined by the number of [4]_____ (electron groups), where each group of electrons is either a bond (single, double, or triple) or an unshared pair of electrons in the valence shell of the central atom. The [5]_____ or arrangement of the groups of *electrons* around the central atom determines the [6]_____, the arrangement of the *atoms* around the central atom. The electronic geometry and the resulting molecular geometry will often determine whether a molecule is a [7]_____, having a dipole, or a [8]_____, in which the electrons are arranged evenly.

Complementary to the VSEPR theory is the [9]_____, which describes *how* the bonding occurs. According to this theory, different kinds of atomic orbitals (*s*, *p*, or *d*) can be mixed, undergoing [10]_____, to produce [11]_____. These hybrid orbitals have names and shapes that are combinations of the atomic orbitals from which they are composed, such as sp, sp^2, sp^3, sp^3d, and sp^3d^2.

The five electronic geometries, the number of electron groups that produce them, and their characteristic bond angles are:

1. [12]_____,...2 groups of electrons, 180° angle (along a straight line)

2. [13]_____,....................3 groups of electrons, 120° angles (a triangle in one plane)

3. [14]_____,...............................4 groups of electrons, 109.5° angles (a four-sided three-dimensional figure)

4. [15]_____,...5 groups of electrons, 90°, 120°, and 180° angles (two triangular-based pyramids joined at their bases, one upside down)

5. [16]_____,..............................6 groups of electrons, 90° and 180° angles (pointing to the six corners of a regular eight-sided three-dimensional figure)

If all of the groups of electrons around the central atom are used to bond other atoms to the central atom, then the molecular geometry is the same as the electronic geometry. However, if the central atom has one or more unshared pairs of electrons (lone pairs), the molecular geometry is described differently. When the central atom has 4 electron groups, but 1 is an unshared electron pair, the remaining 3 groups of electrons bond the central atom to 3 other atoms, and the four atoms form the shape of a shallow pyramid with a triangular base. This shape, described as [17]_____, has bond angles of 109.5° (or close to that), and should not be confused with trigonal planar, which has bond angles of 120°. When the central atom has 4 electron groups, but 2 of them are unshared electron pairs, the remaining 2 groups of electrons bond the central atom to 2 other atoms, with a bond angle of 109.5° (or somewhat less). This shape is described as [18]_____, the same term applied to molecules or ions in which the central atom has 3 electron groups, but 1 of them is an unshared electron pair. The remaining 2 groups of electrons bond the central atom to 2 other atoms, but the bond angle is 120° (or slightly less).

All single bonds between atoms are [19]_____, bonds that result from the head-on overlap of atomic orbitals. Double and triple bonds have one sigma bond and one or two [20]_____, respectively, bonds that result from the side-by-side overlap of p orbitals.

Preliminary Test

These preliminary test questions are quite basic. They are intended to check your understanding of the main ideas and terminology of the chapter. Practice many additional Exercises at the end of Chapter 8 in the textbook.

True-False

Mark each statement as true (T) or false (F).

_____ 1. The number of electron groups on the central atom in H_2O is 4.
_____ 2. The number of electron groups on the central atom in CCl_4 is 4.
_____ 3. The number of electron groups on the central atom in NF_3 is 4.
_____ 4. Because all atoms satisfy the octet rule in their bonding, the number of electron groups on the central atom is always 4 or some multiple of 4.
_____ 5. The number of electron groups is equal to S in the formula $S = N - A$.
_____ 6. All molecules with four electron groups about a central atom have tetrahedral electronic geometry.
_____ 7. All molecules with four electron groups about a central atom have tetrahedral molecular geometry.
_____ 8. A molecule of the type AB_5 (where A is the central atom) could have trigonal bipyramidal molecular geometry.
_____ 9. A molecule of the type AB_5 (where A is the central atom) must have trigonal bipyramidal molecular geometry.
_____ 10. In the molecule BF_3, all F—B—F bond angles are equal.
_____ 11. In the molecule CF_4, all F—C—F bond angles are equal.
_____ 12. In the molecule NF_3, all F—N—F bond angles are equal.
_____ 13. All molecules with tetrahedral electronic geometry must have bond angles equal to 109.5°.
_____ 14. In the molecule NF_3, all F—N—F bond angles are equal to 109.5°.
_____ 15. In the molecule PF_5, all F—P—F bond angles are equal.
_____ 16. A molecule that contains polar bonds may be polar.
_____ 17. A molecule that contains polar bonds must be polar.
_____ 18. A three-atom molecule with linear molecular geometry must be nonpolar.

_____ 19. A three-atom molecule with linear molecular geometry must be polar.
_____ 20. A tetrahedral molecule must be polar.
_____ 21. A tetrahedral molecule must be nonpolar.

Short Answer

Answer with a word, a phrase, a formula, or a number with units as necessary.

1. The theory that accounts for the shapes of molecules by considering the repulsion of the outermost electron pairs on a central atom is called the _____ theory.
2. The theory that describes the formation of covalent bonds as being due to the overlap of the atomic orbitals on two atoms is called the _____ theory.
3. The mixing or rearrangement of individual "pure" atomic orbitals to form new orbitals with different spatial orientations is called _____.
4. The angle between two sp hybrid orbitals on the same atom is _____.
5. The angle between two sp^2 hybrid orbitals on the same atom is _____.
6. The angle between two sp^3 hybrid orbitals on the same atom is _____.
7. The electronic geometry of the central atom in the compound HCN is _____; the hybrid orbitals on the central atom in this compound are _____ orbitals; the molecular shape is described as _____.
8. The electronic geometry of the central atom in the compound PCl_3 is _____; the hybrid orbitals on the central atom in this compound are _____ orbitals; the molecular shape is described as _____.
9. The electronic geometry of the central atom in the compound BCl_3 is _____; the hybrid orbitals on the central atom in this compound are _____ orbitals; the molecular shape is described as _____.
10. The electronic geometry of the central atom in the compound H_2O is _____; the hybrid orbitals on the central atom in this compound are _____ orbitals; the molecular shape is described as _____.
11. The electronic geometry of the central atom in the ion SO_4^{2-} is _____; the hybrid orbitals on the central atom in this ion are _____ orbitals; the ionic shape is described as _____.
12. The electronic geometry of the central atom in the ion AlF_6^{3-} is _____; the hybrid orbitals on the central atom in this ion are _____ orbitals; the ionic shape is described as _____.

Multiple Choice

_____ 1. Which one of the following is a polar molecule?
   (a) CO_2 (a linear molecule)   (b) BF_3 (a trigonal planar molecule)
   (c) CH_4 (a tetrahedral molecule)   (d) NH_3 (a pyramidal molecule)   (e) H_2

122

_____ 2. Which one of the following statements regarding sp^3 hybrid orbitals is *not* correct?
    (a) The result of sp^3 hybridization is a set of four orbitals with tetrahedral geometry.
    (b) The result of sp^3 hybridization is a set of four orbitals, one of which is a lower energy, because one of them came from an s orbital.
    (c) The angle between two sp^3 hybrid orbitals on an atom is about 109.5°.
    (d) If an atom has four electrons in a set of sp^3 hybrid orbitals, the lowest energy configuration would have all four electrons unpaired.
    (e) The result of sp^3 hybridization is four equivalent orbitals.

_____ 3. The main reason for describing the bonding around nitrogen in ammonia, NH_3, in terms of sp^3 hybrid orbitals on nitrogen is that
    (a) the angle between two N—H bonds is closer to 109.5° than to any other ideal value.
    (b) the angle between two N—H bonds is closer to 90° than to any other ideal value.
    (c) the angle between two N—H bonds is closer to 120° than to any other ideal value.
    (d) the hydrogen atoms attain the filled $1s^2$ configuration by bond formation.
    (e) the ground state configuration of nitrogen has four unpaired electrons.

_____ 4. Which of the following contains at least one atom that uses sp^3 hybridization in its bonding?
    (a) H_2O (b) NH_3 (c) CH_4
    (d) All of the molecules listed.   (e) None of the molecules listed.

_____ 5. Which one of the following contains at least one atom whose bonding may be described in terms of sp^2 hybridization?
    (a) H_2O (b) NH_3 (c) C_2H_2 (d) C_2H_4 (e) H_2

_____ 6. In the compound hexachloroethane, C_2Cl_6, the angle between two C—Cl bonds on the same carbon atom is closest to which of the following values?
    (a) 60°   (b) 90°   (c) 109.5°   (d) 120°   (e) 180°

_____ 7. The geometry of a molecule of silane, SiH_4, can be described as
    (a) trigonal planar.   (b) octahedral.   (c) tetrahedral.   (d) linear.   (e) angular.

_____ 8. The shape of a molecule of phosphorus trichloride, PCl_3, can be described as
    (a) tetrahedral.   (b) trigonal planar.   (c) trigonal bipyramidal
    (d) trigonal pyramidal.   (e) angular.

_____ 9. The hybrid orbitals on the central atom in PCl_3 are
    (a) sp.   (b) sp^2.   (c) sp^3.   (d) sp^3d.   (e) sp^3d^2.

_____ 10. The shape of a molecule of phosphorus pentachloride, PCl_5, can be described as
    (a) tetrahedral.   (b) trigonal planar.   (c) pyramidal.
    (d) trigonal bipyramidal.   (e) angular.

_____ 11. The hybrid orbitals on the central atom in PCl_5 are
    (a) sp.   (b) sp^2.   (c) sp^3.   (d) sp^3d.   (e) sp^3d^2.

_____ 12. In the compound SF_6, what is the arrangement of fluorine atoms around the sulfur atom, and what hybrid orbitals of the sulfur atom are used in bonding?
    (a) hexagonal, sp^3 (b) tetrahedral, sp^3 (c) octahedral, sp^3d
    (d) octahedral, sp^3d^2 (e) trigonal planar, sp^2

_____ 13. In the compound CH_3Cl, the bonding orbitals of the carbon atom are
    (a) *p* orbitals.   (b) *sp* hybrid orbitals.   (c) sp^2 hybrid orbitals.
    (d) sp^3 hybrid orbitals.   (e) sp^4 hybrid orbitals.

_____ 14. The compound tetrachloroethylene, C_2Cl_4, has a carbon-carbon double bond. The angle between two C—Cl bonds on the same C atom should be closest to which of the following?
    (a) 60°   (b) 90°   (c) 109.5°   (d) 120°   (e) 180°

_____ 15. If an atom is attached to two other atoms and has no unshared valence-shell electron pairs, then the molecule's shape would be described as
    (a) angular.   (b) linear.   (c) tetrahedral.   (d) trigonal planar.   (e) octahedral.

_____ 16. The water molecule, H_2O, is angular rather than linear, because
    (a) the two hydrogen atoms attract one another.
    (b) the bonding involves *s* orbitals of H.
    (c) the bonding involves sp^2 orbitals of O.
    (d) the bonding involves sp^3 orbitals of O.
    (e) the *s* orbital of one H atom overlaps with the *s* orbital of the other H atom.

_____ 17. Acetylene, C_2H_2, contains a carbon-carbon triple bond. Which one of the following statements about this molecule is *not* true?
    (a) The four atoms are bonded together in a rectangular arrangement.
    (b) Each carbon-hydrogen bond is a single bond.
    (c) Each carbon atom is *sp* hybridized.
    (d) The bonding around each carbon atom is linear.
    (e) The atoms are all in the same straight line.

_____ 18. A triple bond includes
    (a) one σ and two π bonds.   (b) two π and one σ bonds.   (c) three σ bonds.
    (d) three π bonds.   (e) None of the preceding answers is correct.

_____ 19. The following five statements describe differences between a single bond and a double bond. Which one of these statements is *incorrect*?
    (a) A double bond is the sharing of four electrons by two atoms, while a single bond is the sharing of two electrons by the two atoms.
    (b) Atoms connected by a double bond are more difficult to rotate with respect to each other than atoms connected by a single bond.
    (c) A double bond usually consists of a σ bond and a π bond, while a single bond is usually a π bond.
    (d) In a single bond, the region of highest density of bonding electrons is directly between the two atomic centers, while in a double bond another region of high density of bonding electrons also exists on two sides of the line between the atomic centers.
    (e) A double bond is stronger (harder to break) than a single bond between the same two atoms.

_____ 20. A π bond can be formed by the overlap of
   (a) two s orbitals on different atoms.   (b) two p orbitals on the same atom.
   (c) two p orbitals on different atoms, these p orbitals lying side-by-side in the same plane.
   (d) two p orbitals on different atoms, these p orbitals lying end-to-end in the same plane.
   (e) two p orbitals on different atoms, these p orbitals lying side-by-side in planes that are perpendicular to one another.

_____ 21. In order to understand the shape of H_2CO, we describe the hybridization of carbon (the central atom) as sp^2. In order to fully describe the double bond, we postulate that the hybridization of the oxygen atom is
   (a) sp.   (b) sp^2.   (c) sp^3.   (d) sp^3d.   (e) sp^3d^2.

Answers to Some Important Terms in This Chapter

1. valence shell
2. central atom
3. valence shell electron pair repulsion theory
4. regions of high electron density
5. electronic geometry
6. molecular geometry
7. polar molecule
8. nonpolar molecule
9. valence bond theory
10. hybridization
11. hybrid orbitals
12. linear
13. trigonal planar
14. tetrahedral
15. trigonal bipyramidal
16. octahedral
17. trigonal pyramidal
18. angular
19. sigma (σ) bonds
20. pi (π) bonds

Answers to Preliminary Test

Do not just look up answers. *Think about the reasons for the answers.*

True-False

1. True. For all questions like this, you first need to draw the Lewis dot formula for the molecule or ion. Review this topic in Sections 7-5 through 7-9.
2. True
3. True
4. False. This is false for two reasons: (1) Not all atoms obey the octet rule in their covalent bonding. Be familiar with the main types of limitations (Section 7-8). (2) Even if the octet rule is satisfied, the bonding may involve double or triple bonds, which are each counted as only *one* electron group.
5. False. This is false for two reasons: (1) We only count the electron groups on one central atom at a time, and the formula $S = N - A$ refers to an entire molecule or ion. (2) Even if there is only one central atom, the term S in this formula tells the total number of shared electrons. These may be shared in single, double, or triple bonds, each of which counts as only *one* electron group.

6. True

7. False. Such a molecule may have the central atom bonded to fewer than four other atoms, in which case the electronic geometry and the molecular geometry are different. For examples, study the discussions of NH_3 in Section 8-8 and H_2O in Section 8-9.

8. True. This can occur when all of the electron groups are used to bond the central atom to other atoms. An example is PF_5, discussed in Section 8-11.

9. False. This could occur only if all of the electron groups were bonding the central atom to other atoms, but this is not necessary. Suppose there were *six* electron groups (octahedral electronic geometry), with one of these an unshared electron pair. IF_5 or SF_5^- are examples of such species.

10. True. 120°

11. True. 109.5°

12. True. 102.1°. (Why is the angle *not* 120° or 109.5°?)

13. False. See the discussions in Sections 8-8 and 8-9.

14. False. The repulsion of the bonded pairs by the lone pairs causes the bond angles F—N—F to be less than the ideal tetrahedral value of 109.5°.

15. False. Even though there are no lone pairs to account for, the angles in a trigonal bipyramidal arrangement are not all equal. Six are 90° (axial-equatorial), three are 120° (equatorial-equatorial), and one is 180° (axial-axial).

16. True

17. False. If any dipoles due to bond polarity are arranged so that their effects cancel, the molecule would still be nonpolar. A simple example is CO_2.

18. False. HCN is an example of such a molecule that is polar.

19. False. CO_2 is an example of such a molecule that is nonpolar.

20. False. If all four atoms (B) attached to the central atom (A) in a tetrahedral species AB_4 are the same, then the molecule or ion is nonpolar. Examples of this are CH_4, CF_4, and SO_4^{2-}.

21. False. If one (or more) of the atoms is different from the others, the molecule *could* still be polar. An example is CH_3Cl.

Short Answer

1. valence shell electron pair repulsion. This is often referred to as the VSEPR theory. You should remember what those initials stand for, because it will help you to remember the main idea of the theory.

2. valence bond

3. hybridization

4. 180°

5. 120°

6. 109.5°

7. linear; *sp*; linear

8. tetrahedral; sp^3; pyramidal

9. trigonal planar; sp^2; trigonal planar

10. tetrahedral; sp^3; angular

126

11. tetrahedral; sp^3; tetrahedral
12. octahedral; sp^3d^2; octahedral

Multiple Choice

1. (d). NH_3 has one unshared electron pair on the central atom.
2. (b). The four hybrid orbitals are equivalent in energy, each one being ¼ s and ¾ p.
3. (a). Nitrogen has four electron groups, resulting in a tetrahedral electronic geometry. The resulting angles of 109.5° are consistent with sp^3 hybridization.
4. (d). In all three molecules, the central atom has four electron groups, resulting in a tetrahedral electronic geometry, which is consistent with sp^3 hybridization.
5. (d). C_2H_4 has a double bond between the two carbon atoms. Consequently, each carbon atom has three electron groups, resulting in a trigonal planar electronic geometry, which is consistent with sp^2 hybridization.
6. (c). In C_2Cl_6 the bond between the two carbon atoms is a single bond, so each carbon atom has four electron groups, resulting in a tetrahedral electronic geometry, with characteristic angles of 109.5°.
7. (c). The central atom in SiH_4 has four electron groups, resulting in a tetrahedral electronic geometry. Since there are no unshared electron pairs, the molecular geometry is also tetrahedral.
8. (d). The central atom in PCl_3 has four electron groups, resulting in a tetrahedral electronic geometry. There is one unshared electron pair, so the molecular geometry is trigonal pyramidal.
9. (c). The central atom in PCl_3 has four electron groups, resulting in a tetrahedral electronic geometry, which is consistent with sp^3 hybridization.
10. (d). The central atom in PCl_5 has five electron groups, resulting in a trigonal bipyramidal electronic geometry. Since there are no unshared electron pairs, the molecular geometry is also trigonal bipyramidal.
11. (d). The central atom in PCl_5 has five electron groups, resulting in a trigonal bipyramidal electronic geometry, which is consistent with sp^3d hybridization.
12. (d). The central atom in SF_6 has six electron groups, resulting in an octahedral electronic geometry. Since there are no unshared electron pairs, the molecular geometry is also octahedral. Octahedral geometry is consistent with sp^3d^2 hybridization.
13. (d). The central atom in CH_3Cl has four electron groups, resulting in a tetrahedral electronic geometry, which is consistent with sp^3 hybridization.
14. (d). Each carbon atom has three electron groups, resulting in a trigonal planar electronic geometry, with characteristic angles of 120°.
15. (b). An atom bonded to two other atoms and no unshared valence-shell electron pairs has two electron groups, resulting in linear electronic and molecular geometry.
16. (d). The central atom in H_2O has four electron groups, two bonding electron pairs and two unshared electron pairs, resulting in tetrahedral electronic geometry, which is consistent with sp^3 hybridization. sp^3 hybrid orbitals are separated from each other by angles of 109.5°, not 180°.

17. (a). Each carbon atom in C_2H_2 has two electron groups, resulting in a linear electronic geometry. All four atoms are along a straight line.

18. (a). One of the bonds between two atoms is always a σ bond. Double and triple bonds have one or two π bonds, respectively.

19. (c). A single bond is a σ bond.

20. (c)

21. (b). The oxygen atom has three electron groups, the double bond with the carbon atom and two unshared electron pairs. Therefore, the electronic geometry around the oxygen atom is trigonal planar, which is consistent with sp^2 hybridization.

9 Molecular Orbitals in Chemical Bonding

Chapter Summary

In Chapter 8, you learned that valence bond (VB) theory describes covalent bonding as the result of the overlap of orbitals that remain on or between the two atoms involved. You also saw that we could use VSEPR theory to describe the molecular geometry as the result of repulsions among valence shell electron pairs on the atoms. In Chapter 9 you learn about an alternative view of covalent bonding, the **molecular orbital (MO) theory**. Recall that in VB theory, we proposed that the various orbitals on a *single* atom could mix and recombine (hybridize) to form hybrid orbitals. These have different energies and shapes from the "pure" or isolated atom orbitals. In the MO theory, the orbitals that are mixed are on *different* atoms, and the term "hybridization" is not used. The resulting **molecular orbitals** are no longer the "property" of just one atom, but extend over at least the pair of atoms involved, or in some cases over the entire molecule.

For many molecules, either approach gives a good description of the molecule. The VB approach has the advantage of being easily visualized. The MO theory has an advantage with respect to quantitative calculations of energies and geometry, descriptions of electron distributions, and correlation with magnetic properties. In other cases, the MO theory can give a correct description of some molecular property (for example, the paramagnetism of O_2) when the VB approach fails.

When two atomic orbitals from different atoms combine, they produce two molecular orbitals. Of these molecular orbitals, one (the **bonding orbital**) is at lower energy than the original orbitals while the other (the **antibonding orbital**) is at higher energy than the pure atomic orbitals. The superscript star (*) identifies an orbital that is antibonding, while the absence of the star means that the orbital is bonding. Section 9-1 describes the various ways in which atomic orbitals can combine to give molecular orbitals. The relations between the shapes of the atomic orbitals and those of the bonding and antibonding orbitals are described. Be sure that you understand the notation used to describe molecular orbitals. The notation **sigma (σ)** indicates that the molecular orbital is cylindrically symmetrical about a line through the internuclear axis. The notation **pi (π)** means that the molecular orbital has mirror symmetry across a plane through the two nuclei. *Subscripts* indicate the type of atomic orbitals that combined to give this molecular orbital—*s* if both were *s* orbitals, *sp* if one was *s* and one was *p*, and *p* if both were *p*. The only way (at least using only *s* and *p* orbitals as we do in this chapter) that a π molecular orbital can be formed is by *side-on* overlap of *p* orbitals on two different atoms.

Section 9-2 extends the description of the molecular orbitals to include the relative energies of the various molecular orbitals. It is important to remember the **order** of the energy levels for the various homonuclear diatomic molecules of the first- and second-period elements (Figure 9-

129

5). (Remember that for O_2, F_2, and Ne_2 and their ions, the order of energies of the σ_p and the π_p orbitals is reversed from the order of these orbitals for the other molecules and ions.) Fill the orbitals with the correct number of electrons for the molecule, following the Aufbau procedure, Hund's Rule, and the Pauli Principle. You will recognize that this is the same procedure for molecules that was used earlier with isolated atoms (Section 4-18). The procedure for finding the molecular orbital description of a molecule or ion is summarized on pages 333 and 334. Once we have placed the correct number of electrons into the set of molecular orbitals, we can describe the stability of the resulting arrangement using the **bond order**, defined in Section 9-3. The higher the bond order, the more stable we predict the molecule or ion to be, and the shorter the **bond length** and the greater the **bond energy**.

Section 9-4 covers in detail the neutral homonuclear diatomic molecules of the elements of the first and second periods. A careful study of this section will help you learn to apply the MO theory and to become familiar with the notation used. As you will see in this section, the MO theory can also be used to describe bonding in diatomic cations and anions by removing or adding the appropriate number of electrons. For many heteronuclear diatomic molecules, the general features of the molecular orbital diagrams already studied can be adjusted to take into account the electronegativity difference between the atoms. This is discussed in Section 9-5, with NO (a slightly polar molecule) and HF (a very polar molecule) as examples.

The molecular orbital theory is also very useful in discussions of molecules and ions in which electron delocalization is an important feature of the bonding. (These are species that the valence bond theory describes with resonance structures.) Applications of this approach to the carbonate ion (CO_3^{2-}) and the benzene molecule (C_6H_6), are discussed in Section 9-6.

Study Goals

As in other chapters, some typical textbook Exercises related to each Study Goal are indicated.

Introduction to Chapter 9 and Section 9-1
1. Distinguish between the main concepts of the valence bond (VB) theory and the molecular orbital (MO) theory. *Work Exercises 1 through 5 and 13.*
2. Distinguish between molecular orbitals and atomic orbitals. *Work Exercises 3 through 5 and 12.*

Sections 9-1
3. Describe each of the various molecular orbitals discussed in the chapter. Understand how each of these orbitals is related to the "pure" atomic orbitals from which it is formed. *Work Exercises 5 through 7, 10, 12, and 13.*
4. Distinguish between sigma (σ) and pi (π) molecular orbitals. Distinguish between bonding and antibonding molecular orbitals. *Work Exercises 6, 7, 10, 12, and 13.*

Section 9-2
5. Remember the order of energies of the molecular orbitals for simple homonuclear diatomic molecules and ions. Construct molecular orbital diagrams for these molecules and ions. *Work Exercise 8.*

Sections 9-3 and 9-4

6. Apply the concepts of molecular orbital theory to determine bond orders and to predict relative stabilities of simple homonuclear diatomic molecules and ions. *Work Exercises 14 through 28.*

Section 9-5

7. Remember the order of energies of the molecular orbitals for simple heteronuclear diatomic molecules and ions. Relate the appearance of molecular orbital diagrams to polarities of bonds. Construct molecular orbital diagrams for these molecules and ions. *Work Exercises 29 through 36.*

Section 9-6

8. Sketch three-dimensional representations of molecular orbitals in some molecules and ions for which delocalization is important. *Work Exercise 37 and 38.*

General

9. Recognize and answer questions of various types about molecular orbitals. *Work **Mixed Exercises** 39 through 41.*

10. Answer conceptual questions based on this chapter. *Work **Conceptual Exercises** 42 through 50.*

11. Relate what you know from earlier chapters to the new ideas in this chapter. *Work **Building Your Knowledge** exercises 51 and 52.*

12. Exercises at the end of chapter direct you to sources outside the textbook for information to use in solving them. *Work **Beyond the Textbook** exercises 53 through 55.*

Some Important Terms in This Chapter

Some important new terms from Chapter 9 are listed below. Fill the blanks in the following paragraphs with terms from the list. Each term is only used once. Check the Key Terms list and the chapter reading. Study other new terms, and review terms from preceding chapters if necessary.

antibonding orbital	bonding orbital	nonbonding orbital
atomic orbitals	delocalization	pi orbital
bond order	molecular orbitals	sigma orbital

According to valence bond theory, bonds result from the overlap of orbitals of individual atoms; that is, [1]_____. But molecular orbital theory proposes that atomic orbitals combine to form [2]_____ (MOs), which allow electrons to belong to the whole molecule. The overlap of two atomic orbitals always produces two molecular orbitals: a [3]_____, which is of less energy (more stable) than the atomic orbitals, and an [4]_____, which is of greater energy (less stable) than the original atomic orbitals. Based up on their shape, each of these molecular orbitals may be classified as either a [5]_____, which is cylindrically symmetrical about the internuclear axis, or a [6]_____, for which the region of electron sharing is on opposite sides of and parallel to the internuclear axis. Sigma (σ) molecular orbitals result from the head-on overlap of atomic orbitals (*s* or *p*), and pi (*π*) molecular orbitals result from the side-on overlap of *p* atomic orbitals.

The stability of a bond, the bond strength, the bond energy, and the bond length ultimately depend upon the [7]_____, which the difference between the number of bonding electrons and the number of antibonding electrons, divided by two.

A molecular orbital derived only from an atomic orbital of one atom, a [8]_____, lends neither stability nor instability to a molecule or ion.

Valence bond theory represents some molecules and ions by resonance structures. Molecular orbital theory describes these species more precisely by the formation of a set of molecular orbitals that extend over more than two atoms, which is known as [9]_____.

Preliminary Test

As in other chapters, this test will check your understanding of basic concepts and their applications. Practice *many* of the additional textbook exercises, including those indicated with the Study Goals.

Short Answer

1. The maximum number of electrons that can occupy one molecular orbital is _____.
2. When two atomic orbitals combine, the number of molecular orbitals formed is _____.
3. When four atomic orbitals combine, the number of molecular orbitals formed is _____.
4. Head-on overlap of two *p* orbitals results in formation of the molecular orbitals _____ and _____.
5. Side-on overlap of two *p* orbitals results in formation of the molecular orbitals _____ and _____.
6. Molecular orbitals that are at lower energy than the atomic orbitals from which they are formed are called _____ orbitals.
7. Molecular orbitals that, when occupied, increase the stability of the bonding between the two atoms are called _____ orbitals.
8. The quantity that uses the relative numbers of electrons in bonding and antibonding orbitals to predict the stability of bonds is called the _____.
9. The definition of bond order is _____.
10. For a molecule to be stable, its bond order must be _____.
11. The amount of energy necessary to break a mole of bonds is called the _____.
12. Generally, the more stable a bond, the _____ is the bond order, the _____ is the bond length, and the _____ is the bond energy. (Answer with words like greater, smaller, longer, shorter, etc.)
13. The order of energies of the molecular orbitals of the homonuclear diatomic molecules and ions of the first- and second-period elements (atomic numbers up to and including that of N) is

_____.

14. The order of energies of the molecular orbitals of the homonuclear diatomic molecules and ions of the second-period elements (atomic numbers greater than that of N) is _____.

15. A molecule or ion with an odd number of electrons must be _____.

16. In order for a molecule to be paramagnetic, it must have _____ electrons.

17. Of the homonuclear diatomic molecules whose electron distributions are shown in Table 9-1 in the text, the ones that are paramagnetic are _____.

18. The magnetic properties of F_2 would be described as _____.

19. Of the first- and second-row homonuclear diatomic molecules, the most stable one is predicted to be _____.

20. The electron configuration of Li_2 is _____.

21. The bond order of Li_2 is _____; thus, we predict that the molecule of Li_2 would be _____.

22. For heteronuclear diatomic molecules, the atomic orbitals of the more electronegative element are _____ in energy than the corresponding orbitals of the less electronegative element.

23. According to the molecular orbital theory, we predict that the order of stability of NO, NO^-, and NO^+ would be (most stable to least stable) _____.

Multiple Choice

_____ 1. When two *s* orbitals combine to form molecular orbitals, the orbitals formed are

(a) one σ_s and one σ_s^*.   (b) two σ_s.   (c) two σ_s^*.

(d) one σ_s and one π_s.   (e) one π_s and one σ_s^*.

_____ 2. When an *s* orbital and a *p* orbital combine to form molecular orbitals, the orbitals formed are

(a) one σ_s and one σ_p.   (b) one π_s and one π_p.   (c) one σ_{sp} and one σ_{sp}^*.

(d) one π_{sp} and one π_{sp}^*.   (e) one σ_{sp} and one π_{sp}^*.

_____ 3. When two *p* orbitals overlap head-on to form molecular orbitals, the orbitals formed are

(a) one σ_s and one σ_p.   (b) one π_s and one π_p.   (c) one σ_p and one σ_p^*.

(d) one π_p and one π_p^*.   (e) one σ_{sp} and one π_{sp}^*.

_____ 4. When two *p* orbitals overlap side-on to form molecular orbitals, the orbitals formed are

(a) one σ_s and one σ_p.   (b) one π_s and one π_p.   (c) one σ_p and one σ_p^*.

(d) one π_p and one π_p^*.   (e) one σ_{sp} and one π_{sp}^*.

_____ 5. Which one of the following describes a diatomic molecule with bond order equal to zero?
   (a) More electrons are in bonding than in antibonding orbitals.
   (b) More electrons are in antibonding than in bonding orbitals.
   (c) No electrons are in bonding orbitals.
   (d) No electrons are in antibonding orbitals.
   (e) Equal numbers of electrons are in bonding and in antibonding orbitals.

_____ 6. Which one of the following molecules or ions would have a bond order equal to 2½ ?
   (a) He_2 (b) O_2^- (c) NO   (d) O_2 (e) Li_2^+

_____ 7. Which of the following molecules would be predicted to be the most stable?
   (a) H_2 (b) Li_2 (c) B_2 (d) Be_2 (e) C_2

For questions 8 through 10, use the molecular orbital diagram in Figure 9-5(b) of the text. Put 16 electrons into this diagram.

_____ 8. What is the bond order of the molecule or ion described?
   (a) 1   (b) 1½   (c) 2   (d) 2½   (e) 3

_____ 9. Which of the following statements is true about this molecule or ion?
   (a) It is diamagnetic.
   (b) It has a dipole moment.
   (c) It is a homonuclear diatomic species.
   (d) It is a heteronuclear diatomic species.
   (e) None of the preceding statements is true.

_____ 10. The MO diagram best describes which of the following molecules or ions?
   (a) NO^- (b) O_2 (c) CO^- (d) NO^+ (e) CN^-

Answers to Some Important Terms in This Chapter

1. atomic orbitals
2. molecular orbitals
3. bonding orbital
4. antibonding orbital
5. sigma orbital
6. pi orbital
7. bond order
8. nonbonding orbital
9. delocalization

Answers to Preliminary Test

Do not just look up answers. *Think about the reasons for the answers.*

Short Answer

1. two. Just as for pure atomic orbitals, a molecular orbital is a region in space that can hold a maximum of two electrons, and then only if they have opposite spin (that is, are paired).

2. two. The number of molecular orbitals formed is always the same as the number of atomic orbitals used to form them.

3. four. See the answer to Question 2.

4. σ_p and σ_p^*. When two orbitals combine, one of the resultant molecular orbitals is bonding and the other is antibonding. These are σ orbitals because they have cylindrical symmetry about the internuclear line. See Figure 9-3.

5. π_p and π_p^*. See Figure 9-4.

6. bonding

7. bonding

8. bond order. See Section 9-3.

9. $$\frac{\text{(number of electrons in bonding orbitals)} - \text{(number of electrons in antibonding orbitals)}}{2}$$
See Section 9-3.

10. greater than 0. See Section 9-3.

11. bond energy. See Section 9-3.

12. greater; shorter; greater. See Section 9-3.

13. $\sigma_{1s}\ \sigma_{1s}^*\ \sigma_{2s}\ \sigma_{2s}^*\ \pi_{2p_y}\ \pi_{2p_z}\ \sigma_{2p}\ \pi_{2p_y}^*\ \pi_{2p_z}^*\ \sigma_{2p}^*$. See Figure 9-5(a). You should know this order of orbitals.

14. $\sigma_{1s}\ \sigma_{1s}^*\ \sigma_{2s}\ \sigma_{2s}^*\ \sigma_{2p}\ \pi_{2p_y}\ \pi_{2p_z}\ \pi_{2p_y}^*\ \pi_{2p_z}^*\ \sigma_{2p}^*$. See Figure 9-5(b). You should know this order of orbitals.

15. paramagnetic. Remember that the presence of unpaired electrons leads to paramagnetism. If an atom or molecule has an odd number of electrons, they cannot all be paired.

16. unpaired

17. B_2 and O_2. Only these have unpaired electrons.

18. diamagnetic. All electrons are paired.

19. N_2. This molecule has the highest bond order.

20. $\sigma_{1s}^2\ \sigma_{1s}^{*2}\ \sigma_{2s}^2$.

21. 1; stable. The bond order is calculated as $\frac{4-2}{2} = 1$.

22. lower. One way to remember this is to recall that the electrons are drawn preferentially to the more electronegative element.

23. NO^+, NO, NO^-. To figure this out, use the MO energy level diagram. You could use either diagram in Figure 9-5, and lower the O side, because O has the lower electronegativity. Then put in the appropriate number of electrons and determine the bond order. NO has 15 electrons and a bond order of 2½; NO^+ has 14 electrons and a bond order of 3 (same as N_2); NO^- has 16 electrons and a bond order of 2 (same as O_2). Higher bond order means higher stability.

Multiple Choice

1. (a). See Figure 9-2 and the discussion in Section 9-1.
2. (c). See Figure 9-8 and the discussion in Section 9-5.
3. (c). See Figure 9-3 and the discussion in Section 9-1.
4. (d). See Figure 9-4 and the discussion in Section 9-1.
5. (e). Suppose x = the number of bonding electrons = the number of antibonding electrons.

    Then, bond order = $\frac{x - x}{2} = 0$

6. (c). Be sure you know how to fill the orbitals and then calculate bond order. See Figure 9-7.

7. (e). The bond orders are $H_2 = 1$, $Li_2 = 1$, $B_2 = 1$, $Be_2 = 0$, $C_2 = 2$. C_2 has the highest bond order, so it is predicted to be the most stable of the molecules listed here. This would be true in the high-temperature gas phase. However, at room temperature, elemental carbon exists in other forms far more stable than C_2, such as graphite and diamond.

8. (c). $\frac{10 - 6}{2} = 2$

9. (c). We can tell because the atomic orbitals of the two atoms are at the same energies. Remember that for heteronuclear diatomic molecules, the atomic orbitals of the more electronegative element are lower in energy. See Figure 9-7 for such a diagram.

10. (b). Count the number of electrons in each of the species listed. Only O_2 has 16 electrons and is also homonuclear.

10 Reactions in Aqueous Solutions I: Acids, Bases, and Salts

Chapter Summary

This chapter continues and broadens the descriptive chemistry associated with acids, bases, and salts introduced in Chapter 6. Some important occurrences of acids, bases, and salts and practical uses of them are discussed in the introduction to Chapter 10. The general properties of aqueous solutions of protonic acids and soluble bases are summarized in Section 10-1. The list of properties presented there emphasizes several ways in which acids and bases are opposite in their behavior. You have already learned in Chapter 6 about the **Arrhenius** view of acids and bases (Section 10-2). In that view, acids are substances that produce H^+ in aqueous solution, whereas bases act to produce OH^- ions. [We now know that H^+ exists in aqueous solution as the **hydrated hydrogen ion**, usually represented as the **hydronium ion, H_3O^+** (Section 10-3).] The most significant overall feature of acids and bases, common to all of the studies in this chapter, is the essentially *opposite nature* of these two classes of compounds. The most important aspect of their opposite nature is their ability to neutralize one another. One of the drawbacks of the Arrhenius approach is that it does not emphasize sufficiently the "oppositeness" of acids and bases. Much of Chapter 10 is devoted to other ways of looking at acids and bases. Each of these "theories" about acids, bases, and salts is a different way of systematically describing the behavior of these substances. The blue/red color coding for acids/bases, first used in Chapter 6 of the textbook, is continued in this chapter.

One of the most useful descriptions of acid-base behavior is the **Brønsted-Lowry theory** (Section 10-4). The theory is used in this chapter to describe a broad range of reactions. An alternative view of acid-base behavior, the **Lewis theory**, is also presented. The chapter closes with some quantitative considerations of acid-base reactions in water solutions.

According to the Brønsted-Lowry view of acid-base behavior (Section 10-4), an acid is a substance that **donates H^+** (referred to as a proton, even though it is hydrated in aqueous solution), and a base is a substance that **accepts H^+**. Thus, we see that this approach emphasizes the opposite nature of these two classes of substances. Pay close attention to the terminology used to describe behavior and properties so these terms will help you later in the chapter. Reactions are reversible, so we can write many acid-base reactions in the general form

$$\text{acid}_1 + \text{base}_2 \rightleftharpoons \text{acid}_2 + \text{base}_1.$$

When any Brønsted-Lowry acid gives up H^+, it becomes capable of accepting H^+, that is, of acting as a base. Thus, we identify **conjugate acid-base pairs** of substances—a pair of substances that differ by an H^+. Such apparently diverse reactions as the **autoionization** of water (Section 10-5), the dissolving of hydrogen chloride in water to form hydrochloric acid, and the dissolving of ammonia in water to give a basic solution may all be described in this way. Learning to recognize conjugate acid-base pairs will be helpful in your study of this chapter. As a memory aid, the text uses rectangles to indicate one acid-base (blue-red) pair and ovals for the other pair (see page 351). We also learn in Section 10-4 that the stronger an acid is, the weaker is its conjugate base. For any base to accept H^+, it must have available an unshared electron pair. This will be a basis for a different acid-base classification later in the chapter.

Amphoterism (Section 10-6) is the ability of a substance to act either as an acid or as a base. It will be helpful to remember the formulas of the **amphoteric hydroxides** and the complex ions to which they are related (Table 10-1).

One of our continuing goals in chemistry, beyond merely learning and classifying the reactions, is to understand trends in properties in terms of electronic and structural features of the molecules—in other words, in terms of the bonding. Section 10-7 emphasizes such an understanding. First, we correlate relative strengths of **binary** protonic acids with their bonding properties. This section relates trends in strengths of acids (and of course the reverse trends in strengths of their conjugate bases) to such features as electronegativity differences and bond strengths. The important notion of the **leveling** effect is introduced in this section. We also can correlate the acid strengths of **ternary acids** with the bonding features of the acid molecules: structure, oxidation state of the central element, position in the periodic table, etc. Learn the generalizations presented in this section regarding trends in acid strength.

Section 10-8 deals in a systematic way with the neutralization reactions between acids and bases. You should review Section 6-9 before beginning your study of this section. You must also familiarize yourself with the writing of **total ionic equations** and **net ionic equations** (Sections 6-2 and 6-9) to represent reactions in aqueous solution. A review of the general rules regarding solubilities of inorganic compounds in water (Section 6-1, part 5 and Table 6-4) will assist you in understanding this section. As you study the neutralization reactions in Section 10-8, you will see that the useful forms of the equations (that is, which aspects of the reaction that we wish to emphasize) depend on several factors:

(1) the strength of the acid involved,
(2) the strength and solubility of the base involved,
(3) the solubility of the salt in water, and
(4) the ionic or covalent nature of the soluble substances.

Be sure you see how these factors affect the ways in which we write the chemical equation.

When the salts formed by neutralization of acid and base contain no remaining unreacted H^+ or OH^-, they are referred to as **normal salts**. But if less than a stoichiometric amount of base reacts with a **polyprotic acid** (more than one H^+ per molecule), the product is an **acidic salt** (i.e., it still has ionizable H^+). Similarly, a **basic salt** (still has ionizable OH^-) is formed when an acid partially neutralizes a **polyhydroxy base** (more than one OH^- per formula unit of base). The acidic and basic salts are the subject of Section 10-9.

As we observed earlier, a Brønsted-Lowry base must have a lone pair available for bond formation with the H^+ that it accepts. The generalization of this behavior leads to the **Lewis theory** (Section 10-10), the broadest of the acid-base theories. In this theory, any substance that supplies an electron pair for bond formation, whether to H^+ or to any other species, is classified as a base; the substance that accepts a share in this lone pair is then classified as an acid. The Lewis theory includes all acid-base behavior previously discussed, but it extends the acid-base terminology to include reactions that do not involve H^+. This view of acid-base chemistry helps to systematize additional types of reactions, such as those involving coordination compounds (Chapter 25).

Several methods of preparing common acids are discussed in Section 10-11.

Study Goals

Be sure to review the ideas and skills from indicated sections of earlier chapters. Several of the study goals in this chapter take much practice, so be sure you answer many of the Exercises in the text. Be systematic in your study.

Section 10-1 (Review Section 6-1.2 and 6-1.4)
1. Review the fundamental ideas of acids and bases. Memorize the lists of the common strong acids (Table 6-1) and the common strong bases (Table 6-3). Classify common acids as strong or weak; classify common bases as strong, insoluble, or weak. *Work Exercises 1 and 3 through 5.*

Section 10-1
2. Be familiar with the characteristics of acids, bases, and their solutions. *Work Exercises 1, 25 through 27, 86, and 87.*

Section 10-2
3. Define acid, base, and neutralization in terms of the Arrhenius theory. Give specific examples with formulas. *Work Exercise 2.*

Section 10-3
4. Be familiar with the description of the hydrated hydrogen ion. *Work Exercises 6 through 8.*

Section 10-4
5. Understand the Brønsted-Lowry acid-base theory. Be able to identify the Brønsted-Lowry acids and bases in reactions. *Work Exercises 9 through 16.*
6. Understand and apply the Brønsted-Lowry terminology of conjugate acid-base pairs and their relative strengths. *Work Exercises 17 through 23.*

Section 10-5
8. Understand the important aspects of the autoionization of water. *Work Exercises 13 and 100.*

Section 10-6
9. Know the identity, properties, and characteristic reactions of the amphoteric hydroxides. Correlate amphoteric behavior of metal hydroxides with the relative strengths of (a) metal-oxygen and (b) oxygen-hydrogen bonds associated with coordinated water molecules. *Work Exercises 28, 31, and 63.*

Section 10-7
7. Explain trends in acid strengths of the binary acids. Understand the correlations of strengths of ternary acids with oxidation number of the central atom, with position of the central atom in the periodic table, and with number of oxygen atoms. *Work Exercises 31 through 45 and 64.*

Review Section 6-1, part 5, and Table 6-4
10. Review the solubility guidelines for ionic substances in water.

Section 10-8

11. Write balanced molecular equations, total ionic equations, and net ionic equations for neutralization reactions of acids and bases to form salts. Name all compounds involved in these equations. *Work Exercises 46 through 56.*

Section 10-9

12. Tell what is meant by each of the following terms: (a) normal salt, (b) acidic salt, and (c) basic salt. Give examples of each and be able to tell by what kinds of reactions each type of salt could be formed. *Work Exercises 57 through 66.*

Section 10-10

13. Understand the Lewis acid-base theory. Be able to recognize Lewis acid-base reactions and classify their acids and bases. *Work Exercises 67 through 74.*

Sections 10-11

14. Write equations illustrating the preparations of binary acids and of ternary acids. *Work Exercises 75 through 77.*

General

15. Recognize and answer various types of questions about acids, bases, and salts. *Work **Mixed Exercises**, 78 through 85.*

16. Answer conceptual questions based on this chapter. *Work **Conceptual Exercises** 86 through 99.*

17. Apply concepts and skills from earlier chapters to the ideas of this chapter. *Work **Building Your Knowledge** exercises, 100 through 102.*

18. Exercises at the end of chapter direct you to sources outside the textbook for information to use in solving them. Work **Beyond the Textbook** exercises 103 through 105.

Some Important Terms in This Chapter

Some important new terms from Chapter 10 are listed below. Fill the blanks in the following paragraphs with terms from the list. Each term is only used once. Check the Key Terms list and the chapter reading. Study other new terms, and review terms from preceding chapters if necessary.

acidic salt	Brønsted-Lowry acid	Lewis base
amphoterism	Brønsted-Lowry base	net ionic equation
Arrhenius acid	conjugate acid-base pair	neutralization
Arrhenius base	formula unit equation	normal salt
autoionization	ionization	salt
basic salt	leveling effect	ternary acid
binary acid	Lewis acid	total ionic equation

Acids and bases have long interested chemists. Their reactions fascinate; their concepts perplex. Consequently, over time chemists have developed several theories to explain the behavior of acids and bases. They formulated each theory with a certain set of information about acids and bases and a particular purpose in mind. Our study of these theories is not concerned with which is "right," but to uncover which is the best to use for a particular purpose, a specific way of using acids and bases.

According to the theory developed by Svante Arrhenius, a substance that includes hydrogen in its formula and produces hydrogen ions, H^+, (actually hydronium ions, H_3O^+) when dissolved in water is defined as an [1]_____. Conversely, a substance that includes the hydroxyl group (OH) in its formula and produces hydroxide ions, OH^-, when dissolved in water is an [2]_____. Therefore, in the Arrhenius theory, the combination of hydrogen ions, H^+, with hydroxide ions, OH^-, to form water molecules, H_2O, is [3]_____. The other product produced in an Arrhenius acid-base neutralization is an ionic compound, called a [4]_____. When an acid, a base, and a salt are represented by their neutral formula units, the equation written for the neutralization reaction is the [5]_____. However, this form of the equation may be somewhat misleading, since strong acids and bases and most soluble salts separate into ions when dissolved in water. Writing all the ions produced among both the reactants and the products yields the [6]_____. Often this equation will include spectator ions, which are the same on both sides of the equation. Since they are unchanged in the reaction, they can be subtracted from both sides of the equation, leaving the [7]_____. For any strong acid reacting with a strong base, producing a soluble salt, this net ionic equation is always the same: $H^+ + OH^- \rightarrow H_2O$.

According to the theory developed by J. N. Brønsted and T. M. Lowry, a substance that donates a proton (H^+) is a [8]_____, and a substance that accepts a proton is a [9]_____. Furthermore, when a substance donates a proton (an acid), it becomes a species (molecule or ion) that can then accept a proton (a base) to reproduce the original substance. Therefore, the two species are linked by the proton lost and gained, and constitute a [10]_____.

Since a proton has a positive charge, the loss of a proton from a molecule converts the molecule into a negative ion, a process known as [11]_____. Similarly, when a molecule gains a proton, it undergoes ionization, forming a positive ion. Dissolving in water is not required. However, one of the advantages of the Brønsted-Lowry acid-base theory is its ability to explain the separation of a water molecule into a hydrogen ion, H^+, and a hydroxide ion, OH^-. The simplified form of the equation for this reaction is the reverse of the net ionic equation for acid-base neutralization:

$$H_2O \rightarrow H^+ + OH^-$$

A more complete representation of this reaction shows that two molecules of water are involved:

$$2H_2O \rightarrow H_3O^+ + OH^-$$

A proton transfers from one water molecule to another, producing a hydronium ion and a hydroxide ion. Since the water molecules, in effect, ionize each other, this reaction is referred to as the [12]_____ of water. One of the water molecules donates the proton, so it is functioning as an acid, and the other water molecule accepts the proton, so it is function as a base. Thus, water is one of a number of compounds that can react as either an acid or a base, a property known as [13]_____. If any acid stronger than the hydronium ion dissolves in water, it reacts with water to produce hydronium ions. If any base stronger than the hydroxide ion dissolves in water, it reacts with water to produce hydroxide ions. Thus, the strongest acid left in water is the hydronium ion, and the strongest base left in water is the hydroxide ion. Any stronger acid or base is reduced to the level of these two ions. This property of water is known as the [14]_____.

Acids can be categorized by the number of elements in the molecule. An acid composed of just two elements, one of which would have to be hydrogen, is [15]_____. An acid composed of three elements (generally hydrogen, oxygen, and some other element), is [16]_____.

Salts can be categorized by the extent of the neutralization reaction that could be used to produce them. If the acid and base have been completely neutralized, so that there are no ionizable H atoms or OH groups in the salt, the salt is a [17]_____. When a polyprotic acid reacts with a base, all of the protons may be neutralized to produce a normal salt, or only a portion of them may react, producing an [18]_____. (Don't let this term mislead you though—solutions of many acidic salts are actually basic; more on this in Section 18-8.) Similarly, when a polyhydroxy base reacts with an acid, all of the OH groups may be neutralized to produce a normal salt, or only a portion of them may react, producing a [19]_____. (Many basic salts form acidic solutions; see Section 18-9.)

G. N. Lewis proposed the most widely applicable of the classic acid-base theories. According to this theory, any species that can accept a share in a pair of electrons is a [20]_____, and any species that can "donate" a share in a pair of electrons is a [21]_____. No protons are required.

Preliminary Test

As in other chapters, this test will check your understanding of basic concepts and types of calculations. Be sure to practice *many* of the additional textbook Exercises, including those indicated with the Study Goals.

Short Answer

Answer with a word, a phrase, a formula, or a number with units as necessary.

1. Write the definitions of acids and bases according to each of the following theories:

Theory	Definition
Arrhenius	(a) acid
	(b) base
Brønsted-Lowry	(c) acid
	(d) base
Lewis	(e) acid
	(f) base

142

2. The essential feature of bases, according to all theories considered here, is that they act to _____ acids.

3. In either the Arrhenius or the Brønsted-Lowry view of aqueous acid-base neutralization reactions, one of the products of the neutralization is often _____. (Give name and formula.)

4. According to the Brønsted-Lowry description of acids and bases, when an acid acts to donate H^+, it is transformed into its _____.

5. The conjugate base of HCl is _____. (name, formula)

6. The conjugate base of acetic acid is _____. (name, formula)

7. The conjugate acid of ClO_4^- is _____. (name, formula)

8. The conjugate acid of HSO_4^- is _____. (name, formula)

9. The conjugate base of HSO_4^- is _____. (name, formula)

10. The conjugate acid of water is _____. (name, formula)

11. The conjugate base of water is _____. (name, formula)

12. The process of water ionizing to produce an equal number of hydrated hydrogen ions (that is, hydronium ions, H_3O^+) and hydroxide ions OH^- is called the _____ of water.

13. A substance that can act as either an acid or a base is termed _____.

14. The stronger an acid, the _____ is its conjugate base.

15. HCl is a stronger acid than HCN, so we know that CN^- must be a _____ base than Cl^-.

16. Even though it is not included in the formal definition, we see that in order for a substance to act as a Brønsted-Lowry base, it must have _____.

17. We can attribute the weaker acidity of HF compared with the other hydrohalic acids to the _____ between H and F than between H and the other halogens.

18. "… the hydronium ion is the strongest acid that can exist in aqueous solution. All acids stronger than H_3O^+(aq) react completely with water to produce H_3O^+(aq) and their conjugate bases." This is often referred to as the _____ of water.

19. Neutralization reactions in aqueous solution produce _____ and (in most cases) _____.

20. Nearly all soluble salts produce _____ when they dissolve in water.

21. The net ionic equation for the reaction of all strong acids with strong bases to form soluble salts and water is _____.

22. An acid that contains more than one acidic (ionizable) hydrogen atom per formula unit is called _____.

23. The most common weak base is ammonia. The net ionic equation for the reaction of a strong acid HX with aqueous ammonia is: _____.

24. The common driving force for neutralization reactions in aqueous solution is the formation of _____; a secondary factor favoring such neutralizations might be the formation of _____.

25. Salts that contain no ionizable H^+ or OH^- ions are referred to as _____ salts; those with excess available H^+ are termed _____ salts, whereas those with remaining OH^- are called _____ salts.

26. An acid that is composed of hydrogen, oxygen, and one other element is referred to as a _____ acid.

27. The four ternary acids of chlorine are (give name and formula for each):
_____, _____,
_____, and _____.

28. The salts of nitric acid contain the ion _____ (formula), and are called _____; the normal salts of sulfurous acid contain the ion _____, and are called _____.

29. For various ternary acids containing the same central element (i.e., the element other then H and O), strength of the acid increases as the _____ of the central atom _____.

30. For ternary acids containing different elements from the same periodic group in the same oxidation state, acid strength increases with _____ of the central element.

31. Due to the formation of complex hydroxyanions, many amphoteric metal hydroxides are soluble in _____.

32. Strong bases, such as the Group 1A metal hydroxides, can often be prepared by the reaction of the metallic oxide or just _____ with water.

33. Many ternary acids can be prepared by dissolving the appropriate _____ in water.

34. When CO_2 is dissolved in water, as in carbonated drinks, the acid formed is _____ _____. (Give name, formula)

35. In the Lewis acid-base theory, a neutralization reaction always involves _____ _____ bond formation.

Multiple Choice

_____ 1. According to the Brønsted-Lowry concept of acids and bases, which one of the following statements about a base is *not* correct?
(a) A base accepts a hydrogen ion by sharing a previously unshared electron pair with H^+.
(b) A base accepts a hydrogen ion by forming a coordinate covalent bond to H^+.
(c) If a base is weak, its conjugate acid will be a strong acid.
(d) A base can be formed only by ionization of a compound containing a hydroxide group.

_____ 2. Which one of the following could not be a Brønsted-Lowry acid?
(a) H_2O (b) HN_3 (c) H_3O^+ (d) NH_4^+ (e) BF_3

_____ 3. Which one of the following could not act as a Lewis base?
(a) PCl_3 (b) CN^- (c) I^- (d) CH_4 (e) H_2O

_____ 4. When a salt such as sodium acetate is dissolved in water, the acetate ion reacts with water:

$$CH_3COO^-(aq) + H_2O(\ell) \rightleftharpoons CH_3COOH(aq) + OH^-(aq)$$

In this reaction, acetate ion is acting as
(a) an acid.   (b) a base.   (c)   an   amphoteric
substance.
(d) an oxidizing agent.   (e) a reducing agent.

_____ 5. What substance is acting as a Lewis acid in the following reaction?

$$BF_3 + F^- \rightarrow BF_4^-$$

(a) BF_3 (b) F^- (c) BF_4^-
(d) All of substances listed.   (e) None of the substances listed

_____ 6. Cupric ion, Cu^{2+}, is pale blue in water solution. When ammonia is added to the solution, the deep-blue complex ion $[Cu(NH_3)_4]^{2+}$ is formed. In the formation of this complex ion, Cu^{2+} is acting as
(a) a Lewis acid.   (b) a Lewis base.   (c) an Arrhenius acid.
(d) a Brønsted-Lowry acid.   (e) a Brønsted-Lowry base.

_____ 7. In the reaction

$$CN^- + NH_4^+ \rightarrow HCN + NH_3$$

the cyanide ion acts as
(a) a base.   (b) an acid.   (c) an oxidizing agent
(d) a reducing agent.   (e) an amphoteric substance.

_____ 8. The following describe properties of substances. Which one is not a typical acid property?
(a) It has a sour taste.
(b) It reacts with metal oxides to form salts and water.
(c) It reacts with acids to form salts and water.
(d) Its aqueous solutions conduct an electrical current.
(e) It reacts with active metals to liberate H_2.

_____ 9. In water solution, HCN is only very slightly ionized:

$$HCN(aq) + H_2O(\ell) \rightleftharpoons H_3O^+(aq) + CN^-(aq)$$

This observation shows that CN^- is a _____ than is water.
(a) stronger acid   (b) stronger base   (c) weaker acid   (d) weaker base

_____ 10. What is the conjugate base of HI?
(a) H^+ (b) HI   (c) I^- (d) HIO_3 (e) H_2I

_____ 11. HBr is a strong acid when dissolved in water. HF is a weak acid when dissolved in water. Which of the following is the strongest base?
(a) H^+ (b) Br^- (c) F^- (d) H_2O
(e) Insufficient information is given to answer the question.

_____ 12. Which one of the following substances produces an acidic solution in water?
(a) SO_3 (b) NH_3 (c) NaO   (d) $Ca(OH)_2$ (e) LiH

_____ 13. Which one of the following is an example of an acidic salt?
(a) KNO_3 (b) $KHSO_4$ (c) H_3PO_4 (d) $Mg(OH)_2$ (e) $Al(OH)Cl_2$

_____ 14. Which one of the following is an example of a basic salt?

145

(a) KNO_3 (b) $KHSO_4$ (c) H_3PO_4 (d) $Mg(OH)_2$ (e) $Al(OH)Cl_2$

____ 15. Which one of the following is an example of a normal salt?
(a) KNO_3 (b) $KHSO_4$ (c) H_3PO_4 (d) $Mg(OH)_2$ (e) $Al(OH)Cl_2$

____ 16. Each of the following ions forms a stable hydroxide. For which one of the ions would the hydroxide *not* be amphoteric?
(a) Sn^{2+} (b) Sn^{4+} (c) Cr^{3+} (d) Be^{2+} (e) Ba^{2+}

____ 17. In which one of the following pairs is the order of acid strength *incorrect* ?
(a) $HF < HCl$ (b) $H_2SeO_4 < H_2SO_4$ (c) $HNO_2 < HBO_3$
(d) $H_2SO_4 < H_2SO_3$ (e) $HCl < HI$

Answers to Some Important Terms in This Chapter

1. Arrhenius acid
2. Arrhenius base
3. neutralization
4. salt
5. formula unit equation
6. total ionic equation
7. net ionic equation

8. Brønsted-Lowry acid
9. Brønsted-Lowry base
10. conjugate acid-base pair
11. ionization
12. autoionization
13. amphoterism
14. leveling effect

15. binary acid
16. ternary acid
17. normal salt
18. acidic salt
19. basic salt
20. Lewis acid
21. Lewis base

Answers to Preliminary Test

Do not just look up answers. *Think about the reasons for the answers.*

Short Answer

1. (a) acts to produce H^+ (or H_3O^+)   (b) acts to produce OH^-
   (c) a proton (H^+) donor   (d) a proton (H^+) acceptor
   (e) accepts a share in an electron pair   (f) makes available a share in an electron pair
2. neutralize
3. water, H_2O
4. conjugate base
5. chloride, Cl^-
6. acetate, CH_3COO^-. Perhaps you should review the names, formulas, and anions (here called conjugate bases) of common acids, Tables 6-1 and 6-2.
7. perchloric acid, $HClO_4$
8. sulfuric acid, H_2SO_4

9. sulfate, SO_4^{2-}. Notice that some species can have *both* a conjugate acid and a conjugate base. These are often (but not always) anions that contain ionizable hydrogen.

10. hydronium, H_3O^+

11. hydroxide, OH^-

12. autoionization

13. amphoteric

14. weaker

15. stronger

16. an unshared electron pair

17. greater electronegativity difference. See Section 10-7.

18. leveling effect. See Section 10-7.

19. a salt, water

20. ions

21. $H^+(aq) + OH^-(aq) \rightarrow H_2O(\ell)$. See Section 10-8.

22. polyprotic

23. $H^+(aq) + NH_3(aq) \rightarrow NH_4^+(aq)$

24. nonionized water molecules; an insoluble salt

25. normal; acidic; basic

26. ternary

27. perchloric acid, $HClO_4$; chloric acid, $HClO_3$; chlorous acid, $HClO_2$; hypochlorous acid, $HClO$

28. NO_3^- ; nitrates; SO_3^{2-} ; sulfites

29. oxidation state; increases. Notice that this also corresponds to an increasing number of oxygen atoms. For example, the order of increasing acid strengths of the *ternary* acids of chlorine is $HClO < HClO_2 < HClO_3 < HClO_4$. Be sure that you remember that this applies only when comparing *ternary* acids of the same *element*.

30. increasing electronegativity

31. bases

32. the metallic element. Some of these reactions occur at dangerously explosive rates.

33. nonmetal oxide. See Section 10-11.

34. carbonic acid, H_2CO_3

35. coordinate covalent. Be sure that you know what this term means. It refers to the formation of a covalent bond by one atom supplying both electrons for the bond.

Multiple Choice

1. (d). The Brønsted-Lowry concept is not limited by the need for a base to contain a hydroxide group, as Arrhenius bases must.

2. (e). A Brønsted-Lowry acid acts as a source of H^+; BF_3 does not contain H.

3. (d). A Lewis base provides an unshared electron pair for bond formation. Of the compounds listed, only CH_4 has no unshared valence electron pairs.

4. (b). Acetate ion accepts H^+.

5. (a). BF_3 accepts the electron pair from F^- (the base) in the formation of a coordinate covalent bond.

6. (a). Ammonia (the Lewis base) provides electron pairs to Cu^{2+}.

7. (a). In this case, CN^-, as a Brønsted-Lowry base, accepts H^+.

8. (c). It reacts with *bases* to form salts and water.

9. (b). The observation that most of the HCN is still present in undissociated form tells us that CN^- wins the competition for H^+ much more often than does H_2O.

10. (c). $HI \rightarrow H^+ + I^-$ HI loses H^+ to form I^-, its conjugate base.

11. (c). Reason as in Question 9. F^- must be a stronger base than H_2O, which must in turn be a stronger base than Br^-.

12. (a). Understand the relation of acidic oxides to location in the periodic table.

13. (b). $KHSO_4$ contains ionizable hydrogen.

14. (e). $Al(OH)Cl_2$ contains ionizable hydroxide.

15. (a). KNO_3 contains neither ionizable hydrogen nor ionizable hydroxide.

16. (e). $Ba(OH)_2$ is a common strong base.

17. (d). Know the common strong acids, and be able to use the generalizations about strengths of binary and ternary acids discussed in Section 10-7 to predict trends, even if you do not recognize the acids as strong or weak.

11 Reactions in Aqueous Solutions II: Calculations

Chapter Summary

Chapter 11 deals with some calculations about reactions in aqueous solutions. Two general classes of reactions are presented here—acid-base reactions and oxidation-reduction (redox) reactions. These calculations are based entirely on the stoichiometric calculations presented in Chapters 2 and 3. Review those calculations now—especially Sections 3-6 through 3-8.

The first three sections of Chapter 11 deal with some quantitative aspects of acid-base reactions in aqueous solutions. You will see that the terminology of the Brønsted-Lowry and Arrhenius theories (Chapter 10) is used here. The net ionic equation for aqueous acid-base neutralizations is:

$$H^+(aq) + OH^-(aq) \rightarrow H_2O(\ell).$$

The central idea behind neutralization calculations is that equal numbers of moles of H^+ and OH^- react in such reactions, no matter what acid or base is involved. When this condition has been reached, the solution contains water, the anion of the acid, and the cation of the base. If the salt that is formed is insoluble, it precipitates from the reaction mixture as a solid; if it is soluble, it remains in solution as dissolved ions.

The essential criterion for neutralization is that equal numbers of moles of H^+ and OH^- (called **equivalent** amounts of acid and base) are required. But this criterion cannot be applied directly because we do not generally have an instrument to directly measure number of moles of a reactant. Thus, a quantity to express solution concentration that relates *number of moles* (which is hard to measure) to *volume of solution* (which is easy to measure) is quite useful. Molarity is such a quantity.

Section 11-1 uses molarities to describe the amounts of acid and base solutions that are necessary to attain neutralization. Notice that some acids (e.g., HCl or HNO_3) supply 1 mole of H^+ per mole of acid, whereas others (e.g., H_2SO_4) supply 2 moles of H^+ per mole of acid.

Comparable situations exist for bases with different numbers of moles of OH^- per mole of base. Be sure you are thoroughly familiar with the terminology of **titration**, **standard solutions, primary standards,** and **secondary standards** (Section 11-2). In the next section you will learn more about the calculations involved in acid-base neutralizations and in titrations, using the familiar mole method and molarity (Section 11-3).

After you understand the ideas of these sections, get lots of practice by working appropriate problems at the end of the chapter—see Study Goal 4 for suggested exercises. Remember that the basis for these calculations is the following relationship at the **equivalence point** (the point of neutralization):

$$\text{\# eq. of acid} = \text{\# eq. base} \quad \text{or} \quad \text{\# meq. acid} = \text{\# meq. base}$$

149

equivalents of acid = # equivalents base   or   # milliequivalents acid = # milliequivalents base

The rest of this chapter deals with **oxidation-reduction (redox) reactions**. You have learned to use the formalism of oxidation numbers (Section 5-7) to recognize such reactions (Section 6-5). Review these two sections of Chapters 5 and 6.

Many applications of chemical reactions require that we write the equation in **balanced** form. Sometimes redox equations can be difficult to balance. These reactions involve electron transfer, and electrons are neither created nor destroyed in ordinary chemical reactions. These facts lead to the useful principle that the total increase in oxidation numbers must equal the total decrease in oxidation numbers. This, in turn, leads us to a very useful method for balancing redox equations, discussed in Sections 11-4 and 11-5. Be sure that you know how to carry out a completely reliable check on whether the resulting equation is balanced. Be sure you check both mass balance (for *each* element in the equation) and charge balance.

One method that many chemists use is the **half-reaction** method, frequently called the **ion-electron** method (Section 11-4). In this method we formally separate the oxidation (electron loss) and reduction (electron gain) portions of the process into half-reactions. We balance each half-reaction separately and then equalize the number of electrons lost and gained in the two half-reactions. We then add the balanced half-reactions to give the overall balanced equation. As you learn this method, always apply the foolproof check of whether the final equation is balanced. Check also to be sure that no extra electrons remain on either side of the equation. You can see that to apply this method you need to decide which species are changing in oxidation numbers, but the actual values of the oxidation numbers are not used in the balancing process.

Practice first with equations in which all reactants and products are given. This includes all equations given in Exercises 51 through 58 at the end of the text chapter. However, for many reactions in aqueous solutions, the equation as given may not specify all reactants and products. In such a case, it is necessary to add enough H and O to complete the mass balance. Section 11-5 discusses this step, which is accomplished by adding *two* of the three species H^+, OH^-, and H_2O to the appropriate sides of the equation. Use the following reasoning to remember which of these species you should add. (1) In an acidic solution, there is a lot of water (the solvent) and a large amount of H^+, but very little OH^- can be present. Thus, in an acidic solution there is essentially no OH^- to act as a reactant, and it cannot be written on the left-hand side of the equation; likewise OH^- should not be written as a product in acidic solution, because it would immediately be consumed by the excess H^+ to form water. So we conclude that *in acidic solution we can add only H^+ or H_2O, but we never add OH^-*. (2) Similar reasoning (can you go through it?) regarding basic solutions (excess OH^-, very little H^+) leads to the conclusion that *in basic solution we can add only OH^- or H_2O, but we never add H^+*. Exercises 57 and 61 at the end of the text chapter will give you practice in this aspect of balancing redox equations.

The final section of this chapter deals with **redox titrations**. For these, we base our calculations on the quantitative relationships given by the balanced equations that describe oxidation-reduction reactions. As before, we can carry out these calculations with solution concentrations expressed as molarity; in Section 11-6 this approach is illustrated with several examples.

Study Goals

Be sure to review the ideas and skills from indicated sections of earlier chapters. Several of the study goals in this chapter take much practice, so be sure you answer many of the Exercises in the text. Be systematic in your study.

Review Sections 3-6 through 3-8

1. Review earlier calculations involving molarity. *Work Exercises 1 through 7, 11, 16, and 17.*

Section 11-1

2. Using specified concentrations of solutions, be able to calculate:
   (a) whether a specified mixture of acid and base solutions results in complete neutralization and
   (b) the amounts of salts formed and acid or base remaining when an acid and a base solution are mixed.
   Work Exercises 8 through 10, 12 through 15, and 18.

Section 11-2

3. Be familiar with the terminology and procedures of titrations and standardization of solutions of acids and bases. *Work Exercises 19 through 25.*

Section 11-3

4. Using molarity, be able to calculate:
   (a) the concentration of an acid (or base) solution from information about the amount of base (or acid) required to neutralize it (standardization),
   (b) the volume of an acid (or base) solution required to completely neutralize a specified amount of base (or acid), and
   (c) the amount of acid or base in an unknown sample, by titration with a solution of known concentration.
   Work Exercises 26 through 47.

Sections 11-4 and 11-5

5. Balance molecular and net ionic equations for redox reactions by the *ion-electron* method. Be able to add H^+, OH^-, and H_2O, as appropriate, to achieve mass balance of H and O. *Work Exercises 48 through 55, 57 through 61.*

Section 11-6

6. Using molarity, perform stoichiometric calculations involving redox reactions in which one or more of the reactants is in solution. *Work Exercises 62 through 71.*

General

7. Use the concepts of the chapter to recognize and solve a variety of types of questions. *Work **Mixed Exercises** 72 through 82.*

8. Answer conceptual questions based on this chapter. *Work **Conceptual Exercises** 83 through 86.*

9. Apply concepts and skills from earlier chapters to the ideas of this chapter. *Work **Building Your Knowledge** exercises 87 through 98.*

10. Exercises at the end of chapter direct you to sources outside the textbook for information to use in solving them. *Work **Beyond the Textbook** exercises 99 through 102.*

Some Important Terms in This Chapter

Some important new terms from Chapter 11 are listed below. Fill the blanks in the following paragraphs with terms from the list. Each term is only used once. Check the Key Terms list and the chapter reading. Study other new terms, and review terms from preceding chapters if necessary.

end point	oxidation	reduction
equivalence point	primary standard	standard solution
indicator	redox reaction	standardization
molarity	redox titration	titration

The concentration of a solution can be calculated experimentally by measuring the amount of it that will react with a given amount of another solution of known concentration. This is the process of [1]_____. Acid-base titrations often involve colorless reactants and products, so a few drops of a substance, an [2]_____, is added to the solution. This substance indicates the completion of the reaction by changing color. By changing color, the indicator signals when to end the titration, the [3]_____. If the indicator has been properly chosen, this end point will be essentially the same as the [4]_____, the point in the titration at which equivalent amounts of the two reactants have combined. Since it is the standard by which we determine the unknown concentration, the solution of known concentration is called the [5]_____. The process by which the concentration of the standard solution is determined is known as [6]_____. The substance that is used to standardize the standard solution is the [7]_____. The concentrations of the solutions can be expressed in units of moles of solute per liter of solution (or mmols of solute per milliliter of solution), known as [8]_____.

An increase in oxidation number is called [9]_____. A decrease in oxidation number is called [10]_____. A reaction involving both oxidation and reduction is called an oxidation-reduction reaction, abbreviated [11]_____. Quantitatively analyzing the amounts of substances that oxidize or reduce in a chemical reaction is called [12]_____.

Preliminary Test

As in other chapters, this test will check your understanding of basic concepts and types of calculations. Be sure to practice *many* of the additional textbook Exercises, including those indicated with the Study Goals.

Short Answer

Answer with a word, a phrase, a formula, or a number with units as necessary.

1. One millimole of HCl would neutralize exactly _____ millimole(s) of KOH.
2. One millimole of HCl would neutralize exactly _____ millimole(s) of Ba(OH)$_2$.
3. In terms of concentration and volume of solution, the number of millimoles of solute can be calculated as _____.
4. A solution whose concentration is accurately known is called a _____ solution.
5. The process by which we determine the volume of a solution required to react with a specific amount of a substance is called _____.
6. A substance that is of sufficient purity and ease of handling that the number of moles present may be determined by direct weighing is referred to as a _____.
7. For an oxidation-reduction reaction to be balanced: (a) _____ and _____ must occur simultaneously, (b) The same number of _____ of each element must appear among both the reactants and the products, and (c) The sums of the _____ on the reactants and products sides of the equation must equal each other.
8. When we need additional hydrogen and oxygen to balance the equation for a redox reaction that is carried out in acidic solution, we may add _____ or _____ to either side of the equation as needed, but we must not add _____.
9. The "ion-electron" method for balancing redox equations is based on the idea that the _____ by all species being _____ must equal the _____ by all species being _____.
10. In applying the ion-electron method we formally separate the oxidation and the reduction processes, and write these as two separate _____, which we later recombine after we balance each one separately.

Multiple Choice

_____ 1. What is the molarity of the salt produced in the reaction of 50.0 mL of 0.0400 M HBr with 100. mL of 0.0300 M Ca(OH)$_2$?
   (a) 0.0150 M (b) 0.00667 M (c) 0.133 M (d) 0.0670 M (e) 0.075 M

_____ 2. What volume of 0.0750 M Ba(OH)$_2$ solution will react exactly with 125 mL of 0.0350 M HCl?
   (a) 29.2 mL   (b) 42.0 mL,   (c) 230 mL   (d) 536 mL   (e) 53.6 mL

_____ 3. Calculate the molarity of a sulfuric acid solution if 40.0 mL of this H$_2$SO$_4$ solution reacts with 0.212 g of Na$_2$CO$_3$. One mole of Na$_2$CO$_3$ = 106 g.
   (a) 0.025 M (b) 0.050 M (c) 0.075 M (d) 0.100 M (e) 0.090 M

_____ 4. A solution of H$_2$SO$_4$ contains 9.11 mg of H$_2$SO$_4$ per mL. What is the molarity of the solution?
   (a) 0.0464 M (b) 0.0929 M (c) 0.186 M (d) 0.372 M (e) 0.684 M

_____ 5. Calculate the volume of 2.00 M H_2SO_4 required to prepare 200. mL of 0.0500 M H_2SO_4.

(a) 5.0 mL   (b) 7.5 mL   (c) 10 mL   (d) 12.5 mL   (e) 15 mL

_____ 6. What is the molarity of 35.0 mL of a solution of $Ba(OH)_2$ that exactly neutralizes 10.0 mL of 0.150 M H_2SO_4?

(a) 0.0429 M (b) 0.150 M (c) 0.0525 M (d) 0.214 M (e) 0.107 M

_____ 7. Calculate the molarity of a solution of nitric acid, HNO_3, if 25.0 mL of the acid solution neutralizes 0.246 g of sodium carbonate, Na_2CO_3, a primary standard.

$$2HNO_3 + Na_2CO_3 \rightarrow 2NaNO_3 + CO_2 + H_2O$$

(a) 2.09 M (b) 0.00464 M (c) 0.371 M (d) 0. 186 M (e) 0.0928 M

_____ 8. A 30.00-mL sample of a solution of barium hydroxide, $Ba(OH)_2$, reacts with 0.4136 g of potassium hydrogen phthalate (KHP), $KHC_6H_4(COO)_2$ (formula weight of KHP = 204.2).

$$Ba(OH)_2 + 2KHP \rightarrow BaP + K_2P + 2H_2O$$

Calculate the molarity of the $Ba(OH)_2$ solution.

(a) 0.001013 M (b) 1.723 M (c) 0.06752 M (d) 0.03377 M (e) 0.01688 M

_____ 9. Suppose 100.00 mL of a potassium hydroxide, KOH, solution requires 50.00 mL of a 0.1000 M solution of benzoic acid, C_6H_5COOH, for neutralization.

$$KOH + C_6H_5COOH \rightarrow C_6H_5COOK + H_2O$$

What is the concentration of the base solution?

(a) 0.0250 M (b) 0.0500 M (c) 0.100 M (d) 0.200 M (e) 0.400 M

_____ 10. The titration of 25.0 mL of an oxalic acid solution, $H_2C_2O_4$, requires 28.3 mL of standard 0.305 M sodium hydroxide solution for complete neutralization.

$$H_2C_2O_4 + 2NaOH \rightarrow Na_2C_2O_4 + 2H_2O$$

What is the molarity of the oxalic acid solution?

(a) 0.691 M (b) 0.108 M (c) 0.135 M (d) 0.345 M (e) 0.173 M

_____ 11. Aspirin is acetylsalicylic acid, $HC_9H_7O_4$, a monoprotic acid. However, an aspirin tablet may contain other inactive ingredients. Suppose a certain aspirin tablet having a mass of 0.400 g required 15.25 mL of 0.115 M sodium hydroxide solution, NaOH, for complete neutralization. Calculate the percentage of aspirin in the tablet.

(a) 100.%   (b) 78.8%   (c) 28.8%   (d) 22.8%   (e) 2.43%

_____ 12. When we complete and balance the equation

$$SO_3^{2-} + MnO_4^- \rightarrow SO_4^{2-} + Mn^{2+} \qquad \text{(acidic solution)}$$

the coefficients of SO_3^{2-} and Mn^{2+} are, respectively, _____ and _____.

(a) 5 and 3   (b) 2 and 6   (c) 5 and 2   (d) 6 and 5   (e) 3 and 1

_____ 13. When we complete and balance the equation

$$Cr_2O_7^{2-} + H_2S \rightarrow Cr^{3+} + S \qquad \text{(acidic solution)}$$

the coefficients of $Cr_2O_7^{2-}$ and S are, respectively, _____ and _____.

(a) 1 and 3   (b) 3 and 2   (c) 2 and 7   (d) 1 and 6   (e) 10 and 4

_____ 14. When we complete and balance the equation

$$NO_3^- + I_2 \rightarrow NO + IO_3^- \qquad \text{(acidic solution)}$$

the coefficients in front of I_2 and NO are, respectively, _____ and _____.

(a) 3 and 6   (b) 3 and 10   (c) 10 and 3   (d) 1 and 2   (e) 4 and 7)

_____ 15. Chlorous acid, $HClO_2$, oxidizes sulfur dioxide, SO_2, to sulfate ion, SO_4^{2-}, and is reduced to chloride ion, Cl^-, in acidic solution. Write the balanced net ionic equation for the reaction. (H^+ or H_2O may be added as necessary.)

(a) $HClO_2 + SO_2 + H_2O \rightarrow Cl^- + SO_4^{2-} + 2H^+$

(b) $HClO_2 + 2SO_2 + 5H^+ \rightarrow Cl^- + 2SO_4^{2-} + 2H_2O$

(c) $HClO_2 + SO_2 + H^+ \rightarrow Cl^- + SO_4^{2-}$

(d) $HClO_2 + 2H_2O + 2SO_2 \rightarrow Cl^- + 2SO_4^{2-} + 5H^+$

(e) $HClO_2 + SO_2 \rightarrow Cl^- + SO_4^{2-} + H^+$

_____ 16. A compound that has often been used in redox titrations is potassium permanganate, $KMnO_4$. It is a strong oxidizing agent, and, because of its intense purple color, it functions as its own indicator. Write the complete, balanced formula unit equation for the oxidation of iron(II) sulfate, $FeSO_4$, with $KMnO_4$ in an acidic solution (H_2SO_4). The ionic products are Fe^{3+} and Mn^{2+}. Add H^+ or H_2O as needed.

(a) $2FeSO_4(aq) + 2KMnO_4(aq) + 4H_2SO_4(aq) \rightarrow$
$Fe_2(SO_4)_3(aq) + 2MnSO_4(aq) + 4H_2O(\ell) + K_2SO_4(aq)$

(b) $10FeSO_4(aq) + 2KMnO_4(aq) \rightarrow 5Fe_2(SO_4)_3(aq) + 2MnSO_4(aq)$

(c) $2FeSO_4(aq) + 2KMnO_4(aq) + 8H_2SO_4(aq) \rightarrow$
$Fe_2(SO_4)_3(aq) + 2MnSO_4(aq) + 8H_2O(\ell) + K_2SO_4(aq)$

(d) $10FeSO_4(aq) + 2KMnO_4(aq) + 8H_2SO_4(aq) \rightarrow$
$5Fe_2(SO_4)_3(aq) + 2MnSO_4(aq) + 8H_2O(\ell) + K_2SO_4(aq)$

(e) $5Fe^{2+} + MnO_4^- + 8H^+ \rightarrow 5Fe^{3+} + Mn^{2+} + 4H_2O$

Questions 17 and 18 refer to the following reaction:

$$MnO_4^- + 8H^+ + 5Fe^{2+} \rightarrow Mn^{2+} + 5Fe^{3+} + 4H_2O$$

_____ 17. What is the molarity of a solution of $FeSO_4$ if 25.0 mL of it reacts with 38.0 mL of 0.1214 M $KMnO_4$?

(a) 0.185 M (b) 0.399 M (c) 0.0798 M (d) 0.426 M (e) 0.923 M

_____ 18. What volume of 0.100 molar $KMnO_4$ will react with 1.00 g of $FeSO_4$?

(a) 164 mL   (b) 52.6 mL   (c) 26.3 mL   (d) 13.2 mL   (e) 65.8 mL

_____ 19. A solution of sodium thiosulfate, $Na_2S_2O_3$, is 0.1524 M, and 25.00 mL of this solution reacts with 31.58 mL of I_2 solution. Calculate the molarity of the I_2 solution.

$$2Na_2S_2O_3 + I_2 \rightarrow Na_2S_4O_6 + 2NaI$$

What is the molarity of the oxalic acid solution?

(a) 0.1925 M (b) 0.09626 M (c) 0.2413 M (d) 0.06032 M (e) 0.1206 M

> **NOTE**: It takes a great deal of practice to learn to use the material in this chapter. You will need much more practice at balancing redox equations and at working with molarity than the questions in this guide provide. You should work many of the text Exercises suggested in the Study Goals.

Answers to Some Important Terms in This Chapter

1. titration
2. indicator
3. end point
4. equivalence point
5. standard solution
6. standardization
7. primary standard
8. molarity
9. oxidation
10. reduction
11. redox reaction
12. redox titration

Answers to Preliminary Test

Do not just look up answers. *Think about the reasons for the answers.*

Short Answer

1. 1. Each mmol of HCl produces 1 mmol of H^+. Each mmol of KOH produces 1 mmol of OH^-.
2. ½. Each mmol of HCl produces 1 mmol of H^+. Each mmol of $Ba(OH)_2$ produces 2 mmol of OH^-.
3. molarity x milliliters
4. standard
5. titration
6. primary standard
7. (a) oxidation, reduction, (b) atoms, (c) charges
8. H^+; H_2O; OH^-
9. gain of electrons; reduced; loss of electrons; oxidized
10. half-reactions

Multiple Choice

1. (b).

	$2HBr$	$+$	$Ca(OH)_2$	$\rightarrow$	$CaBr_2$	$+ 2H_2O$
Rxn ratio:	2 mmol		1 mmol		1 mmol	2 mmol
Start:	$[50\ mL\ (\frac{0.0400\ mmol}{mL})]$		$[100\ mL\ (\frac{0.0300\ mmol}{mL})]$		0 mmol	
	= 2.00 mmol HCl		= 3.00 mmol $Ca(OH)_2$			
Change:	−2.00 mmol		−1.00 mmol		+1.00 mmol	
After rxn:	0 mmol		2.00 mmol		1.00 mmol	

$$? \ \frac{\text{mmol CaBr}_2}{\text{mL}} = \frac{1.00 \text{ mmol CaBr}_2}{150 \text{ mL}} = 0.00667 \ M$$

2. (a). The balanced equation for the reaction is

$$2HCl + Ba(OH)_2 \rightarrow BaCl_2 + 2H_2O$$
$$2 \text{ mmol} \quad 1 \text{ mmol} \qquad 1 \text{ mmol} \quad 2 \text{ mmol}$$

$$\underline{?} \text{ mL Ba(OH)}_2 =$$

$$125 \text{ mL HCl} \times \frac{0.0350 \text{ mmol HCl}}{1.00 \text{ mL HCl}} \times \frac{1 \text{ mmol Ba(OH)}_2}{2 \text{ mmol HCl}} \times \frac{1.00 \text{ mL Ba(OH)}_2}{0.0750 \text{ mmol Ba(OH)}_2}$$

$$= 29.2 \text{ mL Ba(OH)}_2$$

3. (b). The balanced equation for the reaction is

$$H_2SO_4 + Na_2CO_3 \rightarrow NaSO_4 + CO_2 + H_2O$$
$$1 \text{ mol} \qquad 1 \text{ mol} \qquad 1 \text{ mol} \quad 1 \text{ mol} \quad 1 \text{ mol}$$

$$\underline{?} \ M \text{ soln} = 0.212 \text{ g Na}_2CO_3 \times \frac{1 \text{ mol Na}_2CO_3}{106 \text{ g Na}_2CO_3} \times \frac{1 \text{ mol H}_2SO_4}{1 \text{ mol Na}_2CO_3} \times$$

$$\frac{1}{0.0400 \text{ L H}_2SO_4}$$

$$= 0.050 \ M$$

4. (b). H_2SO_4 has a molecular weight of 98.08.

$$\underline{?} \ M \text{ H}_2SO_4 = 9.11 \text{ mg H}_2SO_4 \times \frac{1 \text{ mmol H}_2SO_4}{98.08 \text{ mg H}_2SO_4} \times \frac{1}{1 \text{ mL H}_2SO_4} = 0.0929 \ M$$

5. (a). $V_1 = \dfrac{V_2 M_2}{M_1} = \dfrac{200 \text{ mL} \times 0.050 \ M}{2.00 \ M} = 5.0 \text{ mL}$

6. (a). The balanced equation for the reaction is

$$H_2SO_4 + Ba(OH)_2 \rightarrow BaSO_4 + 2H_2O$$
$$1 \text{ mmol} \qquad 1 \text{ mmol} \qquad 1 \text{ mmol} \quad 2 \text{ mmol}$$

$$\underline{?} \text{ mmol H}_2SO_4 = 10.0 \text{ mL H}_2SO_4 \text{ soln} \times \frac{0.150 \text{ mmol H}_2SO_4}{1 \text{ mL H}_2SO_4 \text{ soln}} = 1.50 \text{ mmol H}_2SO_4$$

$$\underline{?} \text{ mmol Ba(OH)}_2 = 1.50 \text{ mmol H}_2SO_4 \times \frac{2 \text{ mmol Ba(OH)}_2}{2 \text{ mmol H}_2SO_4} = 1.50 \text{ mmol Ba(OH)}_2$$

$$\frac{\underline{?} \text{ mmol Ba(OH)}_2}{1 \text{ mL Ba(OH)}_2 \text{ soln}} = \frac{1.50 \text{ mmol Ba(OH)}_2}{35.0 \text{ mL Ba(OH)}_2 \text{ soln}} = 0.0429 \ M$$

7. (d). The balanced equation is

$$2HNO_3 + Na_2CO_3 \rightarrow 2NaNO_3 + CO_2 + H_2O$$
$$2 \text{ mol} \qquad 1 \text{ mol} \qquad 2 \text{ mol} \quad 1 \text{ mol} \quad 1 \text{ mol}$$

$$\underline{?} \text{ mol HNO}_3 = 0.246 \text{ g Na}_2CO_3 \times \frac{1 \text{ mol Na}_2CO_3}{106.0 \text{ g Na}_2CO_3} \times \frac{2 \text{ mol HNO}_3}{1 \text{ mol Na}_2CO_3}$$

$$= 0.00464 \text{ mol HNO}_3$$

$$\frac{\underline{?} \text{ mol HNO}_3}{\text{L}} = \frac{0.00464 \text{ mol HNO}_3}{0.0250 \text{ L}} = 0.186 \ M \text{ HNO}_3$$

8. (d). The balanced equation is

$$Ba(OH)_2 + 2KHP \rightarrow BaP + K_2P + 2H_2O$$

$$\text{1 mol} \qquad \text{2 mol} \qquad \text{1 mol} \quad \text{1 mol} \quad \text{2 mol}$$

$$\underline{?}\ \text{mol NaOH} = 0.4136\ \text{g KHP} \times \frac{1\ \text{mol KHP}}{204.2\ \text{g KHP}} \times \frac{1\ \text{mol Ba(OH)}_2}{2\ \text{mol KHP}}$$

$$= 0.001013\ \text{mol Ba(OH)}_2$$

$$\frac{\underline{?}\ \text{mol Ba(OH)}_2}{\text{L}} = \frac{0.001013\ \text{mol Ba(OH)}_2}{0.03000\ \text{L}} = 0.03377\ M\ \text{Ba(OH)}_2$$

9. (b). $\underline{?}$ mmol $C_6H_5COOH = 50.00\ \text{mL}\ C_6H_5COOH\ \text{soln} \times \dfrac{0.1000\ \text{mmol}\ C_6H_5COOH}{1\ \text{mL}\ C_6H_5COOH\ \text{soln}}$

$$= 5.000\ \text{mmol}\ C_6H_5COOH$$

$$\underline{?}\ \text{mmol KOH} = 5.000\ \text{mmol}\ C_6H_5COOH \times \frac{1\ \text{mmol KOH}}{1\ \text{mmol}\ C_6H_5COOH} = 5.000\ \text{mmol}$$

KOH

$$\frac{\underline{?}\ \text{mmol KOH}}{1\ \text{mL KOH soln}} = \frac{5.000\ \text{mmol KOH}}{100.00\ \text{mL KOH soln}} = 0.05000\ M$$

10. (e). The balanced equation is

$$H_2C_2O_4 + 2NaOH \rightarrow Na_2C_2O_4 + 2H_2O$$

$$\text{1 mmol} \qquad \text{2 mmol} \qquad \text{1 mmol} \qquad \text{2 mmol}$$

$$\underline{?}\ \text{mmol NaOH} = 28.3\ \text{mL NaOH soln} \times \frac{0.305\ \text{mmol NaOH}}{1\ \text{mL NaOH soln}} = 8.63\ \text{mmol NaOH}$$

$$\underline{?}\ \text{mmol}\ H_2C_2O_4 = 8.63\ \text{mmol NaOH} \times \frac{1\ \text{mmol}\ H_2C_2O_4}{2\ \text{mmol NaOH}} = 4.32\ \text{mmol}\ H_2C_2O_4$$

$$\frac{\underline{?}\ \text{mmol}\ H_2C_2O_4}{\text{mL}\ H_2C_2O_4\ \text{soln}} = \frac{4.32\ \text{mmol}\ H_2C_2O_4}{25.0\ \text{mL}\ H_2C_2O_4\ \text{soln}} = 0.173\ M$$

11. (b). The balanced equation is

$$HC_9H_7O_4 + NaOH \rightarrow NaC_9H_7O_4 + H_2O$$

$$\text{1 mol} \qquad \text{1 mol} \qquad \text{1 mol} \qquad \text{1 mol}$$

$$\underline{?}\ \text{NaOH} = 0.01525\ \text{L} \times \frac{0.115\ \text{mol NaOH}}{\text{L}} = 0.00175\ \text{mol NaOH}$$

$$\underline{?}\ \text{g}\ HC_9H_7O_4 = 0.00175\ \text{mol NaOH} \times \frac{1\ \text{mol}\ HC_9H_7O_4}{1\ \text{mol NaOH}} \times \frac{180.1\ \text{g}\ HC_9H_7O_4}{1\ \text{mol}\ HC_9H_7O_4}$$

$$= 0.315\ \text{g}\ HC_9H_7O_4$$

$$\%\ \text{purity} = \frac{0.315\ \text{g}\ HC_9H_7O_4}{0.400\ \text{g tablet}} \times 100\% = 78.8\%\ \text{pure}\ HC_9H_7O_4$$

12. (c). The balanced equation is

$$5SO_3^{2-} + 6H^+ + 2MnO_4^- \rightarrow 5SO_4^{2-} + 2Mn^{2+} + 3H_2O$$

We must balance H and O by adding H^+ and H_2O, because the reaction takes place in acidic solution. We cannot add OH^- to either side of the equation.

13. (a). The balanced equation (remember that it is in an acidic solution) is

$$Cr_2O_7^{2-} + 3H_2S + 8H^+ \rightarrow 2Cr^{3+} + 3S + 7H_2O$$

14. (b). The balanced equation is

$$10NO_3^- + 3I_2 + 4H^+ \rightarrow 10NO + 6IO_3^- + 2H_2O$$

15. (d). Cl is reduced from 3+ in $HClO_2$ to 1+ in Cl^-.

$$HClO_2 \rightarrow Cl^- \qquad \text{(red. half-rxn)}$$
$$HClO_2 \rightarrow Cl^- + 2H_2O$$
$$3H^+ + HClO_2 \rightarrow Cl^- + 2H_2O$$
$$4e^- + 3H^+ + HClO_2 \rightarrow Cl^- + 2H_2O \qquad \text{(balanced red. half-rxn)}$$

S is oxidized from 4+ in SO_2 to 6+ in SO_4^{2-}.

$$SO_2 \rightarrow SO_4^{2-} \qquad \text{(ox. half-rxn)}$$
$$2H_2O + SO_2 \rightarrow SO_4^{2-}$$
$$2H_2O + SO_2 \rightarrow SO_4^{2-} + 4H^+$$
$$2H_2O + SO_2 \rightarrow SO_4^{2-} + 4H^+ + 2e^- \qquad \text{(balanced ox. half-rxn)}$$

$$1(4e^- + 3H^+ + HClO_2 \rightarrow Cl^- + 2H_2O)$$
$$2(2H_2O + SO_2 \rightarrow SO_4^{2-} + 4H^+ + 2e^-)$$
$$3H^+ + HClO_2 + 4H_2O + 2SO_2 \rightarrow Cl^- + 2H_2O + 2SO_4^{2-} + 8H^+$$
$$HClO_2 + 2H_2O + 2SO_2 \rightarrow Cl^- + 2SO_4^{2-} + 5H^+$$

16. (d). Fe is oxidized from 2+ in Fe^{2+} to 3+ in Fe^{3+}.

$$Fe^{2+} \rightarrow Fe^{3+} \qquad \text{(ox. half-rxn)}$$
$$Fe^{2+} \rightarrow Fe^{3+} + 1e^- \qquad \text{(balanced ox. half-rxn)}$$

Mn is reduced from 7+ in MnO_4^- to 2+ in Mn^{2+}.

$$MnO_4^- \rightarrow Mn^{2+} \qquad \text{(red. half-rxn)}$$
$$MnO_4^- \rightarrow Mn^{2+} + 4H_2O$$
$$8H^+ + MnO_4^- \rightarrow Mn^{2+} + 4H_2O$$
$$5e^- + 8H^+ + MnO_4^- \rightarrow Mn^{2+} + 4H_2O \qquad \text{(balanced red. half-rxn)}$$

$$5(Fe^{2+} \rightarrow Fe^{3+} + 1e^-)$$
$$1(5e^- + 8H^+ + MnO_4^- \rightarrow Mn^{2+} + 4H_2O)$$

$$5Fe^{2+} + MnO_4^- + 8H^+ \rightarrow 5Fe^{3+} + Mn^{2+} + 4H_2O$$

Add the spectator ions, K^+ and SO_4^{2-}.

$$5FeSO_4 + KMnO_4 + 4H_2SO_4 \rightarrow 2\tfrac{1}{2}Fe_2(SO_4)_3 + MnSO_4 + 4H_2O + \tfrac{1}{2}K_2SO_4$$

To eliminate the fraction, multiply the entire equation by 2.

$$10FeSO_4 + 2KMnO_4 + 8H_2SO_4 \rightarrow 5Fe_2(SO_4)_3 + 2MnSO_4 + 8H_2O + K_2SO_4$$

Finally, add the states of matter.

$$10FeSO_4(aq) + 2KMnO_4(aq) + 8H_2SO_4(aq) \rightarrow$$
$$5Fe_2(SO_4)_3(aq) + 2MnSO_4(aq) + 8H_2O(\ell) + K_2SO_4(aq)$$

17. (e). This is the same type of calculation as in Example 11-15.

$$\underline{}\quad M \quad FeSO_4 \quad = \quad 0.0380 \quad L \quad KMnO_4$$

$$\times \frac{0.1214\ M\ MnO_4^-}{1\ L\ KMnO_4} \times \frac{5\ mol\ Fe^{2+}}{1\ mol\ MnO_4^-} \times \frac{1}{0.0250\ L\ FeSO_4}$$

$$= 0.923\ M\ FeSO_4$$

18. (d). $?\ mL\ KMnO_4 = 1.00\ g\ FeSO_4 \times \frac{1\ mol\ FeSO_4}{151.9\ g\ FeSO_4} \times \frac{1\ mol\ MnO_4^-}{5\ mol\ Fe^{2+}} \times \frac{1\ L\ KMnO_4}{0.100\ mol\ MnO_4^-}$

$$= 0.0132\ L = 13.2\ mL$$

19. (d). $?\ mmol\ S_2O_3^{2-} = 25.00\ mL \times \frac{0.1524\ mmol\ S_2O_3^{2-}\ soln}{mL} = 3.810\ mmol\ S_2O_3^{2-}$

$?\ mmol\ I_2 = 3.810\ mmol\ S_2O_3^{2-} \times \frac{1\ mmol\ I_2}{2\ mmol\ S_2O_3^{2-}} = 1.905\ mmol\ I_2$

$?\ \frac{mmol\ I_2}{mL} = \frac{1.905\ mmol\ I_2}{31.58\ mL} = 0.06032\ M\ I_2$

12 Gases and the Kinetic-Molecular Theory

Chapter Summary

For the previous several chapters, you have been studying primarily the bonding, properties, and reactions of specific substances. In this chapter, you begin a thorough study of some aspects of the behavior of matter that are more dependent on the physical state of the substances than on their chemical identity. In this chapter and the next you will study pure substances in the three states of matter—gas, liquid, and solid. In Chapter 14, you will study mixtures of substances. The introduction and first section of Chapter 12 point out some of the characteristics that distinguish the gas, liquid, and solid states.

First you examine the observable **macroscopic behavior** of gases and then relate them to proposed **molecular behavior**. Two examples of this approach are seen in Section 12-1. (1) The observation that gases have low density tells us that in gases the molecules are much further apart than in liquids or solids, and (2) the ability of gases to fill any container in which they are placed and to diffuse and mix tells us that in gases the molecules are in rapid motion. Sections 12-2 through 12-12, you will examine measurable properties of gases and the observed relationships among these properties—such quantities as **density**, **pressure**, **volume**, amount of gas **(number of moles)**, and **temperature**. Then in Section 12-13 you will interpret these relationships in terms of the kinetic-molecular theory. Most of your study of this chapter is made easier by the observation that all gases behave the same under many conditions. This is referred to as **ideal gas** behavior. In Section 12-15, you will study some conditions for which this ideal behavior is not observed.

Your study begins with the properties pressure (P), volume (V), number of moles (n), and temperature (T) and the ways in which these are related for typical gases. (Of these properties, pressure is the one for which students often have the least prior understanding. Section 12-3 introduces the concept of pressure and some ways of measuring it.) One useful way to interrelate such properties is to keep two of these variables constant and then to see how the other two are related. Many of the relationships in this chapter are valid only under certain conditions, so you *must* keep in mind the limitations of each one. As you study these relationships, you will see the various ways we represent them: (1) **In words**. You should be able to make a verbal statement of each relationship. It is more important to understand any statement than to just memorize it. Be certain that you know what is meant when you use words like "proportional," "inversely," etc. (2) **In mathematical relationships**. To use quantitative relationships effectively, you must be able to understand and use the equations that describe them. This includes using the relationships to solve problems. (3) **In graphs**. This pictorial way of representing mathematical relationships can be a great help to your understanding, remembering, and applying the relations.

In Section 12-4 you see the relationship between pressure and volume for a fixed amount of gas (constant n) at a fixed temperature (constant T). This relationship is known as **Boyle's Law**. This law can be represented in various ways mathematically, some more useful than others. One use of Boyle's Law is to relate some initial set of conditions (i.e., values) of P and V to a final set of conditions of these two variables.

One expression that you will find very useful is

$$P_1V_1 = P_2V_2 \qquad \text{(constant } n, T)$$

Again, remember the set of conditions to which this equation is limited. The relation $P_1V_1 = P_2V_2$ must *not* be expected to hold if the amount of gas changes (n not constant) or if the temperature changes (T not constant).

Experimentally, we also find that the relationship

$$\frac{V_1}{T_1} = \frac{V_2}{T_2} \qquad \text{(constant } n, P)$$

known as **Charles's Law** (Section 12-5), describes the relation of volume and temperature of a gas, so long as we keep the amount of gas and the pressure of the gas constant. The relation takes this simple form only if the temperature is expressed on the **absolute** or **Kelvin** scale. Be sure that you understand the rationale behind establishing this scale and its relation to other temperature scales. Keep in mind that the symbol T in chemistry always refers to the temperature measured on the Kelvin scale. The establishment of a set of conditions of T and P as **standard temperature and pressure (STP)**, as discussed in Section 12-6, helps us to compare different gases easily.

In Section 12-7, you see how Boyle's and Charles's Laws are combined into the **Combined Gas Law equation**,

$$\frac{P_1V_1}{T_1} = \frac{P_2V_2}{T_2} \qquad \text{(constant } n)$$

This equation is convenient to use when all of the quantities P, V, and T change. A common use of this equation is to adjust the volume of a gas at some arbitrary set of conditions to STP.

At the same temperature and pressure, equal volumes of all ideal gases contain the same numbers of molecules (Avogadro's Law, Section 12-8). Thus, the volume occupied by one mole of the gas at STP (the **standard molar volume**) is the same for all gases. Another statement of Avogadro's Law is

$$\frac{V_1}{n_1} = \frac{V_2}{n_2} \qquad \text{(constant } T, P)$$

This equation is the basis for calculations relating volumes and amounts of gas at constant T and P. One mole of each different gas would have a different mass, so different gases have different densities, even at the same conditions.

Each of the gas laws you have studied up to now applies only at certain conditions and covers relationships between (or among) only two or three properties at a time. They are useful primarily to determine changes in one quantity if we know changes in another or others. In Section 12-9, all of these laws are combined into a very useful relation, the **ideal gas equation**:

$$PV = nRT$$

Be sure that you know the significance of the **universal gas constant**, R. Its value does not depend on what gas the equation is being applied to, and it can be expressed in different sets of units (with different numerical values), depending on the units used to describe the other quantities in the equation. Because R is a constant, this equation allows us to predict the value of any one of the four quantities P, V, n, and T, given values for the other three. A summary of the ideal gas laws appears on page 418.

Until now, you have seen the molecular weight, or formula weight, obtained only by adding up the atomic weights of the atoms in the formula. An important use of the ideal gas equation is the *experimental* determination of molecular weight by measurement of gas volumes (Section 12-10). This is the first of several methods you will learn for experimentally determining molecular weights even if the chemical formula is not known. (Watch for other such methods in this and later chapters.) This knowledge lets us find the molecular formulas of unknown substances. This determination requires the measurement of P, V, T, and the mass of the sample. Calculate the number of moles of substance present from the ideal gas equation, and then divide the observed mass by the number of moles to obtain the molecular weight. This approach can also be used to find the molecular weights of **volatile** liquids (those that evaporate easily to form gases).

To this point, all that we have described about gases is independent of the identity of the gas. Thus, it is not surprising that some aspects of the behavior of mixtures of gases (if they do not react) are also quite simply described. **Dalton's Law of Partial Pressures**, Section 12-11, for example, which was based first on direct observation, is consistent with the ideal gas equation. In this section you are introduced to another method for describing the composition of a mixture, by using the dimensionless quantity **mole fraction**. One very useful application of Dalton's Law is the calculation associated with the collection of a gas over water, which exerts a pressure by evaporating.

In your study of stoichiometry in Chapter 3, you saw that the fundamental relationship among amounts of reactants and products is in terms of the numbers of moles. At that stage, the only way you had to "measure" the number of moles was in terms of the mass of the substance. Now you have seen that the volume of a gas at specified conditions of temperature and pressure is a measure of the number of moles present. You now use this as a basis to extend your study of stoichiometry. The volume change that accompanies a reaction is proportional to the changes in numbers of moles of gaseous reactants and products. Using the molecular weight, you can extend these ideas to deal with mass-volume relationships in reactions involving gases (Section 12-12).

In Sections 12-4 through 12-11, you study the observed properties of gases, with no reference to the properties or even the existence of molecules. You see that at the macroscopic level, all gases behave essentially the same. We now describe the molecular level of behavior. Assume that the behavior of gas molecules is essentially independent of their chemical identity. Then see whether this molecular behavior is consistent with the observed macroscopic properties. The result is the **kinetic-molecular theory**, Section 12-13. Not only does this theory provide you with a molecular-level understanding of such quantities as pressure and temperature, but it can also be used to predict additional types of behavior. One such prediction deals with rates of diffusion and effusion of gases. This is the basis for another method for experimentally determining molecular weights of unknown gaseous compounds. Based on how consistent it is with the observed behavior of gases, we now believe that the kinetic-molecular theory gives a good description of how gas molecules act.

Much of the discussion until this point in this chapter has been applicable at high temperature (that is, quite a bit higher than the boiling points of the substances), and at low or moderate pressures. At those conditions, all gases behave alike. However, if the temperature is too low or the pressure is too high, the gas molecules do not behave independently of one another. The gases then do not behave ideally, and the amount of this deviation from ideal behavior is different for different gases. We can take two approaches to describe such "**real gases**" (Section 12-15). One is to develop equations that "fit" the observed properties of gases better at low T and high P than do the various ideal equations studied so far. One of these is the **van der Waals' equation**. This equation contains two constants, a and b, that are different for every gas to which the equation is applied. The second approach, now at the molecular level, is to describe how molecular size and intermolecular attractions can affect molecular properties at low T and high P.

Study Goals

Remember to answer many of the Exercises suggested at the end of the text chapter. Be systematic in your solutions to numerical problems, writing down all steps and reasoning.

Sections 12-1 and 12-2
1. Summarize the general physical properties of gases. *Work Exercises 1, 2, and 8.*

Section 12-3
2. Explain what pressure is, and illustrate how we measure it with (a) an inverted-tube mercury barometer and (b) a mercury manometer connected to a container of gas. Convert pressure measurements to various units. *Work Exercises 3 through 7 and 9 through 11.*

Sections 12-4 through 12-6, 12-8, and 12-11
3. State in words and in mathematical relationships the Laws of (a) Boyle, (b) Charles, (c) Avogadro, and (d) Dalton. *Work Exercises 12, 13, 22, 31, 32, 38, and 60.*
4. Perform calculations using the relationships of Study Goal 3 to relate volume changes, pressure changes, temperature changes, and changes in number of moles of gas. *Work Exercises 14 through 18, 25 through 30, 37, 39 through 42, and 61 through 71.*

Section 12-5
5. Explain the basis of the absolute temperature scale. Convert temperatures from other scales to the Kelvin scale. *Work Exercises 19 through 21.*

Section 12-7
6. State the Combined Gas Law equation and understand the relationships it expresses. *Work Exercise 31, 32, and 36.*
7. Given, for a gaseous system, any five out of six variables (P_1, P_2, V_1, V_2, T_1, T_2), be able to solve for the value of the missing sixth variable. *Work Exercises 33 through 35.*

Sections 12-8
8. Know the standard molar volume of an ideal gas. Relate the volume of a gas at a given set of temperature and pressure conditions to the number of moles (and grams) of the gas. *Work Exercises 37 and 39 through 42.*

Section 12-9
9. State the ideal gas equation and its assumptions; derive the ideal gas equation from the combined gas laws. Perform calculations relating the variables in the ideal gas law. Given any three of the variables P, V, n, and T for a sample of ideal gas, be able to solve for the value of the other variable. *Work Exercises 43 through 52.*

Section 12-10

10. Be able to calculate the molecular weight of a pure gaseous substance, given the density of the gas at any condition of temperature and pressure. Use this information to help determine the molecular formula of a gaseous substance. *Work Exercises 53 through 59.*

Section 12-11

11. Given the total pressure and composition of a mixture of gases, be able to solve for the partial pressure of each gas present in the mixture. *Work Exercises 61 through 71.*

12. Given the vapor pressure of water at various temperatures, be able to correct volumes and pressures of wet gases at specified temperatures and barometric pressures for the vapor pressure of water. *Work Exercises 67 and 68.*

Section 12-12

13. Be able to carry out calculations involving (a) volumes of gases involved in chemical reactions, and (b) mass-volume relationships for chemical reactions involving gases. *Work Exercises 72 through 84.*

Section 12-13

14. State the assumptions of the kinetic-molecular theory. Account for the laws listed in Study Goal 3 in terms of the kinetic-molecular theory. *Work Exercises 85 through 92.*

Section 12-14

15. Determine the relative rates of diffusion and effusion of gases from their molecular weights. *Work Exercise 97.*

Section 12-15

16. Distinguish between real gases and ideal gases. Relate the van der Waals' equation to the ideal gas equation. Relate the van der Waals' constants *a* and *b* to deviations of molecular properties of gases from the assumptions of the kinetic-molecular theory. *Work Exercises 93 through 96.*

17. Be able to use the van der Waals' equation to carry out calculations relating *n*, *P*, *V*, and *T* for real gases. Understand both when and why these results differ significantly from those predicted by the ideal gas equation. *Work Exercises 98 through 101.*

General

18. Use your understanding of this chapter to recognize and answer a variety of types of questions. *Work* **Mixed Exercises** *102 through 117.*

19. Answer conceptual questions based on this chapter. *Work* **Conceptual Exercises** *118 through 131.*

20. Apply concepts and skills from earlier chapters to the ideas of this chapter. *Work* **Building Your Knowledge** *exercises 132 through 139.*

21. Exercises at the end of chapter direct you to sources outside the textbook for information to use in solving them. Work **Beyond the Textbook** *exercises 140 through 142.*

Some Important Terms in This Chapter

Some important new terms from Chapter 12 are listed below. Fill the blanks in the following paragraphs with terms from the list. Each term is only used once. Check the Key Terms list and the chapter reading. Study other new terms, and review terms from preceding chapters if necessary.

absolute zero	ideal gas equation	standard molar volume
diffusion	kinetic-molecular theory	standard temperature and pressure (STP)
effusion	mole fraction	universal gas constant
equation of state	partial pressure	van der Waals equation
fluids	real gas	vapor pressure
ideal gas		

Liquids and gases have one property in common—they flow. Most homes have pipes that allow water to flow into them. Water is a liquid, so it flows. Many homes have pipes that allow natural gas to flow into them. Natural gas is a gas, so it flows. However, homes do not have pipes that allow pizza, for example, to flow into them, because pizza is a solid, so it does *not* flow. Because they flow, liquids and gases are collectively called [1]_____.

Whether a substance is a solid, a liquid, or a gas often depends on the temperature. At high temperatures, many substances are gases. As they cool, these gases eventually condense to liquids. As they cool more, these liquids eventually solidify. There is a minimum temperature at which all atomic motion stops. This is [2]_____, absolutely the coldest temperature possible.

For working with most gases, however, a more moderate temperature and a moderate pressure are desired. Because the volume of a given amount of a gas can vary considerably with changes in temperature and pressure, scientists find it useful to specify sets of standard values for temperature and pressure for comparing different samples of gases. The set of conditions used most often, [3]_____, are 273.15 K and one atmosphere (1 atm). 273.15 K is the freezing point of water (32°F or 0°C), and 1 atm is average sea level pressure. At this temperature and pressure, one mole (6.022×10^{23} molecules) of an ideal gas occupies a volume of 22.4 liters, a value known as the [4]_____. An [5]_____ is a gas that exactly obeys the gas laws. This gas is ideal because: (1) It consists of molecules that are infinitely small, so that the volume of the molecules themselves doesn't take up space in which the molecules can move, and (2) The molecules don't attract each other and "stick together." The equation that combines the values for the pressure (P), the volume (V), the amount or number of moles (n), and the temperature (T) of a gas and is perfectly correct only for an ideal gas is the [6]_____. This equation includes R, a constant of proportionality that is the same amount for all ideal gases, so it is called the [7]_____ _____.

The pressure that each gas in a mixture exerts is called its [8]_____. The composition of a mixture of gases can be expressed by dividing the number of moles of each gas by the total number of moles of all the gases, which expresses the [9]_____ of each gas. Such a mixture of gases results any time an insoluble gas is collected over water. Because the gas is in contact with the surface of the water, its molecules soon become mixed with molecules of water vapor that escape from the liquid, producing a partial pressure of water vapor, called the [10]_____ of water, which is dependent upon the temperature of the water.

The behavior of gases is explained by a theory that imagines gas molecules in rapid motion, the [11]_____. This rapid movement of the molecules allows a gas to escape through a tiny hole, a process known as [12]_____, such as the gradual loss of helium from a latex balloon. It also allows a gas to move into a space or to mix with another gas, a process known as [13]_____, such as the movement of perfume molecules through the air in a room.

Most gases come close to the ideal behavior predicted by the kinetic-molecular theory much of the time. However, a gas of a real substance is a [14]_____, and no real gas behaves like an ideal gas at all times, particularly at high pressures and/or low temperatures. Then the values calculated from the ideal gas equation (pressure, volume, number of moles, or temperature) are significantly different from the real values measured experimentally. In order to be able to calculate values that are closer to the real values, Johannes van der Waals modified the ideal gas equation to adjust for the amount of the volume of the gas container that is taken up by the molecules themselves and for attractive forces between the molecules. The equation he produced is the [15]_____. This equation is called an [16]_____ because it describes a state of matter.

Quotefalls

Fit the letters in each vertical column into the squares directly below them to form words across. The letters do not necessarily go into the squares in the same order in which they are listed. Each letter given is used once. A black square indicates a space. Words starting at the end of a line may continue in the next line. Punctuation in the statement is included in the boxes. When all the letters have been placed in their correct squares, you will be able to read the complete quotation across the diagram from left to right, line by line. Hint: It may be helpful at times to fill in small words, like "the," to use up some letters to help in determining other longer words. Some letters have been "seeded" to help you get started. The solutions are on page 163.

Puzzle #1:

R	O	E	L	T	O	S	A	L	O	W	U	M	T	G	A	C	A	I	P	T	A	N	D	E	P	R	A	P	S	E	P	E	N
B	O	R	T	M	H	E	S	L	O	T	O	A	E	T	E	S	C	N	S	I	I	D	T	B	T	S	E	L	Y	U	F	R	U
P		Y		E	A	S		V	A	L				A	H	O	C	O	P	S	L		I	N	V		R	E	M	S	D	R	A
I	E		I			N			L	F				U	P		E	E		R	Y		E		E	R	I					O	T
			E	'	■						:				■				N					■		T						T	
	,	■						L		■			■					P			■			■			■		D				
	T			S		■			■			■			■			I				■					■	P					
P							■			■				■						E	■			E									.

Puzzle #2:

H	E	R	P	A	O	T	A	L	D	E	P	R	S	F	U	S	E	R	S	I	S	L	T	E	R	E	S	G	U	S	E	S	E	S	I	T
T	H	L	E	T	N	R	O	F	S	I	A	L	A	P	R	E	E	S	S	U	A	E	F	T	T	E	S	S	U	Y	S	A	O	M		X
T	U	E	T	O		R	T	S	I	A	L	A	L	O	G	S	S	A	R	E	E	I	E	R	T	H	P	U	A	M	R	E	M	F		
D	A			R	O	T		I					A	W		S		R	P	E	R	E		A	O	H	D	Y	S				F	S		
	A						'				W	W						R					E			R									:	
																	R			E									U							
T								G										R				T											E		.	

Puzzle #3:

C	H	A	H	S	E	O	O	L	A	L	A	W	O	C	C	U	U	T	E	D	S	C	T	P	Y	R	P	R	U	R	S	I	T	I	I	E
O	M	A	S	R	L	S	F	I	U	S	E	A	B	S	A	C	T	I	R	T	E	M	L	T	E	A	D	E	E	N	U	R	T	E		
N	A	A	R	S	O	F	U	S	T	M	G	A	S	A	I	L	P	C	R	E	B	A	N	R	D	E	P	I	O	R	T	E				
	T	L		V	O	S		S	E					O	O	S	C	I		B	T	Y	E	A	T	F	P	S	E							
		R					'						:					O				T			R		R			S						
,				V								C								O			T					F								
		A													D				C					P												
										L											M					R		.								

Puzzle #4:

N	V	O	A	A	D	A	O	H	S	A	N	A	W	E	E	A	T	H	D	I	N	S	T	E	M	E	O	T	C	M	A	E	R	E	I	O
R	A	L	G	A	T	O	R	S	R	E	S	M	U	R	B	E	R	S	O	F	O	C	A	L	S	T	Y	O	R	G	P	O	I	R	T	U
B	Y	G	A	N	O	R	S	T	E	E	S	S	M	B	E	R	I	O	F	C	V	O	L	U	N	S		P	E	C	U	P	A	T		D
A	E	A	N	G	D		S	P	E	S		U	L			I		F	R	E	O	L	U	N	Y			F	O	O	S	E				
	V					O	'				A			:					N					N								E				
							S						,					D					L					O								
			G		S						M											O		T				P								
N																										G			.							

Puzzle #5:

S	S	Y	A	O	R	S	S	I	N	S	F	A	N	A	W	E	O	C	P	E	W	B	O	L	E	I	N	A	N	D	O	R	R	E
G	A	T	R	E	R	A	T	A	C	S	O	L	S	I	T	E	M	O	C	F	R	G	E	U	N	E	X	T	M	A	S	E	R	M
A	S	U		L	T	A	C	H	O	V	F	R	T	M	D	S	L	E	F	O	M	B	I	R	E	N	P	R	H	S	E	P	D	E
E	S			A	U		T	O	E	O	O	U	U	E	M	P	L	F	A	M	H	T	S	G	R	U			H	V	B	E	S	U
A	C						O	E			T	T	L	M					H	A	A	B	E				E			B	E	L		
		-					A													N				N					V					
	S	:			N											T																		
S			,				V									S							R											
										M							R													M				
		T								M					H											M		.						

168

Preliminary Test

This test will check your understanding of basic concepts and different types of calculation problems. Be sure to answer *many* of the additional textbook Exercises. Those are more challenging than these preliminary test questions, so they will give you additional insight into the meaning and applications of this chapter material. For all Preliminary Test questions in this chapter, assume that the gas is acting ideally, unless you are told otherwise.

True-False

Mark each statement as true (T) or false (F).

_____ 1. Because the noble gases do not easily react to form compounds, they are present in a very high amount in the atmosphere.

_____ 2. All substances are less dense as gases than as liquids.

_____ 3. For any ideal gas, doubling the pressure would always cause its volume to decrease by a factor of two.

_____ 4. For any gas, whether ideal or not, doubling the pressure of a fixed amount of gas at a fixed temperature would cause its volume to decrease by a factor of two.

_____ 5. Doubling the temperature always doubles the volume of a fixed amount of ideal gas at constant pressure.

We have two identical containers. One is filled with hydrogen fluoride gas, HF, and the other is filled with neon gas, Ne. The two containers are at the same temperature and pressure. Questions 6 through 14 all refer to these two samples of gas. The atomic weights to be used for these questions are as follows: H = 1, F = 19, Ne = 20. For Questions 6 through 13, consider the gases to be ideal.

_____ 6. The two gas samples have the same number of molecules.

_____ 7. The two gas samples have the same number of atoms.

_____ 8. The two gas samples have the same total mass of gas.

_____ 9. The two gas samples have the same density.

_____ 10. The molecules in the two gas samples have the same average kinetic energy.

_____ 11. The molecules in the two gas samples have the same average speed.

_____ 12. If the same size hole were opened in each container, the pressure in the neon container would decrease faster than the pressure in the hydrogen fluoride container.

_____ 13. If we were to remove all of the Ne from one box and put it into the other box with the HF, the molecules would mix so as to cause no change in total pressure in the second box.

_____ 14. If the two gases were taken to the same low temperature and the same very high pressure, we would expect HF to behave more ideally than Ne.

Short Answer

1. The pressure of one atmosphere is equal to _____ torr.
2. Charles's Law may be expressed mathematically as _____.
3. The volumes of gases, measured at _____, that are involved in chemical reactions are in the ratios of small whole numbers.
4. The statement "Equal volumes of gases at the same conditions of pressure and temperature contain equal numbers of molecules" is a statement of _____ Law.
5. Tripling the absolute temperature (in K) and doubling the pressure of a given volume of an ideal gas would give a new volume of gas. This new volume could be calculated by multiplying the original volume of the gas by a factor of _____.
6. To cause an ideal gas sample to change from a volume of 1.4 L to 4.2 L at constant temperature, the pressure must be changed to _____ times its initial value.
7. To cause an ideal gas sample to change from a volume of 1.4 L to 4.2 L at constant pressure, the absolute temperature must be changed to _____ times its initial value.
8. The zero of absolute temperature is taken as that temperature at which all gases would exhibit _____. Actually, before this temperature is reached, any real gas would form _____ and then _____.
9. If 1.00 L of an ideal gas at 50.0°C were heated at constant pressure to 100°C, the new volume would be _____ L.
10. For one mole of an ideal gas, PV/T = _____. (The answer is to be a number, with correct units.)
11. The conditions we refer to as "standard temperature and pressure" are _____ and _____.
12. A mixture of H_2 and O_2 is in a 6.00 L vessel. The partial pressure of H_2 in this mixture is 500 torr, and the partial pressure of O_2 is 250 torr. The volume that H_2 would occupy alone at these conditions of temperature and total pressure is _____ L.
13. The mole fractions of H_2 and O_2 in the mixture described in Question 12 are _____ and _____, respectively.
14. The vapor pressure of water at 30°C is 31.8 torr. We collect hydrogen over water at 30°C until the total pressure of wet hydrogen is 1 atm. If we dried the hydrogen at the same temperature, the pressure it would exert would be _____ torr.
15. At STP, 1.00 L of a pure gaseous substance has a mass of 4.00 grams. The molecular weight of this substance is _____.
16. At standard temperature and pressure, 14.007 grams of nitrogen gas (considered ideal) occupies _____ L.
17. A sample of a pure gaseous substance occupying 4.10 L at 760 torr and 227°C is found to weigh 16.4 grams. The density of the substance under these conditions is _____.
18. The molecular weight of the substance in Question 17 is _____.
19. If 1.00 L of gas X weighs 44 times as much as 1.00 L of H_2 at the same temperature and pressure, then the molecular weight of the gas X is _____.
20. One mole of chlorine gas weighs _____ grams, contains _____ molecules, contains _____ atoms, and occupies _____ L at 0°C and 1 atm.

21. In the kinetic-molecular description of gas behavior, pressure is a measure of _____

_____
___.

22. In the kinetic-molecular description of gas behavior, temperature is a measure of _____

_____
___.

23. At the same _____, the molecules of all samples of ideal gases have the same average kinetic energy.

24. A real gas behaves most nearly like an ideal gas at conditions of _____ and _____.

25. In the van der Waals equation describing real (nonideal) gases, the term a is related to the molecular property of _____.

26. The commercial preparation of ammonia is carried out by reaction of nitrogen and hydrogen:

$$N_2(g) + 3H_2(g) \rightarrow 2NH_3(g)$$

To form 1000 L of ammonia gas, we would have to supply _____ L of nitrogen and _____ L of hydrogen, with all gases measured at the same temperature and pressure. (Treat the reaction as if it goes to completion.)

Carbon disulfide, CS_2, burns according to the equation

$$CS_2(\ell) + 3O_2(g) \rightarrow CO_2(g) + 2SO_2(g).$$

The following three questions, 27 through 29, refer to the complete combustion of one mole of CS_2, according to this equation.

27. The volume of O_2, measured at standard temperature and pressure (STP), required for the reaction is _____.

28. The volume of CO_2, measured at STP, that would be produced is _____.

29. The volume of SO_2, measured at STP, that would be produced is _____.

Multiple Choice

_____ 1. Each of the following plots refers to the behavior of a sample of an ideal gas. Which plot is *not* correct?

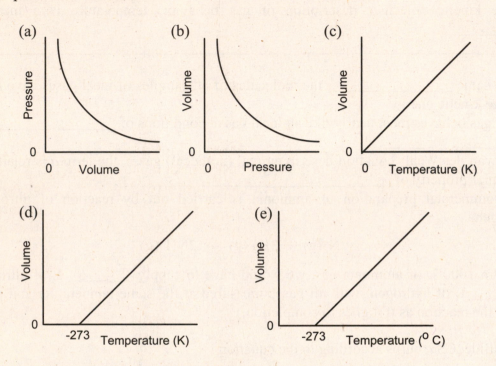

_____ 2. A sample of a gas having a volume of 1 L at 25°C and 1 atm pressure is subjected to an increase in pressure and a decrease in temperature. The density of the gas
(a) decreases. (b) increases. (c) remains the same. (d) becomes zero.
(e) either increases or decreases, depending on the sizes of the pressure and temperature changes.

_____ 3. A sample of a gas having a volume of 1 L at 25°C and 1 atm pressure is subjected to an increase in pressure and an increase in temperature. The density of the gas
(a) decreases. (b) increases. (c) remains the same. (d) becomes zero.
(e) either increases or decreases, depending on the sizes of the pressure and temperature changes.

_____ 4. At the same temperature, the molecules of all gases have the same
(a) average masses. (b) van der Waals constants *a* and *b*.
(c) average kinetic energies. (d) average speeds.
(e) molecular diameters.

_____ 5. How many molecules of an ideal gas are contained in 8.2 L at −73°C and 0.50 atm?
(a) 0.25 (b) 1.5×10^{23} (c) -4.1×10^{23} (d) 7.5×10^{23} (e) 4.2×10^{-25}

_____ 6. One liter of phosphorus gas reacts with six liters of hydrogen gas to form four liters of phosphine gas. All gases are at the same temperature and pressure. What is the formula of phosphine? [Remember that the formula for phosphorus gas is $P_4(g)$.]
(a) P (b) P_2 (c) PH (d) PH_2 (e) PH_3

_____ 7. We place equal weights of O_2 and N_2 in separate containers of equal volume at the same temperature. Which one of the following is true?
   (a) Both flasks contain the same number of molecules.
   (b) The pressure in the nitrogen flask is greater than that in the oxygen flask.
   (c) Molecules in the oxygen flask are moving faster on the average than the ones in the nitrogen flask.
   (d) The nitrogen molecules have a greater average kinetic energy.
   (e) All N_2 molecules have the same energy, and all O_2 molecules have the same energy, but these two energies are not the same.

_____ 8. At 0°C, which one of the following gases would have the lowest average molecular speed?
   (a) NH_3 (b) CO_2 (c) Ar   (d) N_2 (e) H_2

_____ 9. Which one of the following gases would diffuse most rapidly?
   (a) CO_2 (b) O_2 (c) CH_4 (d) He   (e) N_2

_____ 10. Analysis of a gas gives 80.0% C and 20.0% H by mass. A sample of the gas occupying 2.24 L at 0°C and 1 atm is found to weigh 3.0 grams. Which one of the following substances could this gas be?
   (a) methane, CH_4 (b) ethane, C_2H_6 (c) ethylene, C_2H_4
   (d) acetylene, C_2H_4 (e) benzene, C_6H_6

_____ 11. An 1120 mL sample of a pure gaseous compound, measured at 0°C and 1 atm, is found to weigh 2.86 grams. What is the molecular weight of the compound?
   (a) 2.86   (b) 28.6   (c) 5.72   (d) 57.2   (e) 14.3

_____ 12. What volume of pure oxygen gas, O_2, measured at 546 K and 760 torr, is formed by complete dissociation of 0.200 mole of Ag_2O?

$$2Ag_2O(s) \rightarrow 4Ag(s) + O_2(g)$$

   (a) 1.12 L   (b) 2.24 L   (c) 17.92 L   (d) 4.48 L   (e) 8.96 L

_____ 13. How many moles of $KClO_3$ are needed to form 2.8 L of O_2, measured at STP, according to the following equation?

$$2KClO_3(s) \rightarrow 2KCl(s) + 3O_2(g)$$

   (a) 1/12 mole   (b) 1/6 mole   (c) 1/3 mole   (d) 1/4 mole   (e) 1/2 mole

_____ 14. What volume of chlorine gas, measured at 0°C and 1 atm pressure, is needed to react completely with 5.61 grams of KOH in the following equation?

$$Cl_2(g) + 2KOH(aq) \rightarrow KCl(aq) + KClO(aq) + H_2O(\infty)$$

   (a) 0.56 L   (b) 1.12 L   (c) 2.24 L   (d) 4.48 L   (e) 22.4 L

_____ 15. One step in the industrial manufacture of nitric acid involves the production of nitric oxide by the oxidation of ammonia:

$$4NH_3(g) + 5O_2(g) \rightarrow 6H_2O(g) + 4NO(g)$$

We wish to produce 100 L of NO gas. How many liters of oxygen gas, measured at the same temperature and pressure, would be required?
   (a) 100 L   (b) 125 L   (c) 22.4 L   (d) 100×22.4 L
   (e) Cannot be calculated unless we know the values of temperature and pressure.

Answers to Some Important Terms in This Chapter

1. fluids
2. absolute zero
3. standard temperature and pressure (STP)
4. standard molar volume
5. ideal gas
6. ideal gas equation
7. universal gas constant
8. partial pressure
9. mole fraction
10. vapor pressure
11. kinetic-molecular theory
12. effusion
13. diffusion
14. real gas
15. van der Waals equation
16. equation of state

Answers to Quotefalls

Puzzle #1:
Boyle's Law: At constant temperature, the volume occupied by a definite mass of a gas is inversely proportional to the applied pressure.

Puzzle #2:
Dalton's Law of Partial Pressures: The total pressure exerted by a mixture of ideal gases is the sum of the partial pressures of the gases.

Puzzle #3:
Charles's Law: At constant pressure, the volume occupied by a definite mass of a gas is directly proportional to its absolute temperature.

Puzzle #4:
Avogadro's Law: At constant temperature and pressure, the volume occupied by a gas sample is directly proportional to the number of moles of gas.

Puzzle #5:
Gay-Lussac's Law of Combining Volumes: At constant temperature and pressure, the volumes of gases that react or are formed can be expressed as a ratio of simple whole numbers.

Answers to Preliminary Test

True-False

1. False. See Table 12-2. In fact, these are sometimes referred to as the rare gases.
2. True. Gas molecules are very far apart relative to their size, but liquids have very little empty space between molecules.
3. False. This is only true at constant temperature and for a fixed amount of gas.
4. False. Only for ideal gases. We cannot predict in general what the factor would be for other gases.
5. False. This is true only if the temperature is expressed on the absolute (Kelvin) scale.
6. True. See Avogadro's Law, Section 12-8.
7. False. Each molecule of HF has two atoms, while each molecule of Ne is just one atom.

8. True. But only because the molecular weights of these two particular gases are the same. $(1 + 19 = 20)$

9. True. Again, because the molecular weights of these two gases happen to be the same.

10. True. This would be true even if the molecular weights of the gases were not the same.

11. True. Because the molecular weights of the two gases are the same.

12. False. The two types of gas molecules have the same average speed because the molecular weights are the same.

13. False. The pressure would double. The total pressure is the sum of the two partial pressures.

14. False. HF is more polar, so it would behave less ideally than Ne.

Short Answer

1. 760 (by definition)

2. $\dfrac{V_1}{T_1} = \dfrac{V_2}{T_2}$ (constant n, P). This is incorrect without specifying its limitations.

3. the same temperature and pressure

4. Avogadro's

5. $\frac{3}{2}$. Recall that V is directly proportional to T and is inversely proportional to P.

6. $\frac{1}{3}$. V is inversely proportional to P, and $\dfrac{4.2\,\text{L}}{1.4\,\text{L}} = 3$.

7. 3. V is directly proportional to T, and $\dfrac{4.2\,\text{L}}{1.4\,\text{L}} = 3$.

8. zero volume; a liquid; a solid

9. 1.15. Volume is proportional to Kelvin temperature, which changes from 323 K to 373 K. Either multiply old volume by 373/323 or use the mathematical formulation of Charles's Law: $V_1 = 1.00$ L, $T_1 = 323$ K, and $T_2 = 373$ K.

10. 0.0821 $\dfrac{\text{L} \cdot \text{atm}}{\text{K}}$. Other values, with appropriate units, could also be correct.

11. 273.15 K, 1 atm

12. 4.00. Apply Dalton's Law. Because $\frac{2}{3}$ of the total pressure of the mixture is due to H_2, it must account for $\frac{2}{3}$ of the total number of moles. If we removed the O_2 ($\frac{1}{3}$ of the total moles) and adjusted the volume to produce the same total pressure, the moles of H_2 would occupy $\frac{2}{3}$ of the original total volume.

13. $\frac{2}{3}$, $\frac{1}{3}$. You can reason this either from the fraction of total moles or the fraction of total pressure contributed by each gas. Remember that mole fraction is a dimensionless quantity. Notice that the sum of the mole fractions is 1.

14. 728.2. Of the total 760 torr (1 atm) exerted, 31.8 torr is due to water vapor. Only $760 - 31.8 = 728.2$ torr would be due to hydrogen.

15. 89.6. One mole is the amount that would occupy 22.4 L at STP. This would weigh 22.4×4.0 grams, or 89.6 grams.

16. 11.2. Remember that nitrogen is diatomic, so 14.007 grams is only ½ mole. Each mole occupies 22.4 L at STP.

17. 4.0 g/L. $d = \dfrac{m}{V} = \dfrac{16.4\,g}{4.10\,L} = 4.0\,g/L$

18. 164. Apply $PV = nRT$, with $T = 227 + 273 = 500$ K; $P = 760$ torr $= 1$ atm; $V = 4.10$ L. Solving for n, we find that there is 0.100 mole of gas present. This weighs 16.4 grams, so molecular weight $= \dfrac{16.4\,g}{0.10\,mol} = 164$ g/mol.

19. 88. If 1.00 L of gas X weights 44 times as much as 1.00 L of H_2, then 22.4 L of gas X will weight 44 times as much as 22.4 L of H_2, which weighs 2.0 g. 2.0 g × 44 = 88.

20. 71.0 (grams); 6.022×10^{23} (molecules); 1.204×10^{24} (atoms); 22.4 (L). Chlorine gas is diatomic, Cl_2, so 1 mole weighs $2 \times 35.5 = 71.0$ and contains $2 \times 6.022 \times 10^{23} = 1.204 \times 10^{24}$ atoms.

21. collisions of molecules with the walls of the container

22. average molecular kinetic energy

23. temperature

24. high T, low P

25. intermolecular attraction. The term b is related to the size of the molecules.

26. 500; 1500. At the same temperature and pressure, the mole ratio (given in the balanced chemical equation) is the same as the gas volume ratio, 1 : 3 : 2.

27. 67.2 L. The gas volume ratio $CS_2 : O_2$ is 1 : 3. The volume of one mole of CS_2 at STP is 22.4 L, so the volume of the O_2 is 3×22.4 L $= 67.2$ L.

28. 22.4 L. The gas volume ratio $CS_2 : CO_2$ is 1 : 1. The volume of one mole of CS_2 at STP is 22.4 L, so the volume of the CO_2 is 1×22.4 L $= 22.4$ L.

29. 44.8 L. The gas volume ratio $CS_2 : SO_2$ is 1 : 2. The volume of one mole of CS_2 at STP is 22.4 L, so the volume of the SO_2 is 2×22.4 L $= 44.8$ L.

Multiple Choice

1. (d). The volume of an ideal gas extrapolates to 0 at a temperature of 0 K, not –273 K, which is impossible anyway.

2. (b). The density of the gas is directly proportional to the mass, which does not change with changes in pressure or temperature, and inversely proportional to the volume. The volume is inversely proportional to the pressure and directly proportional to the temperature, so increasing the pressure and decreasing the temperature will both decrease the volume, the divisor in the calculation of the density, which will increase the density.

3. (e). An increase in the pressure would decrease the volume, which would increase the density, but an increase in the temperature would increase the volume, which would decrease the density.

4. (c). The average kinetic energy of gaseous molecules is directly proportional to the absolute temperature of the sample.

5. (b). Use the ideal gas equation to determine the number of moles.

$$n = \frac{PV}{RT} = \frac{(0.50\,atm)(8.2\,L)}{\left(\dfrac{0.0821\,L \cdot atm}{mol \cdot K}\right)(-73^\circ C + 273)} = 0.25\,mol$$

Remember that there are 6.0×10^{23} molecules/mole (Avogadro's number).
(0.25 mol)(6.0 × 10^{23} molecules/mole) = 1.5 × 10^{23} molecules

176

6. (e). 1 phosphorus gas + 6 hydrogen gas → 4 phosphine gas, or $1P_4(g) + 6H_2 →$ $4P_xH_y$

To balance the equation, x must be 1 and y must be 3 ($6 × 2 = 4 × 3$), so phosphine is PH_3.

7. (b). Because the atomic weight of oxygen is greater than the atomic weight of nitrogen, equal weights of O_2 and N_2 results in more molecules of nitrogen than oxygen. Equal volumes of O_2 and N_2 at the same temperature and pressure would contain the same number of molecules. For a larger number of molecules of N_2 to occupy the same volume at the same temperature, the pressure must be greater.

8. (b). All have the same average kinetic energy; CO_2 has the highest molecular weight, so it would be moving slowest at the same kinetic energy.

9. (d). All have the same average kinetic energy; He has the smallest molecular weight, so it would be moving most rapidly at the same kinetic energy, and, therefore, would diffuse most rapidly.

10. (b). You can answer this by determining the simplest formula, CH_2. (See Chapter 2.) The gas volume data can be used to find that the molecular weight is 30 g/mol, or twice the simplest formula weight.

Of 100.0 g of the gas, 80.0% or 80.0 g is carbon and 20.0% or 20.0 g is hydrogen.

Element	Mass of Element	Moles of Element (divide mass by AW)	Divide by Smallest	Simplest Formula
C	80.0 g	$\dfrac{80.0\,g}{12.01\,g/mol} = 6.66\ mol$	$\dfrac{6.66\,mol}{6.66\,mol} = 1.00\ C$	CH_3
H	20.0 g	$\dfrac{20.0\,g}{1.008\,g/mol} = 19.8\ mol$	$\dfrac{19.8\,mol}{6.66\,mol} = 2.98\ H$	

The volume at STP is 2.24 L. $\dfrac{2.24\,L}{22.4\,L} = 0.100\ mol$, and $\dfrac{3.0\,g}{0.100\,mol} = 30.0\ g/mol$

The formula weight of CH_3 is $12.0 + 3 × 1.0 = 15.0$, and $\dfrac{30.0}{15.0} = 2.0$

The molecular formula is $2(CH_3)$ or C_2H_6.

11. (d). The volume at STP is 1120 mL = 1.120 L.

$\dfrac{1.120\,L}{22.4\,L} = 0.0500\ mol$, and $\dfrac{2.86\,g}{0.0500\,mol} = 57.2\ g/mol$

12. (d). The mole ratio of $Ag_2O : O_2$ is 2 : 1, so 0.200 mol of Ag_2O produces 0.100 mol of O_2.

0.100 mol occupies 0.100 mol × 22.4 L $= 2.24$ L at STP $× \dfrac{546\ K}{273\ K} = 4.48$ L at 546 K,

or $V = \dfrac{nRT}{P} = \dfrac{(0.100\ mol)\left(\dfrac{0.0821\ L \cdot atm}{mol \cdot K}\right)(546\ K)}{1\ atm} = 4.48\ L$

13. (a). 2.8 L (at STP) is $\dfrac{2.8\,L}{22.4\,L/mol} = \frac{1}{8}$ mol O_2. Since the mole ratio of $KClO_3 : O_2$ is 2 : 3, this would require $\frac{1}{12}$ mole $KClO_3$.

14. (b). $\underline{?}\,\text{L Cl}_2 = 5.61\,\text{g KOH} \times \dfrac{1\,\text{mol KOH}}{56.1\,\text{g KOH}} \times \dfrac{1\,\text{mol Cl}_2}{2\,\text{mol KOH}} \times \dfrac{22.4\,\text{L Cl}_2}{1.00\,\text{mol Cl}_2} = 1.12\,\text{L Cl}_2$

15. (b). The mole ratio (and volume ratio at constant T and P) of NO to O_2 is 4 : 5, so 100 L of NO requires $100\,\text{L} \times \frac{5}{4} = 125\,\text{L}$.

13 Liquids and Solids

Chapter Summary

In Chapter 13, you will take the same approach to the study of the two **condensed phases** or states of matter, liquids and solids, that you did for gases in Chapter 12. You will learn about the observed properties of liquids and solids. You will also describe those properties in terms of the behavior of the molecules that make up the liquids and solids, using the kinetic-molecular theory. Then you will learn about qualitative and quantitative aspects of changes among these three states of matter. In molecular terms, the most important difference between gases versus liquids and solids is that in liquids and solids the molecules are so close together that they can interact quite strongly, whereas in gases they are so far apart that their interactions are very slight. You will use the kinetic-molecular theory to understand differences in properties among these three states of matter. These properties were mentioned briefly in Section 12-1 and are listed in Table 13-1.

As you learn about solids and liquids in this chapter, you will see that we have several goals in mind. One is to understand the properties of liquids by describing what the molecules are doing and how they interact. Another is to gain a similar understanding of properties of solids relative to their molecular and ionic interactions. A third aspect of your study deals with changes between phases—liquid $\rightleftharpoons$ gas, solid $\rightleftharpoons$ liquid, and solid $\rightleftharpoons$ gas. In discussing this last aspect, either descriptively or quantitatively, you will use your understanding of molecular properties in the individual states. We conclude from a study of their observed properties (Chapter 12) that in gases the molecules are far apart and have very little attraction for one another. In this chapter, you will see that properties such as high density, surface tension, and viscosity demonstrates that in liquids the molecules are so close together that their attractions for one another are very strong and are quite important in determining properties. Evaporation and boiling both involve a substance changing from a liquid to a gas. To understand these processes, you must consider the amount of energy required to overcome the attractive forces in the liquid and convert it to a gas.

You can extend the ideas of the kinetic-molecular theory to apply them to liquids and solids (Section 13-1); this provides the basis for the discussion in Chapter 13. The properties of liquids and solids depend on **intermolecular** forces of attraction. These forces may be classified (Section 13-2) as **ion-ion interactions**, **dipole-dipole interactions**, **hydrogen bonding**, and **dispersion forces**. These different types of attractive forces vary considerably in their strengths. They have a great influence on melting and boiling points, heats of fusion and vaporization, and many other physical properties of substances. Table 13-6 summarizes these effects. An understanding of these attractive forces will also aid in understanding properties of solutions, discussed in Chapter 14. In the next few sections, you see the relation of observed macroscopic properties of a liquid such as **viscosity** (Section 13-3), **surface tension** (Section 13-4), and **capillary action** (Section 13-5) to the intermolecular attractions.

The next several sections are concerned primarily with a description of changes between liquid and gas. When the kinetic energy of a molecule near the surface of a liquid is high enough to overcome the attraction of its neighbors, it can **evaporate** or **vaporize**; conversely, gas molecules with low enough kinetic energy can **condense** (Section 13-6). If the container is closed, the two processes eventually occur at equal rates, and the **vapor pressure** (Section 13-7) characteristic of the liquid at that temperature is established. The ideas of **dynamic equilibrium** and **LeChatelier's Principle** presented here will be central to your understanding of many equilibrium situations, including chemical reactions—pay close attention to these important concepts. Section 13-8 deals with a related phenomenon (**boiling**) and the use of differences in the boiling points of substances to separate components of a liquid mixture, the process of **distillation.**

You now begin a quantitative consideration of some of the energy changes involved in heating substances and in their phase changes. How can one understand the energy change in molecular terms? Recall that increasing the temperature of a substance means that the average kinetic energy of its molecules increases. To raise the temperature of a substance, energy must be added, often in the form of heat (we speak of "heating the substance"). Likewise, for a liquid to evaporate, the kinetic energy of its molecules must be raised sufficiently to overcome the attractive forces between molecules. Again, this is done by putting in heat. In Section 13-9, you learn about several kinds of calculations of heat transfer involving liquids. Before beginning that section, you should review the units and measurement of heat, Section 1-13. The amount of heat required to raise the temperature of a substance by a certain amount is expressed as the **specific heat** or the **molar heat capacity** of the substance; the quantity of heat needed to cause vaporization of a liquid is the **heat of vaporization,** which is often expressed on a molar basis. The **Clausius-Clapeyron equation** describes how the vapor pressure of a liquid changes with temperature. One of its important uses is to predict how the boiling point of a liquid changes when the external pressure is changed.

In the same way, the process of melting (and the reverse process, freezing) involves a change of state, this time between liquid and solid. When these two processes are in equilibrium, the substance is at its **melting point** (Section 13-10). Again, this equilibrium is a dynamic one. In molecular terms, melting involves putting in enough heat energy for the molecules or ions to overcome the strong attractions that hold them in their fixed positions in the solid. Terms analogous to those you have already seen are used to describe the heat transfer involving solids—the **specific heat** and **molar heat capacity** of a solid, as well as its **heat of fusion,** again often on a molar basis (Section 13-11). The corresponding transitions from solid to gas (**sublimation**) and gas to solid (**deposition**) and the related idea of the vapor pressure of solids are discussed in Section 13-12.

Much data regarding phase transitions for any pure substance can be summarized in its **phase diagram,** Section 13-13. To help interpret this diagram for various substances, remember that all points along any horizontal line in the diagram are at the same pressure, while all points along any vertical line are at the same temperature.

Solids can be either **amorphous** (lacking internal regularity) or **crystalline** (Section 13-14). Most solids are crystalline, characterized by a regularly repeating pattern of molecules or ions, somewhat like a three-dimensional wallpaper pattern. Because of the regularity of crystalline solids, you can describe and understand them in a systematic way. Much of our information about the structures of solids has been obtained by the technique of **X-ray diffraction** (Enrichment, page 476). The scattering of X-rays (a very short-wavelength form of light) by the regularly repeating arrays of atoms leads to interference of the rays. Analysis of the directions and intensities of these scattered rays has allowed us to determine the three-dimensional structures of many solids. This method has provided one of our major sources of information about the three-dimensional structures of molecules and polyatomic ions (Chapter 8) that make up many solids.

Section 13-15 discusses the various patterns by which atoms, ions, or molecules are arranged in crystals. The main point here is that even though solids consist of many different kinds of particles, the requirement that they are regularly arranged in a repeating pattern in the crystal means that only a few types of *arrangements* are possible. These are the seven crystal systems, given in Table 13-9. This lists the restrictions on the shapes of the possible box (**unit cell**) by which the structure repeats. You should realize that this is just an imaginary box that describes the size and shape of the repeating unit; the repeating unit itself consists of atoms, molecules, or ions, arranged in a particular way.

We often classify crystalline solids according to the types of particles making up the arrangement and the type and strength of the bonding or interactions among them. These categories are summarized in Table 13-10 and discussed in detail with examples in Section 13-16:

(1) In **metallic solids,** metal atoms occupy lattice sites, with valence electrons free to move throughout the array of atoms.
(2) **Ionic solids** consist of extended arrays of ions, with very strong attractions for one another. These strong attractions are due to the arrangement in which each positive ion has negative ions as its nearest neighbors, and vice versa. As seen in Chapters 2 and 7, these substances do not consist of molecules. Knowing the size and shape of the repetitive pattern of such substances can often give us information about ionic sizes, as seen in the various examples of this section.
(3) In **molecular solids,** the fundamental particle of the crystal is a molecule. Each molecule contains strong covalent bonds, but the molecules interact in the solid only weakly, giving these solids their characteristic properties.
(4) **Covalent solids** are those in which atoms are covalently bonded together over many unit cells. Examples of these are diamond, graphite, quartz, and many minerals.

The **band theory** of metals (Section 13-17) describes the properties of metallic solids in terms of their electronic structures. The "Chemistry in Use" essay in this chapter discusses the important technological applications of solids in semiconductors.

Study Goals

There are many important ideas and skills in this chapter. To help you to organize your study, concentrate on them in the following groups. Answer as many of the text Exercises as you can.

Chapter Introduction
1. Compare and contrast the properties of gases, liquids, and solids. *Work Exercise 18.*

Section 13-1

2. Apply the kinetic-molecular theory to explain the similarities and differences in properties of the three states of matter. *Work Exercises 1, 2, 21, 75, and 76.*

Section 13-2

3. Describe and illustrate each of the following kinds of intermolecular attractions: (a) ion-ion interactions, (b) dipole-dipole interactions, (c) hydrogen bonding, and (d) dispersion forces. *Work Exercises 1 through 4, 9 through 12, 20, 26, 29, and 38.*

Sections 13-3 through 13-8

4. Describe and illustrate the following terms concerning liquids: (a) viscosity, (b) surface tension, (c) cohesive and adhesive forces, (d) capillary action, (e) evaporation and condensation, (f) vapor pressure, (g) boiling point, and (h) distillation. *Work Exercises 15, 23, 27, 28, 31, 35 through 37, and 39 through 42.*

Sections 13-2 through 13-8

5. Understand the significance of the types of intermolecular forces in Study Goal 3 in determining properties listed in Study Goal 4. *Work Exercises 3 through 11, 13, 14, 16, 17, 30, and 32 through 34.*

Section 13-9

6. Use the Clausius-Clapeyron equation to relate vapor pressure, temperature, and heat of vaporization. *Work Exercises 57 through 64.*

Sections 13-10 through 13-12

7. Describe and illustrate terms concerning solids: (a) melting point, (b) fusion and solidification, and (c) sublimation and deposition. Understand the significance of the types of intermolecular forces in Study Goal 3 in determining these properties of solids. *Work Exercises 39, 46, and 76 through 83.*

Sections 13-6, 13-9, 13-10, and 13-12

8. Write equations for phase transitions.

Sections 13-7, 13-9, and 13-11

9. Use the kinetic-molecular theory to describe the heat flow that accompanies changes of temperature and phase transitions. *Work Exercise 45.*

Sections 13-9 and 13-11

10. Given the appropriate heat capacities or heats of fusion, vaporization, or sublimation of a substance, be able to relate energy (heat) gain or loss to temperature changes and phase transitions of the substance. *Work Exercises 43 and 47 through 56.*

Section 13-13

11. Interpret phase diagrams. *Work Exercises 65 through 74.*

Section 13-14

12. Distinguish between amorphous and crystalline solids. Describe a glass as a supercooled liquid, as distinguished from a true solid. *Work Exercise 75.*

Enrichment section

13. Illustrate the ideas underlying X-ray diffraction by crystals. Carry out some X-ray diffraction calculations. *Work Exercises 98 through 101.*

Section 13-15

14. Characterize the seven crystal systems by names and by patterns of axial lengths and interaxial angles.

15. Given a picture or description of a unit cell and its contents, determine the numbers of atoms, ions, or molecules per unit cell. *Work Exercises 85 and 94.*

16. Describe and distinguish between the close-packed crystal structures. Sketch the three kinds of cubic lattices. *Work Exercises 84 and 86.*

Section 13-16

17. Classify solids into one of the following four categories and summarize the distinguishing characteristics of these classes: (a) metallic solids, (b) ionic solids, (c) covalent solids, and (d) molecular solids. *Work Exercises 77 through 83.*

18. Perform calculations to relate unit cell data to such quantities as atomic and ionic radii and volumes, densities of solids, and the number of particles per unit cell. *Work Exercises 86 through 101.*

Section 13-17

19. Describe metallic bonding in terms of band theory. Use band theory to describe electrical conductors, nonconductors (insulators), and semiconductors. *Work Exercises 102 through 104.*

General

20. Recognize and answer types of questions based on the concepts in this chapter. *Work* **Mixed Exercises** *105 through 118.*

21. Answer conceptual questions based on this chapter. *Work* **Conceptual Exercises** *119 through 129.*

22. Apply concepts and skills from earlier chapters to the ideas of this chapter. *Work* **Building Your Knowledge** *exercises 130 through 136.*

23. Exercises at the end of chapter direct you to sources outside the textbook for information to use in solving them. Work **Beyond the Textbook** *exercises 137 through 140.*

Some Important Terms in This Chapter

Chapter 13 has a very large number of new terms, many of which will be used in later chapters. This is *only a partial list*, so be sure you study others as well. Fill the blanks in the following paragraphs with terms from the list. Each term is only used once. Check the Key Terms list and the chapter reading. Study other new terms, and review terms from preceding chapters if necessary.

band	dynamic equilibrium	LeChatelier's Principle
band gap	forbidden zone	metallic bonding
boiling point	heat of fusion	metallic solid
Clausius-Clapeyron equation	heat of vaporization	molar heat of fusion
conduction band	hydrogen bonding	molar heat of vaporization
covalent solid	intermolecular forces	molecular solid
crystalline solid	intramolecular forces	normal boiling point
dipole-dipole interaction	ionic solid	vapor pressure
dispersion forces	ion-ion interaction	

The prefix *intra* means "within," and the prefix *inter* means "between" or "among." Therefore, attractive forces within molecules or polyatomic ions, which strongly hold the atoms together, are [1]_____, and attractive forces between molecules, which allow the molecules to adhere, are [2]_____. Intermolecular forces are generally much weaker than intramolecular forces, but without them, there would be no liquids or solids. Positive and negative ions attract each other in ionic compounds, a force known as [3]_____, which can be thought of as both *inter*molecular and *intra*molecular. Polar molecules have a dipole, an uneven distribution of the electron charge, so that one end of the molecule is slightly positive and the other end is slightly negative. These "partial charges" attract each other, much like ions do, which results in an attractive force known as [4]_____. One kind of dipole-dipole interaction is so specific in the elements involved and so much stronger than other polar attractions that it belongs to its own category. This attraction occurs when a hydrogen atom that is bonded to a very electronegative atom (fluorine, oxygen, or nitrogen) is attracted to another one of these very electronegative atoms, producing [5]_____. Nonpolar atoms or molecules, too, can be attracted to other particles, if their electron clouds are briefly distorted to produce a temporary dipole. This distortion may be cause by a passing cation or anion, a polar molecule, or even another nonpolar particle. The resulting attractive forces may be called dipole-induced dipole forces, London forces, or [6]_____.

The relationships between the liquid and vapor states, the solid and liquid states, and the solid and gaseous states (in the case of sublimation) often involve a situation in which the two opposing processes that change each state into the other both occur at the same time and, eventually, at the same rate. Then, everything *seems* to stop (static), but, actually, the processes continue to occur (dynamic), with the number of particles changing one way being equal to the number changing the other way. The system is said to have achieved a [7]_____. A French chemist observed a pattern that can be used to predict the effects of changes to this equilibrium, a concept called [8]_____. For example, the partial pressure of vapor molecules above the surface of a liquid at equilibrium at a given temperature is the [9]_____, and the increase in the vapor pressure of any liquid with increasing temperature can be explained by consideration of LeChatelier's Principle (Chapter 17). Eventually, the vapor pressure becomes equal to the atmospheric pressure, and bubbles of the vapor of the substance form throughout the liquid, rising to the surface. The liquid is said to be boiling, and the temperature at which this occurs is called the [10]_____. Because the atmospheric pressure over the liquid affects this temperature, any comparison of boiling points requires that they be measured at the same atmospheric pressure. The temperature at which the vapor pressure of the liquid becomes equal to one atmosphere (760 torr); that is, the temperature at which the liquid boils at standard (normal?) pressure, is the [11]_____.

Heat energy is required to change a liquid to its vapor, even *at* the boiling point, just to break the intermolecular forces and separate the particles. If the amount of heat required to produce the vapor is measured per gram, it is called simply the [12]_____, but if it is measured per *mole*, it is called the [13]_____, represented by the symbol ΔH_{vap}. The change in the vapor pressure when the temperature is changed can be calculated from the molar heat of vaporization, using the mathematical expression called the [14]_____.

Similarly, heat energy is required to change a solid to a liquid, even *at* the melting point, just to break enough of the intermolecular forces to allow the particles to move over and around each other. If the amount of heat required to fuse (melt) is measured per gram, it is called simply the [15]_____, but if it is measured per *mole*, it is called the [16]_____, represented by the symbol ΔH_{fus}.

Most solids have well-defined, sharp melting temperatures and shatter to produce fragments with flat faces and the same angles between faces. They do this because the particles that compose them are in very regular positions within the solid, with bonds of the same length and strength holding the particles together. This type of solid is called a [17]_____. There are at least four types of crystalline solids, depending on the type of bonding:

   (a) bonds between metal atoms produces a [18]_____,
   (b) bonds between ions produces an [19]_____,
   (c) weaker intermolecular attractive forces (dipole-dipole interactions, hydrogen bonds, and dispersion forces) between molecules produces a [20]_____, and
   (d) networks of strong covalent bonds between atoms produces a [21]_____.

The bonding in metals is called [22]_____. The interaction of atomic orbitals from a large number of metals atoms produces a series of very closely spaced molecular orbitals, which form a [23]_____ of orbitals that belongs to the whole crystal. A band within which, or into which, electrons must move to allow electrical conduction is called a [24]_____. In semiconductors and insulators the filled band of molecular orbitals and the empty band do not overlap, but are separated by an energy difference called the [25]_____. *Insulators* have such a large band gap that electrons cannot jump from the filled band to the empty band, so this energy gap is called a [26]_____.

Preliminary Test

As in other chapters, this test will check your understanding of basic concepts and types of calculations. Be sure to practice *many* of the additional textbook Exercises indicated with the Study Goals.

Short Answer

1. The much lower compressibility of solids and liquids compared with gases tells us that in these condensed phases, the molecules are _____ than in gases.
2. Forces between separate particles of a substance are referred to as _____ forces, whereas those within molecules are called _____ forces.
3. Forces between molecules within a liquid are called _____ forces.
4. Forces between molecules of a liquid and its container are called _____ forces.

5. Because the forces acting on molecules at the surface of a liquid are not equal in all directions, the liquid exhibits a property known as _____.

6. The movement of a liquid up a tube of small diameter is called _____; this occurs when the _____ exceed the _____.

7. The stronger the intermolecular forces in a liquid, the _____ is its vapor pressure, the _____ is its heat of vaporization, and the _____ is its boiling point.

8. A high viscosity observed for a liquid tells us that the intermolecular forces in the liquid are _____.

9. A method of separating the components of a mixture of liquids with sufficient different boiling points is called _____ .

10. When a liquid is evaporating exactly as fast as its vapor is condensing, a state of _____ is said to exist; the pressure exerted by the gas at this condition is said to be the _____ of the liquid.

11. For a pure substance at any pressure, the temperature at which the vapor pressure is equal to the applied pressure is called the _____ of the liquid. If the pressure is 1 atm, this temperature is termed the _____.

12. The vapor pressure of a liquid at 25°C is 200 torr. If the atmospheric pressure over the liquid is lowered to 200 torr at 25°C, the liquid will _____.

13. Substance A has greater tendency to evaporate at a given temperature than substance B does. We describe substance A as being more _____ than substance B. We also expect that the _____ of A would be higher and that its _____ would be lower.

14. The amount of heat that is required to melt one gram of a substance is called the _____ of the substance. The amount of heat required to melt one mole of the substance would be called the _____ of the substance.

15. For any substance at a given temperature, the molar heat of vaporization must equal the _____.

16. In order to calculate the molar heat capacity of a substance, we would have to _____ its specific heat by its _____.

17. Because of the highly polar nature of an H–F bond, we should expect the molar heat of vaporization of HF to be considerably _____ than that of F_2.

18. Even though both are gases are room temperature, we would expect the temperature at which CH_4 condenses to a liquid to be _____ than that for NH_3.

19. The abnormally high boiling and melting points of water compared with those of H_2S, H_2Se, and H_2Te are due to the presence of _____ in water.

20. Consider the substances $CaCl_2$, PH_3, N_2, H_2, and NH_3. The order of increasing boiling points of these substances would be _____.

21. The two substances CO and N_2 have the same molecular weight, 28. It is observed that the molar heat of fusion of carbon monoxide is higher than that of nitrogen. The reason for this is the _____ of the CO molecules.

22. Hydrogen bonding interactions represent a special, unusually strong form of _____.

23. The temperature at which the solid and liquid phases of a substance are in equilibrium with one another is called the _____ of the substance.
24. The set of conditions, at a particular pressure, at which the solid and gaseous states of a substance can exist in equilibrium, is called the _____ of the substance.
25. The set of conditions at which a substance can exist simultaneously as solid, liquid, and gas in equilibrium is called the _____ of the substance.
26. A solid that shows no order (or only limited order) on the molecular scale is called _____.
27. The melting points of ionic crystals are usually _____ than those of molecular crystals.
28. Two examples of substances that exist as covalent crystals are _____ and _____.
29. A method that is used for the study of atomic arrangements in solids is _____ _____.
30. In a single layer of spheres of equal size in a close-packed arrangement, the number of nearest neighbors of each sphere is _____.
31. In a three-dimensionally close-packed arrangement of spheres of equal size, each sphere has _____ nearest neighbors.
32. In the NaCl structure, each Cl^- ion is surrounded by _____ nearest neighbors, each of which is a _____.
33. Two types of defects in the structures of crystalline solids that affect the properties of the solids are _____ and _____.

Multiple Choice

_____ 1. A drop of liquid has a spherical shape due to the property of
    (a) surface tension.   (b) capillary action.   (c) viscosity.
    (d) vapor pressure.   (e) close packing.

_____ 2. As a liquid is heated, its vapor pressure
    (a) does not change.   (b) increases.   (c) decreases.   (d) disappears.
    (e) may increase or decrease, depending on the liquid.

_____ 3. When the external pressure is decreased, the boiling point of a liquid
    (a) does not change.   (b) increases.   (c) decreases.   (d) disappears.
    (e) may increase or decrease, depending on the liquid.

_____ 4. If energy, in the form of heat, is added to a pure liquid substance at its boiling point while keeping the pressure constant, the
    (a) temperature will increase.   (b) temperature will decrease.
    (c) temperature will remain constant until all of the liquid has vaporized.
    (d) temperature of the vapor will be greater than that of the liquid.
    (e) temperature of the liquid will be greater than that of the vapor.

_____ 5. As we increase the temperature of a liquid, its properties change.
Which one of the following would *not* be an expected change in the properties of a liquid as we increase its temperature?
(a) decrease in viscosity   (b) decrease in density
(c) increase in surface tension   (d) increase in vapor pressure
(e) increase in tendency to evaporate

_____ 6. The vapor pressure of all liquids
(a) is the same at 100°C.
(b) is the same at their freezing points.
(c) increases with volume of liquid present.
(d) decreases with the increasing volume of the container.
(e) increases with temperature.

_____ 7. On a relative basis, the weaker the intermolecular forces in a substance,
(a) the larger its heat of vaporization.   (b) the more it deviates from the ideal gas law.
(c) the greater its vapor pressure.   (d) the larger its molar heat capacity as a liquid.
(e) the higher its melting point.

_____ 8. It is found that 600 joules of heat are required to melt 15 grams of a substance.
What is the heat of fusion of this substance?

(a) 600×15 J/g   (b) $\dfrac{600 \times 15}{1000}$ kJ/g   (c) $\dfrac{600 \times 1000}{15}$ kJ/g   (d) $600/15$ J/g

(e) cannot be answered without knowing the specific heat of the substance.

_____ 9. Substances have properties that are related to their structures.
Which one of the following statements regarding properties of substance is *not* expected to be correct?
(a) Molten KBr should be a good conductor of electricity.
(b) Diamond should have a high melting point.
(c) Solid sodium should be a good conductor of electricity.
(d) Solid CO_2 should have a low melting point.
(e) Solid K_2SO_4 should sublime readily.

_____ 10. Which of the following, in the solid state, would be an example of a molecular solid?
(a) carbon dioxide   (b) diamond   (c) barium fluoride
(d) iron   (e) none of the preceding

_____ 11. Which of the following, in the solid state, would be an example of a covalent solid?
(a) carbon dioxide   (b) diamond   (c) barium fluoride
(d) iron   (e) none of the preceding

_____ 12. Which of the following, in the solid state, would be an example of an ionic solid?
(a) carbon dioxide   (b) diamond   (c) barium fluoride
(d) iron   (e) none of the preceding

_____ 13. Which of the following, in the solid state, would be an example of a metallic solid?
(a) carbon dioxide   (b) diamond   (c) barium fluoride
(d) iron   (e) none of the preceding

_____ 14. The number of nearest neighbors of each atom in the face-centered cubic crystalline arrangement is
(a) 4.   (b) 6.   (c) 8.   (d) 12.   (e) 16.

_____ 15. The unit cell constants for a particular crystalline compound were determined by X-ray diffraction to be $a = 12.52$ Å, $b = 5.67$ Å, $c = 12.88$ Å, $\alpha = 90.8°$, $\beta = 98.2°$, $\gamma = 101.3°$. What is its crystal system?
(a) cubic   (b) orthorhombic   (c) tetragonal   (d) monoclinic   (e) triclinic

_____ 16. A hypothetical metal, atomic weight 60, forms simple cubic crystals. The nearest distance between centers of adjacent atoms is 2.0 Å. What is the density of the crystalline material?
(a) 12. g/cm^3 (b) 2.5 g/cm^3 (c) 62 g/cm^3 (d) 120 g/cm^3 (e) 1.2 g/cm^3

Questions 17 through 20 refer to the following phase diagram, which is for a hypothetical substance.

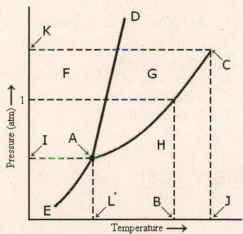

_____ 17. Where would you find the conditions necessary for the existence of a solid as a single phase?
(a) region F   (b) region G   (c) region H   (d) along line AE   (e) along line AC

_____ 18. The normal boiling temperature of the substance is indicated in the diagram by which of the following letters?
(a) A   (b) B   (c) C   (d) J   (e) L

_____ 19. In the diagram, the point A is the
(a) triple point.   (b) critical point.   (c) normal freezing point.
(d) normal boiling point.   (e) point of maximum density.

_____ 20. Which of the following represents a pressure below which the substance cannot exist as a liquid?
(a) K   (b) B   (c) L   (d) D   (e) I

_____ 21. Which of the following is the band theory explanation for the conduction of electricity by metals?
   (a) The occupied band of molecular orbitals overlaps a vacant conduction band of lower energy.
   (b) The forbidden zone becomes polarized.
   (c) Addition of electron-rich impurities by doping causes the conduction.
   (d) The conduction band is fully occupied.
   (e) The occupied band is either partially filled or overlaps a conduction band of higher energy.

Answers to Some Important Terms in This Chapter

1. intramolecular forces
2. intermolecular forces
3. ion-ion interaction
4. dipole-dipole interaction
5. hydrogen bonding
6. dispersion forces
7. dynamic equilibrium
8. LeChatelier's Principle
9. vapor pressure
10. boiling point
11. normal boiling point
12. heat of vaporization
13. molar heat of vaporization
14. Clausius-Clapeyron equation
15. heat of fusion
16. molar heat of fusion
17. crystalline solid
18. metallic solid
19. ionic solid
20. molecular solid
21. covalent solid
22. metallic bonding
23. band
24. conduction band
25. band gap
26. forbidden zone

Answers to Preliminary Test

Short Answer

1. much closer together
2. intermolecular; intramolecular
3. cohesive
4. adhesive
5. surface tension
6. capillary action; adhesive forces; cohesive forces
7. lower, higher, higher
8. stronger
9. distillation
10. dynamic equilibrium; vapor pressure
11. boiling point; normal boiling point
12. boil
13. volatile; vapor pressure; boiling point
14. heat of fusion; molar heat of fusion
15. molar heat of condensation
16. multiply; formula weight (or molecular weight)
17. higher

18. lower. NH_3 is polar, with strong intermolecular attractions due to hydrogen bonding. Remember that the temperature at which the gas condenses to a liquid is the same at that at which the liquid boils. Thus, you can apply the reasoning of Section 13-2 to questions such as this.
19. hydrogen bonding
20. $H_2 < N_2 < PH_3 < NH_3 < CaCl_2$. Reason as follows: $CaCl_2$ is ionic, so it has the highest boiling point. NH_3 and PH_3 are both polar covalent molecules, but NH_3 is more polar and can undergo hydrogen bonding, so it boils at a higher temperature than PH_3. Both H_2 and N_2 are nonpolar. N_2 is larger and has more electrons, so it boils at a higher temperature than H_2, due to its larger dispersion forces. See Section 13-2.
21. polar nature (or polarity). We have to supply more energy to overcome the dipole–dipole interactions between CO molecules.
22. dipole–dipole interaction
23. melting point. Note that the melting point of a substance changes with pressure much less than the boiling point. Can you see that in the phase diagrams of Figure 13-17?
24. sublimation point
25. triple point
26. amorphous
27. higher
28. Several answers would be correct here. Among these are diamond, graphite, quartz, and many minerals.
29. X-ray diffraction
30. 6
31. 12
32. 6; Na^+ ion
33. vacancies; substitutions of different types of atoms

Multiple Choice

1. (a). A sphere has the least possible surface area.
2. (b). As the temperature increases, the rate of evaporation increases.
3. (c). A liquid boils at the temperature at which its vapor pressure equals the external pressure.
4. (c). All of the added heat is used to vaporize the liquid.
5. (c). In general, at higher temperature, the molecules are less able to attract one another. This *would* cause the results listed in (a), (b), (d), and (e) to occur, along with a *decrease* in surface tension.
6. (e). As the temperature is increased, the rate of evaporation increases.
7. (c). Weaker forces between molecules would make it easy to evaporate (*lower* heat of vaporization, *higher* vapor pressure), would make it obey the ideal gas law better, would make it easier to move the molecules past one another in the liquid (smaller molar heat capacity), and would make it easier to melt.
8. (d). The heat of fusion is the amount of heat (J) required to melt a solid *per* gram (g).

9. (e). KBr and K_2SO_4 are both ionic solids, diamond is a covalent solid, Na is a metallic solid, and CO_2 is a molecular solid. The characteristics of various types of solids are summarized in Table 13-10.

10. (a). Carbon dioxide, consisting of two nonmetals, forms molecules.

11. (b). Carbon atoms in diamond form a network of covalent bonds.

12. (c). Barium, a metal, forms an ionic compound with fluorine, a nonmetal.

13. (d). Iron is a metallic element.

14. (d). In close-packed structures, such as face-centered cubic, the coordination number is 12.

15. (e). This unit cell has edges of three different lengths with three different angles between them, none of which is 90°.

16. (a). This problem is similar to Examples 13-8 and 13-9, except that this simple cubic cell contains only one metal atom (eight corners, each with $\frac{1}{8}$ atom).

$$\underline{?}\ \text{g metal per unit cell} = \frac{1\,\text{metal atom}}{\text{unit cell}} \times \frac{1\,\text{mol metal}}{6.022 \times 10^{23}\,\text{atoms}} \times \frac{60\,\text{g metal}}{1\,\text{mol metal}}$$

$$= 9.96 \times 10^{-23}\ \text{g metal/unit cell}$$

$$V_{\text{unit cell}} = (2.0\ \text{Å})^3 = 8.0\ \text{Å}^3 \times \left(\frac{10^{-8}\ \text{cm}}{\text{Å}}\right)^3 = 8.0 \times 10^{-24}\ \text{cm}^3/\text{unit cell}$$

$$\text{Density} = \frac{9.96 \times 10^{-23}\ \text{g metal/unit cell}}{8.0 \times 10^{-24}\ \text{cm}^3/\text{unit cell}} = 12.\ \text{g/cm}^3$$

17. (a). Region F is solid, region G is liquid, and region H is gas. Line AE is the sublimation curve, and line AC is the vapor pressure curve.

18. (b). Point A is the triple point, point C is the critical point, point J is the critical temperature, and point L is the temperature of the triple point.

19. (a)

20. (e). Point K is the critical pressure, above which the substance cannot exist as a liquid. Points B and L are temperatures (0 pressure).

21. (e)

14 Solutions

Chapter Summary

Much of your previous study of substances and their properties has been of pure elements and compounds. But substances usually occur in mixtures, so you must consider how they behave in the presence of other substances. In Chapter 14, you will study solutions and their properties. The chapter introduction stresses the importance of solutions in some common processes. It also reminds you of some of the terminology used to refer to the components of the solution (you should also review Section 3-6). Sections 14-1 through 14-7 describe the ways in which substances interact to form solutions. In general terms, a solution can be in any phase—solid, liquid, or gas. Most of your study in this chapter will concern liquid solutions.

Throughout this chapter, we continue our general theme of describing macroscopic properties as the result of properties and interactions of molecules and ions. In Section 14-1, we introduce the important idea that a process is favored (1) by a **decrease in energy** and (2) by an **increase in randomness or disorder**. In Chapter 15, you will consider these ideas in more detail. To understand the dissolution process, you must think about the various kinds of attractions that occur on the molecular level: solute-solute attractions, solvent-solvent attractions, and solute-solvent attractions. The stronger the solute-solute attractions and the stronger the solvent-solvent attractions are, the more each kind of species tends to self-associate, and the less **soluble** or **miscible** the substances are. Conversely, very strong solvent-solute attractions can favor the dissolution process. For instance, the strong electrostatic attractions in an ionic solid and the strong attractions due to hydrogen bonding in water can be overcome by the even stronger interactions (**hydration**) between ions and water. As a result, many ionic substances are quite soluble in water. By contrast, even though the molecules of a nonpolar hydrocarbon such as benzene have relatively weak attractions for one another, they are not able to **solvate** ions very well. These strong solute-solute attractions, weak solvent-solvent attractions, and very weak solvent-solute attractions generally make ionic substances quite insoluble in benzene and other hydrocarbons. These ideas will help you to organize your study of the dissolution of solids, liquids, and gases in liquids (Sections 14-2 through 14-4). As you study these sections, observe the factors that help to determine the strengths of the interactions—such features as charge density of ions, hydrogen bonding in liquids, dipole-dipole interactions, relative sizes of solute and solvent particles, and the tendency of some covalent substances to ionize in solution.

The **rate of dissolution** of a solid solute in a solvent depends on such factors as surface area, solubility, and temperature; this is opposed by the **rate of crystallization**, the rate at which dissolved solute particles form crystals of undissolved solid (Section 14-5). When these two rates are equal, a state of **dynamic equilibrium** exists, and the solution is **saturated**. Keep in mind that it is incorrect to describe this state of equilibrium as a situation in which "nothing is happening." On the molecular and ionic level, much is happening, but there is *no net change* in the amount of dissolved solute; that is, there is no change at the macroscopic level. This is another application of some ideas regarding equilibrium that we first encountered in Chapter 13.

The effects of temperature and pressure on solubility are summarized and related to molecular-level behavior in Sections 14-6 and 14-7. Pressure changes usually have a significant effect only on the solubility of gases in liquids; this relationship is quantitatively summarized in **Henry's Law**. It is important to understand the effects of temperature and pressure on solubility in terms of **LeChatelier's Principle**, which was briefly introduced in Chapter 13.

Section 14-8 discusses two methods for describing solution concentrations, **molality** and **mole fraction**. You should also review the concentration calculations in Sections 3-6, 3-7, and 12-11. Different methods of expressing concentrations are useful for different applications in chemistry. You should learn all of the methods and practice solving problems associated with each one.

In the next several sections, you will study some solution properties that are closely related to concentration. A number of physical properties of solutions depend on the *number*, not the *kind*, of solute particles. These properties, which apply to solutions in which the *solvent* is volatile and the *solute* is nonvolatile, are called **colligative properties**. The four colligative properties are **vapor pressure lowering, boiling point elevation**, **freezing point depression,** and **osmotic pressur**e. Be sure you understand that vapor pressure, boiling point, and melting point (or freezing point) are *not* colligative properties—*changes* in these (vapor pressure *lowering*, boiling point *elevation*, and freezing point *depression*) are considered colligative properties.

Why are we interested in the colligative properties of solutions? One reason is to be able to *describe* the behavior, both verbally and mathematically, where possible. A second reason is to *understand* this behavior in terms of the interactions among molecules and ions. A third objective of this study is to be able to *use* these properties. Uses discussed in this chapter include separation of mixtures by **fractional distillation** and **reverse osmosis**, alteration of the freezing and boiling points of fluids such as the coolant in automobile radiators, experimental determination of the relative strengths (degrees of ionization) of electrolytes, and methods for the experimental determination of molecular weight. The latter two uses strongly emphasize that the colligative properties depend quantitatively on the number of particles in solution and not on the kind of particles. Sections 14-9 through 14-15 discuss the colligative properties, their interpretation, and their uses.

When both the solvent and the solute are volatile (i.e., able to evaporate, so vapor pressures are not zero), we can apply Raoult's Law (Section 14-9) to *each* component. The composition of the vapor above a solution is always richer in the more volatile component than the liquid solution is. This composition difference is the basis for the separation of components of a solution by simple distillation (Section 13-8) or by fractional distillation (Section 14-10).

The last portion of the chapter, Sections 14-16 through 14-18, discusses **colloids**. These mixtures are, in a sense, intermediate between homogeneous and heterogeneous. In a colloid, the dispersed particles are larger than in a true solution, so the mixture is not entirely homogeneous, but they are too small to settle out, so it is not truly heterogeneous either. These dispersed particles may be collections of molecules or ions, or they may be individual large molecules such as proteins. Table 14-4 summarizes the various types of colloids and important examples of each. Colloids are involved in many problems of biological and ecological significance. Colloids are frequently recognized by their ability to scatter light (the **Tyndall effect**, Section 14-16). They consist of finely divided particles with large total surface area, so their properties are understood in terms of surface phenomena, such as **adsorption** (Section 14-17). The interactions that keep the colloidal particles suspended or dispersed depend on the nature of the colloidal particle. Thus, we often classify colloids as **hydrophilic** or **hydrophobic**, depending on whether or not they interact with water to remain suspended. Examples of these two important classes of colloids are discussed in Section 14-18, which emphasizes some useful characteristics of colloids.

Study Goals

Write your own outline of the ideas presented in each study goal. Exercises at the end of the text chapter are indicated for your study.

Introduction to Chapter 14 and Introduction to Colloids
1. Describe and distinguish the terms in each group:
   (a) solvent, solute, and solution;
   (b) solution and colloid.
   Work Exercises 1, 3, and 91.

Introduction to Chapter 14 and Sections 14-1 through 14-4
2. Give several examples of various kinds of solutions, involving different combinations of solids, liquids, and gases as solvent and solute. *Work Exercise 2.*

Sections 14-1 through 14-4
3. Describe the relative effects on solubility of the following kinds of interactions: (a) solute-solute attractions, (b) solvent-solvent attractions, and (c) solvent-solute attractions. Be able to discuss these effects for solid-liquid, liquid-liquid, and gas-liquid solutions. *Work Exercises 4 through 6 and 13.*

Section 14-2
4. Describe and illustrate the mechanism of dissolution of ionic solids and polar covalent substances in water. *Work Exercises 7 through 12.*
5. Relate the relative magnitudes of crystal lattice energies of ionic solids to their formulas. *Work Exercises 19 through 22.*

Sections 14-1 and 14-6
6. Distinguish between exothermic and endothermic dissolution processes. Know the effects of exothermicity or endothermicity and of an increase in disorder on the spontaneity of the dissolution process. Describe the effects of exothermicity or endothermicity on the temperature dependence of water solubility of a compound. *Work Exercise 6 and 22.*

Section 14-5
7. Distinguish among unsaturated, saturated, and supersaturated solutions. *Work Exercise 16.*

Section 14-7
8. Describe the effect on solubility of changing the pressure when the solute is (a) a gas, (b) a liquid, or (c) a solid. *Work Exercises 14, 17, and 18.*

Section 14-8 (Review Sections 3-6, 3-7, and 12-11)
9. Relate specific amounts of solute dissolved in specific amounts of solvent or solution to concentration expressed as (a) percent by mass, (b) molarity, (c) mole fraction, and (d) molality. *Work Exercises 23 through 35.*

Sections 14-9, 14-11, 14-12, and 14-15

10. Describe the colligative properties and the factors on which they depend. *Work Exercises 36 and 81.*

Section 14-9

11. Use Raoult's Law to describe the vapor pressures of (a) solutions of nonvolatile solutes and (b) solutions in which both components are volatile. *Work Exercises 36 through 45.*

Section 14-10

12. Relate Study Goal 11 to the purification technique of fractional distillation. *Work Exercises 41 through 45.*

Sections 14-11, 14-12, and 14-15

13. Carry out calculations involving boiling point elevation, freezing point depression, and osmotic pressure to relate these physical properties of solutions to the amount of nonvolatile, nonelectrolyte solute dissolved in a given amount of solvent. *Work Exercises 46 through 57, 62, and 84 through 90.*

Sections 14-13 and 14-15

14. Use the colligative properties to determine molecular weights of solutes. *Work Exercises 58 through 61 and 90.*

Section 14-14 (Review Section 6-1)

15. Describe solutions of strong electrolytes, weak electrolytes, and nonelectrolytes, giving specific examples. *Work Exercises 11, 12, and 15.*

Section 14-14

16. Understand how the properties of electrolyte solutions depend on such factors as ion association and degree of ionization. Be able to determine and interpret the van't Hoff factor, *i. Work Exercises 63 through 69.*

17. Carry out calculations concerned with percent ionization and the van't Hoff factor. *Work Exercises 70 through 80, 86, and 88.*

Sections 14-15 and Chemistry in Use: Water Purification and Hemodialysis

18. Describe and illustrate osmosis and reverse osmosis. *Work Exercises 81 through 85 and 87 through 90.*

Introduction to Colloids, Sections 14-16, and 14-17

19. Know the features and properties of colloids. Distinguish among, and give common examples of, the kinds of colloids listed in Table 14-4, such as gels, sols, foams, and so on. *Work Exercises 91 through 93.*

Section 14-18

20. Describe and distinguish between hydrophilic and hydrophobic colloids. *Work Exercise 94.*

21. Describe and illustrate the action of emulsifying agents, soaps, and detergents. *Work Exercises 95 through 97.*

General

22. Recognize and answer questions about solutions and their properties. *Work **Mixed Exercises** 98 through 108.*

23. Answer conceptual questions based on this chapter. *Work **Conceptual Exercises** 109 through 123.*

24. Apply concepts and skills from earlier chapters to the ideas of this chapter. *Work **Building Your Knowledge** exercises 124 through 127.*

25. Exercises at the end of chapter direct you to sources outside the textbook for information to use in solving them. *Work **Beyond the Textbook** exercises 128 through 133.*

Some Important Terms in This Chapter

This is another chapter with *many* important new terms, some of which are listed at the top of the next page. You should study many others, too. Fill the blanks in the following paragraphs with terms from the list. Each term is only used once. Check the Key Terms list and the chapter reading. Study other new terms, and review terms from preceding chapters.

boiling point elevation	freezing point depression	osmosis
coagulation	Henry's Law	osmotic pressure
colligative properties	hydration	percent ionization
colloid	hydrophilic colloids	Raoult's Law
distillation	hydrophobic colloids	saturated solution
emulsifier	molality	solvation
fractional distillation	mole fraction	van't Hoff factor

The process in which water molecules surround and interact with solute ions or molecules is called [1]_____. The more general term applied when any solvent is used is [2]_____. Eventually, dissolved particles are in equilibrium with excess undissolved particles, and the mixture forms a [3]_____. The quantitative relationship between the concentration of a gas in a liquid and the pressure of the gas above the surface of the solution is expressed by the statement known as [4]_____.

Calculations of the properties of a solution based upon the concentration of the solution often require that the concentration be expressed, not in molarity, but in either [5]_____ (the number of moles of solute per *kilogram* of *solvent*) or [6]_____ (the number of moles of a component of a mixture divided by the total number of moles of all components). This requirement is especially true of the properties that depend on the *number* of solute particles in a given amount of solvent, not the *kind*, which are called [7]_____.

The lowering of the vapor pressure of the solvent by a nonvolatile, nonionizing solute is described by [8]_____, which says that the vapor pressure of a solvent in an ideal solution is directly proportional to the mole fraction of the solvent in the solution. Although simple [9]_____ can be used to separate a liquid solution into volatile and nonvolatile components, the separation of two volatile components requires either repeated distillations or [10]_____, which allows the mixture to repeatedly distill within the fractionating column.

Because the vapor pressure of the solution is lower than for the pure solvent, a higher temperature is required to raise the vapor pressure to the atmospheric pressure, allowing the solution to boil. This relationship is known as [11]_____. Similarly, the presence of solute particles *decreases* the temperature at which the solution will begin to freeze, a pattern referred to as [12]_____.

Since colligative properties depend on the number of solute particles in a given amount of solvent, ionic compounds, which dissociate into two or more ions for each formula unit, should have two or more times the effect on these properties. Their effect, however, is not as great as expected because the ions do not move about in a totally independent way—they undergo *association* in solution. The extent of dissociation can be expressed as the ratio of the actual colligative property to the value that would be observed if no dissociation occurred, a value known as the [13]_____, but this value must be compared with the ideal value for the compound, the number of ions produced from each formula unit. A more direct expression of the degree of ionization is the [14]_____, which is calculated by dividing the concentration of the compound that ionizes by the original concentration, times 100.

The process by which solvent molecules pass through a semipermeable membrane from a solution of lower concentration of solute into a solution of higher concentration is [15]_____. The hydrostatic pressure produced on the surface of the semipermeable membrane by osmosis is [16]_____.

A colloidal dispersion, or [17]_____, is a mixture in which solute-like particles are suspended but they do not settle out. Colloids come in two types: [18]_____ _____, which are "water loving" and [19]_____, which are "water hating." Hydrophobic colloids in polar solvents, like water, require the presence of an emulsifying agent, or [20]_____, to coat the particles of the dispersed phase to prevent their [21]_____ into a separate phase.

Preliminary Test

As in other chapters, this test will check your understanding of basic concepts and types of calculations. Be sure to practice *many* of the additional textbook Exercises, including those indicated with the Study Goals.

True-False

Mark each statement as true (T) or false (F).

_____ 1. When one substance dissolves in another, the result is a situation of greater disorder.

_____ 2. All cases of solids dissolving in water are endothermic.

_____ 3. All cases of gases dissolving in water are exothermic.

_____ 4. It is impossible, under any conditions, to dissolve more solute in a solvent than the amount required to saturate it.

_____ 5. Ions are usually dissolved in water without any interaction with the solvent.

_____ 6. In general, the larger the charge and the smaller the size of an ion, the more easily it can be hydrated.

_____ 7. An ionic substance such as KCl is able to dissolve in water because in the solution each K^+ ion can be surrounded closely by several Cl^- ions, while each Cl^- ion has several close K^+ neighbors.

_____ 8. The more negative the lattice energy for a solid substance, the less likely that it is able to dissolve in a specific solvent.

_____ 9. Gases can dissolve in liquids only if they react.

_____ 10. Increasing the pressure usually makes solids more soluble in liquids.

_____ 11. The degree to which an electrolyte ionizes in solution is greater for solutions that are more concentrated.

_____ 12. For a substance to be very soluble in water, it must be a strong electrolyte.

_____ 13. Nitric acid is a weak electrolyte solution.

_____ 14. Strong electrolytes are those that ionize almost completely in dilute aqueous solutions.

_____ 15. A solution described as concentrated would have a higher molarity than a dilute solution of the same substance.

_____ 16. Before we can answer any question regarding a solution, we must decide which component is the solute and which is the solvent.

_____ 17. Any solution that is 20% A dissolved in B has a mole fraction of A less than that of B.

_____ 18. A solution described as a strong acid solution would have a higher molarity than a weak acid solution.

_____ 19. For a given solution, it would probably be easier to measure the freezing point depression than the boiling point elevation.

_____ 20. It is not necessary to use measurements, such as colligative properties, to determine the degree of ionization of $CaCl_2$ in solution, because we can predict from the formula that 1 mole of $CaCl_2$ must dissolve to give 3 moles of solute particles.

_____ 21. All colloid particles are clumps of molecules or ions, just small enough to keep from settling out.

Short Answer

Answer with a word, a phrase, a formula, or a number with units as necessary.

1. A solution is an example of a _____ mixture.

2. When we describe a solution of a little sugar in water, we identify the sugar as the _____ and the water as the _____.

3. The reverse of the dissolution process is called _____.

4. A solution that contains dissolved solute in equilibrium at a given temperature with excess undissolved solute is said to be _____.

5. The ease of the dissolution process depends on two factors: the change in _____ and the change in _____.

6. A process in which the substances involved decrease in energy is called _____.

7. The three types of interactions that are useful to consider in assessing the change in heat content on dissolution are _____, _____ _____, and _____.

8. The dissolution process is favored by _____ solute-solute attractions, by _____ solvent-solvent attractions, and by _____ solute-solvent attractions.

9. The close clustering of solvent molecules around a solute particle is generally called _____; when the solvent is water, it is called _____.

10. Classify each of the following substances as soluble or insoluble in water (for review).

   (a) KNO_3 _____   (k) $Ca(OH)_2$ _____

   (b) $SrCl_2$ _____   (l) $AgNO_3$ _____

(c) SO_2 _____

(m) $Mg(CH_3COO)_2$ _____

(d) $Al(OH)_3$ _____

(n) $Zn(ClO_4)_2$ _____

(e) $(NH_4)_2S$ _____

(o) H_3PO_4 _____

(f) FeS _____

(p) $Ba_3(PO_4)_2$ _____

(g) $CaCO_3$ _____

(q) MgS _____

(h) HBr _____

(r) $(NH_4)_2CO_3$_____

(i) $PbCl_2$ _____

(s) KOH _____

(j) NaSCN _____

(t) Rb_2SO_4_____

11. To have a solution of acetone and water that has a mole fraction of acetone equal to 0.20, we must add _____ moles of acetone to 6.0 moles of water.

12. The lowering of the vapor pressure of a solvent due to the presence of dissolved solute is described by _____ Law, which states in words that _____

_____

___

_____

___

_____

___.

13. The vapor pressure of pure water is always _____ the vapor pressure of an aqueous solution of a nonvolatile solute at the same temperature.

_____

We make up two solutions. Solution A contains 0.10 mole of NaBr dissolved in 1000 grams of water, while solution B contains 0.10 mole of sugar dissolved in 1000 grams of water. The next four questions, 14 through 17, refer to these two solutions.

14. We expect that the vapor pressure of solution A would be _____ that of solution B.

15. We expect that the freezing point of solution A would be _____ that of solution B.

16. We expect that the boiling point of solution A would be _____ that of solution B.

17. If each solution is put in contact through a semipermeable membrane with a sample of pure water, solution A would exhibit _____ osmotic pressure compared to solution B.

_____

At 25°C, the vapor pressure of pure chloroform, $CHCl_3$, is 172.0 torr and that of pure carbon tetrachloride, CCl_4, is 98.3 torr. The formula weights of $CHCl_3$ and CCl_4 are 119.4 and 153.8, respectively. Questions 18 through 20 refer to ideal solutions of these two liquids at 25°C.

18. In a solution containing 40.0 g of $CHCl_3$ and 70.0 g of CCl_4, the mole fraction of $CHCl_3$ is _____ and the mole fraction of CCl_4 is _____.

19. In the solution of Question 18, the vapor pressure of $CHCl_3$ is _____ torr, the vapor pressure of CCl_4 is _____ torr, and the total vapor pressure is _____ torr.

20. In a solution of $CHCl_3$ and CCl_4 with a total vapor pressure of 150.0 torr, the mole fraction of $CHCl_3$ is _____ and the mole fraction of CCl_4 is _____.

---

21. We prepare a solution by dissolving 0.01 mole of $MgCl_2$ in 1.0 kg water and observe the freezing point. We also calculate the freezing point depression, using $\Delta T_b = K_b m$, with $m = 0.01$ molal. We estimate that the ratio of observed freezing point depression to calculated freezing point depression would be about _____.

22. The passage of solvent molecules selectively through a semipermeable membrane is called _____.

23. Liquids diffuse through a membrane from regions of high solvent concentration to regions of low solvent concentration due to a force known as _____.

24. When we distill a mixture of the very volatile substance A and the moderately volatile substance B, the first distillate (liquid obtained by condensing the first vapor to boil off) should be richer in substance _____.

25. Of the four colligative properties of a solution, the one that would give the most easily determined measurement from which to calculate the solute molecular weight would be _____.

26. In order to carry out "reverse osmosis," the pressure that we exert on the solution must _____.

27. The Tyndall effect is the name given to the observation that _____.

28. A fog is an example of a(n) _____.

29. The only combination of phases that cannot lead to formation of a colloid would be _____ dispersed in _____.

30. Colloids consisting of solids or liquids dispersed in a gas medium are called _____.

31. The process by which ions and molecules stick to the surface of a particle is called _____.

32. Colloid particles that readily attract water molecules to their surfaces are called _____.

33. To keep them from coming together to form a separate phase, the particles of a hydrophobic colloid are treated with an agent called a(an) _____.

Multiple Choice

_____ 1. For a gas that does not react chemically with water, the solubility of the gas in water
   (a) increases with increasing gas pressure and decreases with increasing temperature.
   (b) decreases with increasing gas pressure and decreases with increasing temperature.
   (c) increases with increasing gas pressure and increases with increasing temperature.
   (d) decreases with increasing gas pressure and increases with increasing temperature.
   (e) does not change appreciably with change in either gas pressure or temperature.

_____ 2. The formation of tiny bubbles when a beaker of water is mildly heated indicates that
   (a) the boiling point is being approached.
   (b) water is being decomposed into hydrogen and oxygen.
   (c) water is extensively hydrogen bonded.
   (d) air is less soluble in water at higher temperature.
   (e) the lattice energy of water is large.

_____ 3. Hydrogen chloride in the gaseous state is in the form of covalent molecules, yet in aqueous solution, it behaves as a strong electrolyte, conducting electricity. The explanation for this solution behavior is that
   (a) HCl forms ionic crystals in the solid state.
   (b) HCl reacts with water to form ions.
   (c) although it is a covalent substance, the molecules of HCl are polar and stay in solution as polar molecules, conducting electricity.
   (d) there is more than one form of gaseous HCl.
   (e) at lower temperatures, HCl becomes ionic.

_____ 4. Which of the following alcohols would be the _least_ miscible with water?
   (a) hexanol, $CH_3CH_2CH_2CH_2CH_2CH_2OH$ (b) pentanol, $CH_3CH_2CH_2CH_2CH_2OH$
   (c) propanol, $CH_3CH_2CH_2OH$ (d) ethanol, CH_3CH_2OH (e) methanol, CH_3OH

_____ 5. In four of the following groups, the three substances are either all strong electrolytes, all weak electrolytes, or all nonelectrolytes. Which group has one member that is not like the others in the group?
   (a) H_2SO_4, NaOH, HCl   (b) NH_3, CH_3COOH, HCN   (c) NaCl, $Ca(OH)_2$, HNO_3
   (d) O_2, $CO(NH_2)_2$, CH_3OH (e) $BaCl_2$, $NaNO_3$, HF

_____ 6. Hydrogen chloride is observed to be soluble in benzene, although not to the extent that it is in water. We would expect that
   (a) the HCl in benzene would be ionized, as in water.
   (b) the solution of HCl in benzene would be a good conductor of electricity.
   (c) the HCl molecules would be largely intact in benzene.
   (d) the benzene molecules would solvate the HCl molecules very well.
   (e) the limited solubility of HCl in benzene is due to the very strong interaction between benzene molecules.

_____ 7. Four of the following are colligative properties of solutions. Which one is not a colligative property?
    (a) molality   (b) vapor pressure lowering   (c) osmotic pressure
    (d) freezing point depression   (e) boiling point elevation

_____ 8. A 1-molar solution has
    (a) 1 mole of solute per 1 liter of solvent.
    (b) 1 mole of solute per 1 liter of solution.
    (c) 1 mole of solute per 1 kilogram of solvent.
    (d) 1 mole of solute per 1 kilogram of solution.
    (e) 1 mole of solute per 1 mole of solvent.

_____ 9. A 1-molal solution has
    (a) 1 mole of solute per 1 liter of solvent.
    (b) 1 mole of solute per 1 liter of solution.
    (c) 1 mole of solute per 1 kilogram of solvent.
    (d) 1 mole of solute per 1 kilogram of solution.
    (e) 1 mole of solute per 1 mole of solvent.

_____ 10. We dissolve 7.5 grams of urea (a nonelectrolyte, molecular weight = 60) in 500 grams of water. At what temperature would the solution boil? Additional useful data may be found in Table 14-2.
    (a) 100.46°C   (b) 100.13°C   (c) 0.13°C   (d) 99.54°C   (e) 99.87°C

_____ 11. An aqueous solution of compound X is described as "a 10% X solution." Its density is 1.10 g/mL. Which of the following statements about this solution is not true?
    (a) 1 liter of the solution would weigh 1.10 kg.
    (b) 1 liter of the solution would contain 100 grams of compound X.
    (c) 1 kg of the solution would contain 100 grams of compound X.
    (d) The molality of the solution would be greater than the molarity.
    (e) The boiling point of the solution would be higher than 100°C.

_____ 12. How many mL of a 0.320-molar solution of sucrose in benzene (density = 0.8787 g/mL) are needed to have 0.160 moles of sucrose?
    (a) 500 mL   (b) 498 mL   (c) 438 mL   (d) 567 mL   (e) 219 mL

_____ 13. Which of the following aqueous solutions would exhibit the highest boiling point?
    (a) 0.1 m urea   (b) 0.1 m NaCl   (c) 0.1 m $CaCl_2$ (d) 0.1 m HCl
    (e) All would show the same boiling point.

_____ 14. According to Raoult's Law, the lowering of vapor pressure of the solvent in dilute solutions of a nonvolatile solute is proportional to the
    (a) molarity of the solute.   (b) molarity of the solvent.
    (c) mole fraction of the solute.   (d) mole fraction of the solvent.
    (e) None of the preceding answers is correct.

Answers to Some Important Terms in This Chapter

1. hydration	4. Henry's Law	7. colligative properties
2. solvation	5. molality	8. Raoult's Law
3. saturated solution	6. mole fraction	9. distillation

10. fractional distillation
11. boiling point elevation
12. freezing point depression
13. van't Hoff factor
14. percent ionization
15. osmosis
16. osmotic pressure
17. colloid
18. hydrophilic colloids
19. hydrophobic colloids
20. emulsifier
21. coagulation

Answers to Preliminary Test

Do not just look up answers. *Be sure you understand the reasons for the answers.*

True-False

1. True. The two kinds of molecules are mixed in the solution, so they are less ordered.
2. False. Some are exothermic, including those examples listed in Section 14-2.
3. True. See the discussion in Section 14-4.
4. False. Read about supersaturated solutions, Section 14-5.
5. False. Hydration must occur.
6. True. Remember the definition of charge density and know how it relates to hydration of ionic substances, Section 14-2.
7. False. The solubility of KCl in water is due to the ability of water to surround each ion and thus to *prevent* the kind of interaction described.
8. True. Lattice energy is a measure of the attraction of the ions (or molecules) in the solid. The more negative it is, the more strongly these particles are held in the solid.
9. False. If they do react (e.g., HCl, SO_2, CO_2 to a limited extent) they are often much more soluble, but some nonreacting gases are slightly soluble in water (O_2, N_2).
10. False. Changing the pressure usually has no significant effect on the solubilities of either solids or liquids in liquids, but it does greatly alter the solubilities of gases in liquids.
11. False. High concentration promotes "clumping" of ions, rather than each ion existing separately.
12. False. Many substances that are quite soluble in water are either weak electrolytes (e.g., HCN, HNO_2, acetic acid, ammonia) or are nonelectrolytes (e.g., sugar, alcohol, urea).
13. False. Remember that nitric acid is one of the common strong acids (Chapters 6 and 10) and is thus essentially completely ionized.
14. True. But weak electrolytes only partially ionize in dilute aqueous solutions.
15. True
16. False. This terminology is a convenience, but sometimes it is neither easy nor necessary to identify one component as solute.
17. False. This would depend on the molecular weights of A and B. Suppose that A had a much lower molecular weight than B, so that the fewer number of grams were a greater number of moles—then the mole fraction A would be greater than that of B.
18. False. It may or may not. Remember that the terms "strong" and "weak" describe the efficiency of ionization of the solute, not its concentration.
19. True. See Table 14-2. For all the solvents listed (in fact, for all solvents) the value of K_f is greater than that for K_b. Thus, a solution of a given concentration would undergo a bigger

change in freezing point than in boiling point. Another factor is that freezing points are much less sensitive to external atmospheric pressure.

20. False. The actual degree of ionization depends on concentration—the more dilute the solution, the greater the degree of ionization of a particular solute.

21. False. Some are individual macromolecules, such as proteins or polymers.

Short Answer

1. homogeneous
2. solute; solvent
3. crystallization (or precipitation)
4. saturated
5. energy; disorder
6. exothermic
7. solute-solute attractions; solvent-solvent attractions; solute-solvent attractions
8. weak; weak; strong
9. solvation; hydration
10. (a) soluble   (k) soluble
    (b) soluble   (l) soluble
    (c) soluble   (m) soluble
    (d) insoluble   (n) soluble
    (e) soluble   (o) soluble
    (f) insoluble   (p) insoluble
    (g) insoluble   (q) soluble
    (h) soluble   (r) soluble
    (i) insoluble   (s) soluble
    (j) soluble   (t) soluble

    Know and be able to apply the solubility guidelines in Section 6-1.5. You may need to review chemical nomenclature of inorganic compounds, Sections 6-3 and 6-4.

11. 1.5. We want the fraction of the total number of moles that is acetone to be 0.20, so if x = number of moles of acetone, then the total number of moles $(6.0 + x)$ can be calculated as $x = 0.2(6.0 + x)$, or $x = 1.5$.

12. Raoult's; the vapor pressure of a solvent above a solution is equal to the mole fraction of the solvent in the solution times the vapor pressure of the pure solvent.

13. higher than

14. lower than. NaBr dissolves to give ions in solutions, resulting (ideally) in 2 moles of solute particles per mole, while sugar gives only 1 mole of solute particles (molecules) per mole. The *lowering* of the vapor pressure depends on the total concentration of dissolved solute particles.

15. lower than. Freezing point is depressed, again by an amount depending on the total concentration of dissolved solute particles.

16. higher than. Boiling point is elevated, by an amount depending on the total concentration of dissolved solute particles.

17. higher. Osmotic pressure, like all colligative properties, depends on the total concentration of solute particles.

18. 0.424; 0.576

The solution contains $\dfrac{40.0\,g}{119.4\,g/mol}$ = 0.335 mol CHCl$_3$ and $\dfrac{70.0\,g}{153.8\,g/mol}$ = 0.455 mol

CCl$_4$. Calculate the mole fraction of each component as in Example 14-3.

$$X_{CHCl_3} = \frac{\text{no. mol CHCl}_3}{\text{no. mol CHCl}_3 \ + \ \text{no. mol CCl}_4} = \frac{0.335\,mol}{(0.335 + 0.455)\,mol} = 0.424$$

$$X_{CCl_4} = \frac{\text{no. mol CCl}_4}{\text{no. mol CHCl}_3 \ + \ \text{no. mol CCl}_4} = \frac{0.455\,mol}{(0.335 + 0.455)\,mol} = 0.576$$

19. 72.9, 56.6, 129.5. The vapor pressure for each component is equal to its mole fraction in the solution times its vapor pressure when pure.

$P_{CHCl_3} = X_{CHCl_3}P^0_{CHCl_3} = (0.424)(172.0\ \text{torr}) = 72.9\ \text{torr}$

$P_{CCl_4} = X_{CCl_4}P^0_{CCl_4} = (0.576)(98.3\ \text{torr}) = 56.6\ \text{torr}$

The total vapor pressure is the sum of the vapor pressures of the two components.

$P_{total} = P_{CHCl_3} + P_{CCl_4} = 72.9\ \text{torr} + 56.6\ \text{torr} = 129.5\ \text{torr}$

20. 0.701, 0.299. This uses the same ideas as Question 19, except that the mole fractions X_{CHCl_3} and X_{CCl_4} are unknown. We can relate these two unknowns, because their sum must equal 1;

$X_{CHCl_3} + X_{CCl_4} = 1$

This lets us eliminate one variable and solve.

Let $x = X_{CHCl_3}$; then $X_{CCl_4} = 1 - x$

$P_{total} = P_{CHCl_3} + P_{CCl_4} = (x)(172.0\ \text{torr}) + (1-x)(98.3\ \text{torr}) = 150.0\ \text{torr}$

$$172.0\,x\ \text{torr} + 98.3\ \text{torr} - 98.3\,x\ \text{torr} = 150.0\ \text{torr}$$
$$73.7\,x\ \text{torr} = 51.7\ \text{torr}$$
$$x = 0.701 = X_{CHCl_3}$$
$$1 - x = 1 - 0.701 = 0.299 = X_{CCl_4}$$

21. 3. Each mole of MgCl$_2$ dissolves to give 1 mole of Mg^{2+} and 2 moles of Cl$^-$, or 3 moles total of solute particles per mole of MgCl$_2$ dissolved. This would probably be a pretty good prediction for such a dilute solution.

22. osmosis

23. osmotic pressure

24. A. The more volatile substance evaporates more, so the vapor is richer in the more volatile substance.

25. osmotic pressure. See the discussion in Section 14-15.

26. exceed the osmotic pressure of the solution.

27. colloidal particles scatter light.

28. colloid (or liquid aerosol, which is a type of colloid)

29. gases; gases. This is because all gas mixtures are homogeneous.

30. aerosols

31. adsorption. Do you know the difference between adsorption and absorption?

32. hydrophilic

33. emulsifier (or emulsifying agent)

Multiple Choice

1. (a). See Henry's Law (Section 14-7). Also, the dissolution processes in water are exothermic. (See Section 14-4.)
2. (d). The solubility of gases dissolved in water decreases with increased temperature.
3. (b). The ionization involves protonation of a water molecule by HCl to form a hydrated hydrogen ion and a chloride ion.
4. (a). The three alcohols with the fewest carbon atoms, propanol, ethanol, and methanol, are miscible with water in all proportions, but the alcohols with larger nonpolar hydrocarbon ends are much less soluble. Of the alcohols listed, hexanol is the most hydrocarbon-like, and hence least miscible with water.
5. (e). In (a), all substances are strong electrolytes; in (b), all are weak electrolytes; in (c), all are strong electrolytes; in (d), all are nonelectrolytes. In answer (e), $BaCl_2$ and $NaNO_3$ are strong electrolytes but HF is a weak electrolyte. You may need to review the discussion of strong and weak electrolytes in Section 6-1.
6. (c). Benzene, a nonpolar solvent, cannot solvate very well, so the ionization process does not readily occur, as it does in water.
7. (a). Molality is an expression of the concentration.
8. (b). $M = \dfrac{\text{moles of solute}}{\text{liters of solution}}$
9. (c). $m = \dfrac{\text{moles of solute}}{\text{kilograms of solvent}}$
10. (b). $\Delta T_b = K_b m$ Molality (m) is $\dfrac{7.5 \text{ g}}{0.500 \text{ kg solvent}} \times \dfrac{1 \text{ mol}}{60 \text{ g}} = 0.24$ molal.

    From Table 14-2, K_b is 0.512°C/molal, so $\Delta T_b = 0.13°C$. Remember that the boiling point is raised from its normal value by this amount. $T_b = T_b^o + \Delta T_b = 100° + 0.13° = 100.13°C$.
11. (b). One liter would weigh 1100 grams. Compound X would be 10% or 110 grams, of this total.
12. (a). Molarity directly relates moles of solute to volume of solution. Density is not needed to answer this question.
13. (c). Consider the number of dissolved particles in each solution. Urea is a nonelectrolyte.
14. (d). See Section 14-9.

15 Chemical Thermodynamics

Chapter Summary

At many stages in your study of chemistry, you have been concerned with questions such as these: Which is more stable—situation A or situation B? Will a specific chemical reaction or other process occur? Will the gas expand or contract if we change the pressure? Which is more stable at the stated conditions—solid, liquid or gas? Will the substance dissolve in water? Is the dissolution process exothermic or endothermic? What energy changes are associated with the process of melting or vaporization? Varied as they seem, all of these are versions of the same question, "Is a proposed process spontaneous at a specified set of conditions?" **Chemical thermodynamics** is a general approach to answering such questions, based on the study of energy changes that accompany a process. In this chapter, you will also learn how to understand and predict the amount of heat produced or absorbed when a change takes place—the topic called **thermochemistry**.

Section 15-1 reminds you that an **exothermic** change, either physical or chemical, *releases* heat energy to its surroundings, *increasing* the temperature of the surroundings. Conversely, an **endothermic** change *absorbs* heat energy from its surroundings, *decreasing* the temperature of the surroundings. Energy is useful for doing work. In fact, you have probably noticed that many exothermic changes are **spontaneous** (product-favored). However, you must understand that just because a process is exothermic, does not *necessarily* mean that it is spontaneous. Exothermicity favors, but does not guarantee, spontaneity. You will study more on this in Section 15-12. In addition, energy cannot be created out of nothing, an observation included in the **First Law of Thermodynamics**. So, how can work be done? First, recognize that "you can't get something for nothing." Incidentally, notice how the Law of Conservation of Energy has been expanded from the way it was stated in Chapter 1.

Section 15-2 introduces you to several thermodynamic terms, some of which have been used earlier in the text. You will encounter others throughout the chapter. In order to understand the logic of chemical thermodynamics, you must pay close attention to the terms used and interpret them as they are defined in chemistry. In the terminology of thermodynamics, a **process** may be any proposed change in the **system** under observation—a change of state, an expansion or contraction of the sample, the absorption or evolution of heat with or without a change of state, a chemical reaction, and so on.

As you saw above, one factor that helps to make a process spontaneous is a lowering of the energy. In thermodynamics, we express the change in the energy of a sample (the system) as ΔE. If the final energy is greater than the initial energy, ΔE is positive. The total amount of energy in the universe is constant, so in order for the system to increase its energy it must absorb energy from its surroundings.

One very useful comment about notation can be made here. We are using a symbol that will become quite common. Whenever we write Δ(any quantity) we mean the *change* in that quantity.

This is always calculated as: (the final value of the quantity) minus (the initial value of the quantity). This is the idea of a **state function**. This important characteristic of a state function will be used throughout the chapter. (Study the marginal note at the bottom of page 555 for an analogy to help you remember this concept.) All of the Δ values we will see (ΔE, ΔH, and so on) can be considered in this way.

Let us clarify another use of terminology that sometimes confuses students. When we say that some variable is "constant" for a process, we mean only that it has the same value at the end of the process as at the beginning, not that it necessarily had that value throughout the change. For instance, when we talk about a reaction "at constant temperature and pressure," we mean that after the reaction has occurred, we adjust the products to the same temperature and pressure as before the reaction. We do not necessarily mean that the temperature and the pressure were held invariable while the process was occurring.

Use the thermodynamic quantity (or state function) known as the **enthalpy change**, ΔH, to describe the heat change at constant pressure (Section 15-3). A lowering of the enthalpy content of a system (that is, a *negative* value of ΔH) is favorable to spontaneity of the process. The measurements of chemical thermodynamics involve heat changes, so we must have methods for measuring the flow of heat from one body to another. Heat energy can only be given up by one body if it is absorbed by another. We can use the method of **calorimetry** (Section 15-4) to measure the release or absorption of heat by a sample. The type of calorimeter described here, the coffee-cup calorimeter, can be used to measure the heat change at constant pressure (ΔH). Before you study the calculations associated with calorimetry, review specific heat calculations in Section 1-14.

We use a **thermochemical equation** (Section 15-5) to describe the relation between the amounts of reactant and products involved in a reaction and the amount of heat liberated or consumed at constant pressure. Be sure you know how to interpret such equations. Pay special attention to the conventions presented in the middle of page 559, and know how to apply them. If ΔH is evaluated at 1 atm pressure and *thermodynamic* standard temperature (298 K, which is 25°C), its value for the reaction is written as ΔH^0 and is referred to as the standard enthalpy change. (Notice that this temperature, the so-called standard temperature for thermodynamic studies, is different from that used in gas law problems.) Be sure that you understand what is meant by the **standard state** of a substance (Section 15-6). Section 15-7 introduces the idea of a **standard molar enthalpy of formation** of a substance. This is the enthalpy change for the reaction in which *one mole* of the substance is produced from the elements, all reactants and products being at standard-state conditions. Therefore, the standard molar enthalpy of formation of a substance, ΔH_f^0, is just a measure of how much higher or lower the substance is in enthalpy content than are its constituent elements.

According to **Hess's Law** of heat summation (Section 15-8), you can calculate what would be the ΔH (or ΔH^0) for any proposed reaction. To do this, either (1) think of reactions to measure that would total algebraically to the desired reaction, or (2) find tabulations of the ΔH_f^0 values for all substances involved in the reaction. This is a very important idea. You will use the value of ΔH^0 for a proposed reaction to determine whether it is spontaneous, and you can calculate this important quantity without having to carry out the reaction. Another use would be to tell you how much heat would have to be supplied, dissipated, or used for doing work when the reaction occurs.

In earlier chapters **bond energy** was introduced (Section 7-4) and used as a measure of bond strength and stability in covalent substances (Chapter 9). Now you examine how these bond energies are measured experimentally, Section 15-9. As you study this section, keep in mind that the bond energy calculation refers only to the step in which the bonds in gaseous molecules break to form isolated gaseous atoms. Usually we cannot measure this directly. However, you can calculate it (approximately) from first measuring the enthalpy change associated with the formation of bonds from the elements in their standard states, and then correcting, via Hess's Law, for the conversion of elements from their stable forms (solids, diatomic molecules, etc.) to isolated atoms. Many students find diagrams such as Figure 15-5 helpful in understanding bond energy calculations. Once derived, tabulations of bond energies can help you to understand why reactions that involve the breaking of very stable bonds are hard to carry out. They can also be used to estimate ΔH^0_{rxn} for reactions in which all reactants and products are gases.

Section 15-10 discusses the change in internal energy, ΔE, for a system and its relationship to heat and work. We can understand this important relationship as follows. When we put heat into a system it can (1) increase the internal energy of the system, (2) cause the system to do some work (for example, by expanding against an external force), or result in some combination of (1) and (2). Because energy can be neither created nor destroyed, the total increase in energy of the system (ΔE) must be the sum of the amount of heat absorbed by the system (q) and the work done on the system (w). We can express this as $\Delta E = q + w$, a mathematical expression of the First Law of Thermodynamics (Section 15-1). Remember the conventions for when the values of q, w, and ΔE are positive. The most direct way to measure ΔE is in a bomb calorimeter; in this experiment, the volume is held constant ($\Delta V = 0$) so no work is done and the value of ΔE is given directly by the observed heat flow. The relationship between ΔH and ΔE for a reaction is discussed in Section 15-11. You can think of this as the comparison between the amount of heat absorbed at constant pressure (ΔH) and at constant volume (ΔE); the difference is the work done on the system. For a process involving expansion or contraction without reaction, the relationship

$$\Delta H = \Delta E + P\Delta V$$

should be used; when the volume change is due to a chemical reaction, the equation

$$\Delta H = \Delta E + (\Delta n)RT$$

is more convenient. As you work with this relationship, remember that Δn is equal to the change in the number of moles of *gaseous substances* in the balanced chemical equation.

As you saw briefly in Chapter 14, the spontaneity of a process is favored by two factors: (1) a lowering of the energy (exothermicity) and (2) an increase in the disorder or randomness. It may help to recall that rocks roll *downhill* (lower potential energy), not uphill, and that it is easier to *mix* things (scrambling an egg) than to "unmix" them. So far in this chapter you have studied the first factor that favors spontaneity, the exothermicity of the process. But many endothermic reactions are also spontaneous, at least under some conditions. A second favorable factor, an increase in the disorder of the system, is introduced in a general way in Section 15-12. The next three sections deal in more detail with the second favorable factor, an increase in disorder. After an examination (Section 15-13) of the dispersal of energy (being spread over many particles) and matter (particles of a substance expanding or particles of different substances being mixed), we introduce a thermodynamic measure of disorder called **entropy** (Section 15-14). An increase in entropy of the system, a positive ΔS, favors a process being spontaneous. The **Third Law of Thermodynamics** allows us to determine absolute entropies for substances and, by a relationship analogous to Hess's Law, to calculate entropy changes for any proposed process. The **Second Law of Thermodynamics** seems to say that "you can't even break even." Section 15-15 presents general remarks about the more disordered situation being more favorable and investigates the connection between the entropy of the system (ΔS_{sys}) and the entropy of the universe (ΔS_{univ}). The discussion of melting/freezing that accompanies Table 15-6 will help you to clarify these ideas.

In the final sections of the chapter, we are ready to combine the two factors that affect spontaneity into a single measure, the change in Gibbs free energy, ΔG. As seen in Section 15-16, the equation,

$$\Delta G = \Delta H - T\Delta S \qquad \text{(constant } T, P)$$

allows us to formulate simple rules for predicting whether or not a reaction will be spontaneous. A negative value of ΔG means that the reaction is **product-favored** (spontaneous) and *can* occur, whereas a positive value of ΔG means that the reaction is **reactant-favored** (not spontaneous) under the conditions specified. (Realize, however, that just because a reaction is spontaneous does not guarantee that it will occur at a measurable rate.)

Just as for other thermodynamic quantities, we usually carry out the calculation of free energy change with respect to special conditions. One important set of conditions is that all reactants and products be at 1 atm pressure (gases) or concentrations of 1 M (solutions). We designate the free energy change at these special conditions as

$$\Delta G^0 = \Delta H^0 - T\Delta S^0$$

The presence of T in this equation tells us that *some* reactions that are impossible at some temperatures can be made spontaneous at other temperatures (Section 15-17). If the ΔH^0 aspect is favorable (–) and the $T\Delta S^0$ aspect is unfavorable (–), lower temperatures make the unfavorable term less significant and the reaction can become product favored (spontaneous). Conversely, if the ΔH^0 aspect is unfavorable (+) and the $T\Delta S^0$ aspect is favorable (+), higher temperatures make the favorable term more significant and the reaction can become product-favored. We can calculate from this relationship the temperature at which the reaction would become spontaneous, if it could be spontaneous at any temperature. It is very important that you understand the discussion accompanying Table 15-7 in the text. Be sure that you can calculate ΔG^0 either (1) from known or experimentally determined values of ΔH_f^0 and S^0 or (2) from known ΔG_f^0 values. Be sure you remember that the latter method can be used *only* for the temperature for which the ΔG_f^0 values are tabulated—25°C for the values in Appendix K. In the former method, be careful of the units (J vs. kJ). We will see many uses of ΔG^0 in subsequent chapters, including its relation to equilibrium constants (Chapter 17) and electrochemistry (Chapter 21).

There is much detail and considerable subtlety in this chapter. You must proceed carefully through the chapter, being sure that you master the terminology and the calculations of each section before proceeding to the next. Do not be content with just seeing how to do the problems; because they involve only addition, subtraction, and multiplication of numbers, they are easy to calculate. The usual difficulty with chemical thermodynamics at first encounter is in seeing what these calculations mean. Remember that we are dealing with energy changes, calculating the predicted energy changes that will accompany reactions, and trying to use these energy changes to predict whether or not a reaction can go or to understand why it does or does not go at certain conditions. And do not just memorize—*think*!

Arnold Sommerfeld, one of the greatest physicists of the 20th century and supervising professor to Heisenberg and Pauli, among many others, is reported to have said, "Thermodynamics is a funny subject. The first time you go through it, you don't understand it at all. The second time you go through it, you think you understand it, except for one or two small points. The third time you go through it, you know you don't understand it, but by that time you are so used to it, it doesn't bother you anymore."

Study Goals

It may help you to write out a brief summary of the ideas of each study goal, summarizing the applicable parts of your class notes and text readings. Be sure to answer many of the suggested Exercises related to the Study Goals.

Review Section 1-14

1. Be familiar with the calculations and units used to describe heat transfer. *Review Exercises 61 and 66 of Chapter 1.*

Section 15-1

2. Understand the First Law of Thermodynamics and its relation to the Law of Conservation of Energy. *Work Exercises 7 and 8.*

Section 15-2, Spontaneity of Physical and Chemical Changes, and Section 15-12

3. Understand the basic terminology of thermodynamics as presented and used in this chapter (be able to give examples of each): (a) system, (b) surroundings, (c) state of a system, (d) state function, (e), standard conditions, (f) standard state, (g) endothermic process, (h) exothermic process, (i) spontaneous process, and (j) nonspontaneous process. *Work Exercises 1 through 11 and 24.*

Sections 15-3 and 15-6

4. Know what is meant by the enthalpy change of a reaction. Know what is meant by the standard enthalpy change of a reaction. *Work Exercises 12 and 25.*

Section 15-4

5. Be familiar with the measurements and calculations of calorimetry, either at constant pressure or at constant volume. Be able to carry out the calculations, using experimental calorimeter data, to determine enthalpy or energy changes for reactions and processes. *Work Exercises 59 through 71.*

Section 15-5

6. Know how to write and interpret thermochemical equations. *Work Exercises 14 through 23, 28, and 29.*

Section 15-7

7. Know what is meant by the standard molar enthalpy of formation of a substance. *Work Exercises 30 and 31.*

Section 15-8

8. Understand the meaning of Hess's law, and why it works. Know how to carry out the calculations of Hess's Law to determine enthalpy changes for specified reactions. *Work Exercises 11 and 32 through 45.*

Section 15-9

9. Given the enthalpy changes for appropriate reactions or physical changes, be able to calculate average bond energies. Given appropriate bond energies, be able to calculate enthalpy changes for reactions. *Work Exercises 46 through 58.*

Section 15-10

10. Understand the First Law of Thermodynamics in mathematical form. Carry out calculations relating ΔE, q, and w. *Work Exercises 72 through 81.*

Section 15-11

11. Explain the relationship between energy change and enthalpy change. Explain why the use of ΔH is often preferred.

Sections 15-12 and 15-14

12. Explain how ΔH and $T\Delta S$ are related to spontaneity of a reaction. *Work Exercises 105 and 106.*

Section 15-14

13. Understand the interpretation of entropy of a system in terms of microscopic disorder within that system. Be able to predict the sign of ΔS for various kinds of chemical and physical changes. *Work Exercises 86 through 102.*

14. Use tabulated values of absolute entropies of reactants and products to calculate the entropy change for a reaction. *Work Exercises 103 and 104.*

Section 15-15

15. State the Second Law of Thermodynamics and summarize its implications with respect to reaction spontaneity. *Work Exercises 82, 83, and 85.*

Section 15-16

16. Explain what is meant by the Gibbs free energy change for a process. Be able to relate it, both mathematically and descriptively, to enthalpy and entropy changes. *Work Exercises 105 and 106.*

17. Use the standard Gibbs free energy change for a reaction as an indicator of spontaneity. Be able to calculate the standard Gibbs free energy change for a reaction from either:
    (a) standard Gibbs free energies of formation of reactants and products, or
    (b) standard molar enthalpies of formation and standard entropies of reactants and products.
    Work Exercises 107 through 113.

Section 15-17

18. Determine the temperature range of spontaneity of a reaction from tabulated thermodynamic data. Make general statements regarding the spontaneity of a reaction at relatively low and high temperatures based on the signs of ΔH^0 and ΔS^0. *Work Exercises 114 through 125.*

General

19. Be able to recognize and solve a variety of types of problems concerning the material of this chapter. *Work **Mixed Exercises** 127 through 132.*

20. Answer conceptual questions based on this chapter. *Work **Conceptual Exercises** 133 through 139.*

21. Apply concepts and skills from earlier chapters to the ideas of this chapter. *Work* **Building Your Knowledge** *exercises 140 through 146.*

22. Exercises at the end of chapter direct you to sources outside the textbook for information to use in solving them. Work **Beyond the Textbook** *exercises 147 through 152.*

> **NOTE:** After you believe you have mastered the material of this chapter, you should work many of the exercises in the text. This will give you valuable experience in recognizing problems of various kinds as well as applying the concepts and skills you have learned in the chapter.

Some Important Terms in This Chapter

Some important new terms from Chapter 15 are listed at the top of the next page. Fill the blanks in the following paragraphs with terms from the list. Each term is only used once. Check the Key Terms list and the chapter reading. Study other new terms, and review terms from preceding chapters if necessary.

absolute entropy	free energy change	standard molar enthalpy of formation
bond energy	kinetic energy	state function
enthalpy	potential energy	surroundings
enthalpy change	spontaneity	system
entropy	standard enthalpy change	thermochemistry
free energy	standard free energy change	thermodynamics

Because the energy changes that accompany physical and chemical processes usually involve *heat*, the study of these changes is called [1]_____. One of the major areas of thermodynamics is [2]_____, which is concerned with how we observe, measure, and predict energy changes for both physical changes and chemical reactions. Energy can be divided into two types: [3]_____, the energy matter has due to its motion, and [4]_____, the stored energy that a system possesses by virtue of its position or composition.

In thermodynamics, the substances involved in the chemical and physical changes that we are studying are called the [5]_____. Everything in the system's environment constitutes its [6]_____. Each of the properties of a system—such as *P*, *V*, and *T*—is called a [7]_____ because its value depends *only* on the state of the system and not on the way in which the system came to be in that state. One of these state functions is the "heat content" of the substances, called the [8]_____, and represented by the letter *H*. The quantity of heat transferred into or out of a system as it undergoes a chemical or physical change at constant pressure is defined as the [9]_____, ΔH, of the process. The ΔH for a reaction when the specified number of moles of reactants, all at standard states, are converted *completely* to the specified number of moles of products, all at standard states, is called the [10]_____, ΔH^0_{rxn}. The ΔH for the reaction in which *one mole* of a substance in a specified state is formed from its elements in their standard states, is called the [11]_____ _____, ΔH_f^0, of the substance. The standard molar enthalpy of formation can be calculated from the [12]_____ (B.E.), the amount of energy necessary to break *one mole* of bonds in a gaseous covalent substance to form products in the gaseous state at constant temperature and pressure.

The tendency to favor the products in a physical or chemical change, or [13]_____, is the result of a situation in which the products are thermodynamically *more stable* than the reactants under the given conditions. Spontaneity is favored when the change causes a dispersal of energy and matter. [14]_____, S, is a measure of the dispersal of energy and matter in a system. The entropy of a substance at any condition is its [15]_____, also called the standard molar entropy. A change will be spontaneous if it releases useful energy obtainable in the form of work at constant temperature and pressure, a quantity known as Gibbs [16]_____, ΔG. It is the difference in the Gibbs [17]_____, G, of the reactants and products, a quantity that is defined by the equation:

$G = H - TS$. The values of ΔG_f^0 may be used to calculate the [18]_____ _____, ΔG_{rxn}^0, of a reaction at 1 atm and 298 K.

Quotefalls

Fit the letters in each vertical column into the squares directly below them to form words across. The letters do not necessarily go into the squares in the same order in which they are listed. Each letter given is used once. A black square indicates a space. Words starting at the end of a line may continue in the next line. Punctuation in the statement is included in the boxes. When all the letters have been placed in their correct squares, you will be able to read the complete quotation across the diagram from left to right, line by line. Hint: It may be helpful at times to fill in small words, like "the," to use up some letters to help in determining other longer words. The solutions are on page 223.

Puzzle #1:

C	I	R	B	T	N	E	D	W	A	O	O	N	I	V	E	R	F	E	M	A	S	A	E	O	N	A	N	D	N	T	E	E
F	G	Y	S	I	N	L	T	H	E	M	F	U	T	T	E	R	S	O	D	Y	N	T	C	R	C	S	T	A	T	E	N	
R	O	M		I			A				U			N	H		O	M			I	T		M	I		S				H	

Puzzle #2:

P	Y	C	O	F	D	S	H	A	N	U	O	F	V	O	H	S	R	M	I	A	N	G	E	A	I	E	N	T	R	I
N	E	A	O	N		T	P	O	W	T	A	N	E	T	U	S	E	C	H	D	C	R	E	M	S	E	S			O
S			N	Y			L	E			N	I		E	R	E			O	N	Y	N	A			C	S			
						■				■			■															:	■	
	■				■												■							,	■					
		■			■				■									■										.	■	■

Puzzle #3:

T	Y	N	R	E		I	A	P	Z	O	E	O	T	A	T	R	F	B	C	T	N	C	T	Y	S	S	E	R	L	H	N	E	E	K	T	B	S
P	A	I	C	D		L	S	W	U	R	R			H	E	R	A	E	S	O	L	A	M	I	C	Z	A	L	T	I	E			N	U	R	O
T	H		O	F		A				E	F			P	E		M	O	D	Y		U	R	E		T			O					S			
					■				■			■															:	■				■					
		■			■		■					,	■								■												■				
					■			■					■			■									■					■	(	0	■		)	.	■

Puzzle #4:

T	N	E	P	A	O	P	E	B	W	H	A	F	G	H	S	E	I	T	S	U	M	M	A	R	C	O	E	O	S	O	N	S	
S	T	E	H	S	S	R	Y	A	C	H	A	N	Y	E	R	F	O	I	E	A	C	C	F	T	S	T	N	P	N	T	I	E	
H	H	S	S		L	M	L		Y		O	N	H	E	E	A	R	R		O		R	E	A	S	T	I	Y			H	E	
E	E	T			A				W		E	T					T			S		O	U		I		B						
				'		■				■			■					■										:	■				■
								■							■				■		■									■			■
			■					■								■			■							■			■				■
				■			■			■				■							■			■						.	■	■	■

Preliminary Test

True-False

Mark each statement as true (T) or false (F).

____ 1. For a body to have energy it must do some work.
____ 2. Potential energy can be changed into kinetic energy.
____ 3. Kinetic energy can be changed into potential energy.
____ 4. Chemical energy is a form of kinetic energy.
____ 5. The heat capacity of a body is the same as its specific heat.
____ 6. When water at 100°C is mixed with water at 20°C, the final water temperature is the average, 60°C.
____ 7. When we put a 100-g piece of metal at 100°C into 100 g of water at 20°C, the final temperature is 60°C.
____ 8. The reverse of an exothermic process must be endothermic.
____ 9. The term ΔH means $H_{\text{final}} - H_{\text{initial}}$.

_____ 10. The term ΔT means $T_{final} - T_{initial}$.

_____ 11. A process in which the system absorbs heat from its surroundings is called endothermic.

_____ 12. For a process in which the system absorbs heat from its surroundings, the system has a negative ΔH value.

_____ 13. The process described by the thermochemical equation

$$H_2(g) + I_2(g) \rightarrow 2HI(g) + heat$$

is described as exothermic.

_____ 14. The process described by the thermochemical equation

$$H_2(g) + I_2(g) \rightarrow 2HI(g) + heat$$

has a negative value of ΔH.

_____ 15. The thermochemical equation: $C(graphite) + \frac{1}{2}O_2(g) \rightarrow CO(g)$ $\Delta H = -110.5$ kJ/mol

refers to the reaction of one atom of C with one-half molecule of O_2 to form one molecule of CO.

_____ 16. If the temperature of a reaction mixture in a coffee-cup calorimeter increases as the reaction proceeds, the reaction is endothermic, because the mixture is absorbing heat from the surrounding air.

_____ 17. Hess's Law is true because enthalpy is a state function.

_____ 18. The standard molar enthalpy of formation of a substance is the amount of heat one mole of the substance contains at standard conditions.

_____ 19. The enthalpy change at standard conditions for the reaction

$$H_2(g) + I_2(g) \rightarrow 2HI(g)$$

is referred to as the standard molar enthalpy of formation of HI(g).

_____ 20. The enthalpy change at standard conditions for the reaction

$$\frac{1}{2}H_2(g) + \frac{1}{2}I_2(g) \rightarrow HI(g)$$

is referred to as the standard molar enthalpy of formation of HI(g).

_____ 21. The enthalpy change at standard conditions for the reaction

$$CO(g) + \frac{1}{2}O_2(g) \rightarrow CO_2(g)$$

is referred to as the standard molar enthalpy of formation of $CO_2(g)$.

_____ 22. According to the First Law of Thermodynamics, the only way we can increase the internal energy of a system is to heat it.

_____ 23. The advantage of a bomb calorimeter is that the volume is held constant, so that the heat absorbed by the system is equal to the change in internal energy of the system.

_____ 24. A reaction that is exothermic must be spontaneous.

_____ 25. A reaction that is spontaneous must be exothermic.

_____ 26. The reverse of an ordering reaction is a disordering reaction.

_____ 27. The reverse of a spontaneous reaction is a nonspontaneous reaction.

_____ 28. When we say that ΔE is a state function, we mean that the energy change of the system does not depend on its initial or final state.

_____ 29. The enthalpy change, ΔH, of a system is a state function.

_____ 30. The heat, q, absorbed by a system is a state function.

_____ 31. The work, w, done on a system is a state function.

_____ 32. The entropy change, ΔS, of a system is a state function.

_____ 33. The Gibbs free energy change, ΔG, of a system is a state function.

_____ 34. The bond energy of any binary compound (consisting of only two elements) is equal to its ΔH_f^0 value divided by the number of bonds formed per molecule.

_____ 35. The enthalpy change associated with the reaction in which 2 moles of H atoms are formed from 1 mole of H_2 is called the bond energy of the H–H bond (all substances at standard conditions).

_____ 36. The enthalpy change associated with the reaction in which ½ mole of H_2 and ½ mole of Cl_2 are formed from 1 mole of HCl is called the bond energy of the H–Cl bond (all substances at standard conditions).

_____ 37. The enthalpy change associated with the reaction in which 1 mole of H atoms and 1 mole of Cl atoms are formed from 1 mole of HCl is called the bond energy of the H–Cl bond (all substances at standard conditions).

_____ 38. The enthalpy change of a gas phase reaction is the energy required to break all the bonds in reactant molecules minus the energy required to break all the bonds in product molecules.

_____ 39. For the vaporization of 1 mole of water at 1 atm at its normal boiling point, $\Delta E = \Delta H$.

The next five questions, 40 through 44, refer to the following standard reaction:

$$SiH_4(g) + 2O_2(g) \rightarrow SiO_2(s) + 2H_2O(\ell) \qquad \Delta H_{rxn}^0 = -1516 \text{ kJ/mol}$$

_____ 40. The reaction is exothermic.

_____ 41. When this reaction is carried out at constant pressure with 1 mole of SiH_4 and sufficient oxygen, the reaction mixture gives off 1,516,000 joules of heat to its surroundings.

_____ 42. When this reaction is carried out at constant pressure with 2 moles of SiH_4 and sufficient oxygen, the reaction mixture gives off 1,516,000 joules of heat to its surroundings.

_____ 43. When this reaction is carried out at constant pressure, the reaction mixture does work against its surroundings.

_____ 44. When this reaction is carried out at constant volume at room temperature with 1 mole of SiH_4 and sufficient oxygen, the reaction mixture gives off 1,516,000 joules of heat to its surroundings.

_____ 45. The entropy of a substance is greater as a liquid than as a solid.

_____ 46. The entropy of a substance is greater as a gas than as a liquid.

_____ 47. All reactions become more spontaneous at higher temperatures.

Short Answer

1. The sample or portion of the universe that we wish to study or describe in thermodynamics is called the _____.
2. Three examples of state functions of a gaseous sample are _____, _____, and _____.
3. In calculations with the First Law of Thermodynamics, we use the convention that heat absorbed by the system from its surroundings is given a _____ sign.
4. In calculations with the First Law of Thermodynamics, we use the convention that work done on the system by its surroundings is given a _____ sign.
5. If a gas sample absorbs 1,000 joules of heat from its surroundings and does 400 joules of work by expansion against its surroundings, its internal energy must _____ by _____ joules.
6. A reaction in which the total chemical potential energy of the products is greater than the total chemical potential energy of the reactants is said to be _____.
7. Measurements in calorimetry are based on equating heat lost by the sample to heat gained by the surrounding. This is an application of the Law of _____ _____.
8. A pure substance is said to be in its standard state at conditions of _____ and _____.
9. For a change that takes place at constant volume, the heat gained or lost by the system is a measure of _____ for the system.
10. For a change that takes place at constant pressure, the heat gained or lost by the system is a measure of _____ for the system.
11. A statement of Hess's Law in words is that the enthalpy change for a reaction is the same whether it occurs _____ or _____.
12. The enthalpy change associated with the formation of 1 mole of any substance at standard conditions from its elements, also at standard conditions, is called the _____ _____ of the substance.
13. The enthalpy change associated with the reaction in which 1 mole of HBr(g) at standard conditions is formed from the appropriate numbers of moles of hydrogen gas and bromine liquid, all at standard conditions, is called the _____ _____ of HBr.
14. The enthalpy change associated with the reaction in which 1 mole of HBr(g) at standard conditions breaks apart to form the appropriate numbers of moles of hydrogen atoms and bromine atoms is called the _____ of HBr.

The following equations, referred to by letter, are to be used in the next five questions, 15 through 19. Consider all reactants and products to be at 1 atm and 298 K.

A. $C(graphite) + O_2(g) \rightarrow CO_2(g)$
B. $C(g) + O_2(g) \rightarrow CO_2(g)$
C. $C(graphite) + 2O(g) \rightarrow CO_2(g)$
D. $C(g) + 2O(g) \rightarrow CO_2(g)$

    E. $CO(g) + \frac{1}{2}O_2(g) \rightarrow CO_2(g)$

    F. $C(graphite) + \frac{1}{2}O_2(g) \rightarrow CO(g)$

    G. $C(g) + \frac{1}{2}O_2(g) \rightarrow CO(g)$

    H. $C(graphite) + O(g) \rightarrow CO(g)$

    I. $C(g) + O(g) \rightarrow CO(g)$

15. The enthalpy change for Reaction _____ is referred to as the standard molar enthalpy of formation of carbon monoxide.

16. The enthalpy change for the reverse of Reaction _____ is referred to as the bond energy of carbon monoxide.

17. The enthalpy change for Reaction _____ is referred to as the standard molar enthalpy of formation of carbon dioxide.

18. If we subtract the ΔH^0 value for Reaction C from that for Reaction A, the result would be equal to the _____ in O_2.

19. The standard enthalpy change for Reaction _____ could be called the standard molar enthalpy of combustion of carbon monoxide.

---

20. When we reverse a chemical equation, we must multiply its ΔH by _____.

21. The symbol for enthalpy is _____.

22. The symbol for enthalpy change for a process at any conditions is _____.

23. The symbol for enthalpy change for a process in which all substances are at standard conditions is _____.

24. The symbol for standard molar enthalpy of formation of a substance is _____.

25. The energy holding the atoms, molecules, or ions in a crystal in their regular arrangement is called the _____.

26. A reaction in which the total heat content of the products is lower than the total heat content of the reactants is said to be _____.

27. A reaction in which the system absorbs heat from its surroundings is called _____.

28. A reaction in which the system absorbs heat from its surroundings has a value of ΔH that is _____.

29. The standard enthalpy change for the reaction

$$\frac{1}{2}H_2(g) + \frac{1}{2}F_2(g) \rightarrow HF(g)$$

is called the _____ of HF(g).

30. The reaction for which the standard enthalpy change represents the standard molar enthalpy of formation of $NH_3(g)$ is _____.

For each of the changes described in Question 31 through 40, indicate whether the entropy of the system would increase (I), decrease (D), or remain unchanged (U). You should also know which ones would be described as having $\Delta S = +$, 0, or $-$.

31. A liquid is frozen to make a solid. _____
32. A new deck of cards is carried home from the store, its wrapper intact. _____
33. A new deck of cards is shuffled. _____
34. A shuffled deck of cards is rearranged according to suits. _____
35. Salt is dissolved in water. _____
36. Two moles of hydrogen atoms combine to form one mole of hydrogen molecules: $2H(g) \rightarrow H_2(g)$. _____
37. Calcium carbonate decomposes on heating: $CaCO_3(s) \rightarrow CaO(s) + CO_2(g)$. _____
38. Nitrogen gas and hydrogen gas combine to form ammonia gas: $N_2(g) + 3H_2(g) \rightarrow 2NH_3(g)$. _____
39. A sample of ammonia gas, $NH_3(g)$, is carried from New York to London. _____
40. Dry ice, $CO_2(s)$, sublimes to form $CO_2(g)$. _____

---

41. Complete the relationship: $\Delta G = \Delta H$ _____.
42. A reaction in which ΔS is positive tends to become more spontaneous as the temperature _____.

Multiple Choice

_____ 1. If a system absorbs heat and also does work on its surroundings, its energy
  (a) must increase. (b) must decrease. (c) must not change
  (d) may either increase or decrease, depending on the relative amounts of heat absorbed and work done.

_____ 2. A calorimeter is
  (a) a dieting aid.
  (b) a device used to measure transfer of heat energy.
  (c) equal to the molar enthalpy of reaction.
  (d) an indicator of spontaneity.
  (e) only useful in measuring exothermic reactions.

_____ 3. To change 1 gram of ice to 1 gram of liquid water at 0°C, 334 joules of heat are needed. For this process,
  (a) $q = 0$. (b) $\Delta E = 0$. (c) $\Delta H = 334$ joules.
  (d) $\Delta H = -334$ joules. (e) $w = 334$ joules.

_____ 4. All spontaneous reactions
  (a) give off heat. (b) are fast. (c) have $\Delta G < 0$.
  (d) have $\Delta H < 0$. (e) have $\Delta S > 0$.

_____ 5. Given the following standard heats of reaction:

$$S(s) + H_2(g) \rightarrow H_2S(g) \qquad \Delta H^0 = -20.6 \text{ kJ/mol}$$
$$H_2(g) \rightarrow 2H(g) \qquad \Delta H^0 = 436.0 \text{ kJ/mol}$$
$$S(s) \rightarrow S(g) \qquad \Delta H^0 = 278.8 \text{ kJ/mol}$$

The bond energy for the H−S single bond is
  (a) 736 kJ/mol. (b) 507 kJ/mol. (c) 368 kJ/mol
  (d) 239 kJ/mol. (e) 173 kJ/mol.

_____ 6. For a reaction in which gases are neither produced nor consumed, ΔH is _____ ΔE.

    (a) the same as   (b) less than   (c) greater than   (d) unrelated to

_____ 7. For a reaction in which more moles of gas are produced than are consumed, ΔH is _____ ΔE.

    (a) the same as   (b) less than   (c) greater than   (d) unrelated to

_____ 8. A reaction occurring under certain conditions has an enthalpy change of $\Delta H = -50$ kJ/mol. This means that 50 kJ of heat will be liberated when one mole of reaction occurs

    (a) at constant temperature and volume.

    (b) at constant temperature and pressure.

    (c) at constant volume and pressure.

    (d) at constant temperature, volume, and pressure.

    (e) under any conditions.

_____ 9. Is an exothermic process spontaneous?

    (a) Yes, always   (b) No, never   (c) Sometimes

_____ 10. Given the following reactions and their associated enthalpy changes:

$$CH_4(g) \rightarrow C(g) + 4H(g) \qquad \Delta H^0 = \ 1663 \text{ kJ/mol}$$
$$O_2(g) \rightarrow 2O(g) \qquad \Delta H^0 = \ 498 \text{ kJ/mol}$$
$$H_2O(g) \rightarrow 2H(g) + O(g) \qquad \Delta H^0 = \ 927 \text{ kJ/mol}$$
$$CO_2(g) \rightarrow C(g) + 2O(g) \qquad \Delta H^0 = \ 1608 \text{ kJ/mol}$$

Use these values to calculate ΔH^0 for the combustion (burning) of 1 mole of CH_4 to form $H_2O(g)$ and $CO_2(g)$ as described by the reaction

$$CH_4(g) + 2O_2(g) \rightarrow 2H_2O(g) + CO_2(g)$$

    (a) −372 kJ/mol   (b) 4696 kJ/mol   (c) −803 kJ/mol   (d) 6121 kJ/mol   (e)   124 kJ/mol

_____ 11. The value of ΔH^0 for the following reaction, as written, is −2220 kJ/mol. How much heat is produced when 11.0 g of propane gas, C_3H_8, is burned in a constant pressure system?

$$C_3H_8(g) + 5O_2(g) \rightarrow 3CO_2(g) + 4H_2O(\ell)$$

    (a) 555 kJ   (b) 2220 kJ   (c) 50.5 kJ   (d) 24420 kJ   (e) 25.96 kJ

_____ 12. When a liquid boils, which of the following must be true?

    (a) $\Delta S < 0$ (b) $\Delta S > 0$ (c) $\Delta T < 0$ (d) $\Delta T > 0$

_____ 13. Which one of the following statements is correct?

    (a) The standard molar enthalpy of formation of an element is usually positive.

    (b) ΔH can never equal ΔE.

    (c) q and w are state functions.

    (d) $q + w$ is a state function.

    (e) The standard entropy S^0 of an element is zero.

_____ 14. For some reaction carried out at constant atmospheric pressure and at constant temperature of 25°C, it is found that $\Delta H^0 = -38.468$ kJ/mol and $\Delta S^0 = +51.4$ J/mol·K. What is the value of ΔG^0 for this reaction at these conditions?

    (a) −53.8 kJ/mol   (b) −84.5 kJ/mol   (c) 53.8 kJ/mol   (d) 84.5 kJ/mol

_____ 15. Is the reaction referred to in Question 14 spontaneous at the conditions given?
(a) Yes   (b) No   (c) Insufficient information is given to answer.

_____ 16. For the process $H_2O(\ell) \rightarrow H_2O(g)$ at 1 atm and 100°C, the change in enthalpy is
(a) positive.   (b) negative.   (c) zero.

_____ 17. For the process referred to in Question 16, the change in entropy is
(a) positive.   (b) negative.   (c) zero.

_____ 18. For the process referred to in Question 16, the change in Gibbs free energy is
(a) positive.   (b) negative.   (c) zero.

_____ 19. A reaction for which $\Delta H = +$ and $\Delta S = -$
(a) is spontaneous at all temperatures.
(b) is not spontaneous at any temperature.
(c) could become spontaneous at high temperature.
(d) could become spontaneous at low temperature.

_____ 20. A reaction for which $\Delta H = -$ and $\Delta S = -$
(a) is spontaneous at all temperatures.
(b) is not spontaneous at any temperature.
(c) could become spontaneous at high temperature.
(d) could become spontaneous at low temperature.

_____ 21. A reaction for which $\Delta H = +$ and $\Delta S = +$
(a) is spontaneous at all temperatures.
(b) is not spontaneous at any temperature.
(c) could become spontaneous at high temperature.
(d) could become spontaneous at low temperature.

> **NOTE:** This Preliminary Test has not provided you with enough practice at thermodynamic calculations. Be sure that you work *many* problems from the end of Chapter 15 in the text.

Answers to Some Important Terms in This Chapter

1. thermodynamics
2. thermochemistry
3. kinetic energy
4. potential energy
5. system
6. surroundings
7. state function
8. enthalpy
9. enthalpy change
10. standard enthalpy change
11. standard molar enthalpy of formation
12. bond energy
13. spontaneity
14. entropy
15. absolute entropy
16. free energy change
17. free energy
18. standard free energy change

Answers to Quotefalls

Puzzle #1:
First Law of Thermodynamics: The combined amount of matter and energy in the universe is constant.

Puzzle #2:
Second Law of Thermodynamics: In any spontaneous change, entropy of the universe increases.

Puzzle #3:
Third Law of Thermodynamics: The entropy of a pure, perfect crystalline substance is zero at absolute zero (0 K).

Puzzle #4:
Hess's Law of Heat Summation: The enthalpy change for a reaction is the same whether it occurs by one step or by any series of steps.

Answers to Preliminary Test

Do not just look up answers. *Think about the reasons for the answers.*

True-False

For those that are false, it is especially important to think carefully *why* they are false.

1. False. Energy is the *capacity* to do work or transfer heat; it is not necessary that work actually be done or heat actually be transferred.
2. True. Think about a rock, whose higher potential energy at the top of the hill can be changed (partly) into kinetic energy as it rolls down the hill.
3. True. Suppose you throw a ball into the air. As it goes up its kinetic energy decreases; that kinetic energy is transformed into increasing potential energy. After it has reached the top of its travel, the reverse begins to happen.
4. False. Chemical energy is a form of potential energy.
5. False. The heat capacity of a body is the amount of heat required to raise the temperature of that body (whatever its mass) by 1 degree Celsius. The specific heat refers specifically to one gram of the substance of which the body is composed. In the terminology of Chapter 1, heat capacity is an extensive property, whereas specific heat is an intensive property. However, we often cite the *molar heat capacity* of a substance, which is the heat capacity of one mole of that substance.
6. False. Common sense would tell you, for instance, that putting 1 teaspoon of 100°C water into a gallon of 40°C water cannot make the final temperature equal to 60°C! The result depends on the relative masses of water. Review the heat transfer calculations in Section 1-14.

7. False. The temperature change depends on the specific heat of each substance as well as the mass of each. Review the heat transfer calculations in Section 1-14.

8. True. If a reaction releases heat, reversing that reaction will absorb the same amount of heat.

9. True

10. True

11. True. And an exothermic reaction *releases* heat to its surroundings.

12. False. If it absorbs heat from the surrounding, its enthalpy is increasing, so that $H_{final} > H_{initial}$. Then $\Delta H = H_{final} - H_{initial} > 0$.

13. True. Heat is released.

14. True. The reaction releases heat to the surrounding, so its enthalpy is decreasing. $H_{final} < H_{initial}$. Then $\Delta H = H_{final} - H_{initial} < 0$.

15. False. Thermochemical equations always refer explicitly to the numbers of *moles* of substances involved. Be sure you are clear about the distinction between *molecules* and *moles* (Chapter 2).

16. False. The Styrofoam™ in this kind of calorimeter serves as in insulator. Any increase in temperature must come from heat liberated by the reacting substances, so the reaction is exothermic.

17. True. See the discussion in Section 15-8.

18. False. It is the heat content (enthalpy) of 1 mole of the substance at its standard state *compared with* those of the corresponding elements in their standard states. The ΔH_f^0 values of the elements are conventionally taken to be zero, but this does not mean that they "have no heat content."

19. False. The value must refer to the formation of *1 mole* of the substance.

20. False. The reaction must involve all reactants and products *at their standard states*. I_2 is a solid at its standard state.

21. False. All reactants must be *elements* in their standard states.

22. False. $\Delta E = q + w$. Doing work on the system can also increase its entropy.

23. True. No $P\Delta V$ work is done.

24. False. Changes in entropy of the system also affect spontaneity. Think of the example of water freezing (an exothermic process); it is certainly not spontaneous at 10°C, although it is at   −10°C.

25. False. For example, the melting of water (ice) is spontaneous at 10°C, and this is an endothermic process (absorbs energy from the surroundings).

26. True

27. True

28. False. The change in a state function *does* depend on the initial and final states. For instance, think about (1) the heating of a substance at constant volume from the initial state of 25°C to a final state of 50°C, compared to (2) heating of a substance at constant volume from the initial state of 25°C to a final state of 75°C. It is apparent that ΔE is different for these two changes. You may need to review Section 15-2 where the concept of the thermodynamic state of a system is discussed.

29. True

30. False. A system may absorb 15 joules of heat and do 5 joules of work ($q = 15$, $w = −5$), or it may release 11 joules of heat and have 21 joules of work done on it ($q = −11$, $w = 21$).

Either way $\Delta E = 10$ joules. Both processes had the same initial and final states, and ΔE is the same, regardless of the path, but q and w are different for the two changes.

31. False. See the answer to Question 30 above.

32. True. ΔS, like ΔH, ΔE, and ΔG are state functions.

33. True. ΔG, like ΔH, ΔE, and ΔS are state functions.

34. False. The bond energy is ΔH for the reaction in which the *only* thing that happens is the breaking of 1 mole of bonds to form isolated atoms. By definition, the determination of ΔH_f^0 requires the elements to be in their standard states, and for many elements the standard state is not isolated gaseous atoms. Many elements are diatomic, or solid, or liquid, etc. in their standard states.

35. True. This reaction involves *only* the breaking of 1 mole of H–H bonds to form isolated H atoms.

36. False. This reaction involves not only the energy required to break 1 mole of H–Cl bonds but also the energy released in the formation of 1 mole of H–H bonds and 1 mole of Cl–Cl bonds. See the answer to Question 34 above.

37. True

38. True. Even though reactions do not usually proceed by first breaking all bonds in reactants and then forming products from the resultant atoms, we can calculate enthalpy changes as if they did. This is a major advantage of thermodynamic calculations—the answer does not depend on the pathway used to get from initial to final states.

39. False. Gas is formed, so Δn is not equal to zero. In other words, the system does work against its surroundings, so ΔE cannot be equal to ΔH.

40. True. Negative values of ΔH^0 mean that the reaction is exothermic (enthalpy of products is less than that of reactants).

41. True. The ΔH^0 value is, among other things, a measure of the heat transferred at constant pressure.

42. False. Remember that the ΔH^0 value is for the specific amount of material referred to in the equation as written. The burning of 2 moles of SiH_4 gives off twice as much heat as the 1 mole that is referred to in the reaction, or 3,032,000 joules.

43. False. The number of moles of gas decreases so at constant pressure the volume would decrease. Thus, the surroundings (the atmosphere) does work on this system.

44. False. At constant volume, the amount of heat given off is equal to ΔE. Because Δn for this reaction is -3, ΔE is algebraically greater than ΔH for the reaction. Study Section 15-11, which relates ΔE to ΔH.

45. True. Remember that entropy is a measure of disorder. In the liquid state, the particles are more dispersed, and thus at higher entropy.

46. True. In the gaseous state, the particles are more dispersed than in the liquid state.

47. False. Study Section 15-17, with special emphasis on the discussion of Table 15-7 and Examples 15-21 through 15-23. Only reactions with $\Delta S > 0$ (producing more dispersal of energy and matter) become more likely to be spontaneous at higher temperatures.

Short Answer

1. system
2. In this chapter, we have seen many possible answers to this question. Acceptable answers include pressure, volume, temperature, energy, enthalpy, entropy, and Gibbs free energy.
3. positive
4. positive
5. increase; 600. $q = 1000$ J, $w = -400$ J, $\Delta E = q + w = 1000$ J $+ (-400$ J$) = 600$ J
6. endothermic
7. Conservation of Energy
8. 1 atmosphere (actually 1 bar, which is 0.987 atm); 25°C (unless otherwise specified)
9. ΔE
10. ΔH
11. by one step; by a series of steps
12. standard molar enthalpy of formation
13. standard molar enthalpy of formation
14. bond energy
15. F. Notice that in thermodynamics we often write the reactions with fractional coefficients. This just means that we are dealing with some fractional numbers of moles. Do not think about a fraction of a molecule!
16. I
17. A. All substances must be in their standard states. For carbon this is solid graphite and for oxygen this is the diatomic gas.
18. bond energy of the oxygen-oxygen bond
19. E. Combustion is a reaction with oxygen, O_2; in this reaction, one mole of carbon monoxide, CO, reacts with $\frac{1}{2}$ mole of O_2.
20. -1
21. H
22. ΔH
23. ΔH^0
24. ΔH_f^0
25. crystal lattice energy
26. exothermic
27. endothermic
28. positive
29. standard molar enthalpy of formation
30. $\frac{1}{2} N_2(g) + \frac{3}{2} H_2(g) \rightarrow NH_3(g)$. Be sure that you write the reaction so that 1 mole of $NH_3(g)$ is formed. Also, you must always explicitly include the physical state of each substance; in this case, they are all gases.

See the discussion in Section 15-14 as you think about the answers to questions 31 through 40. An increase in entropy is denoted by positive ΔS, a decrease by negative ΔS, and no change by $\Delta S = 0$.

31. D. A solid is more ordered (less disorder).

32. U
33. I. More disorder.
34. D. The cards are more ordered.
35. I. The solid has more order; in the solution the ions are mixed with the water molecules.
36. D. The atoms are paired up in the molecules—more ordered.
37. I. A gas is produced.
38. D. Four molecules of gas change to two molecules of gas.
39. U
40. I. The gas is less ordered than the solid.

41. $-T\Delta S$
42. increases. The term in ΔG that includes entropy change is $-T\Delta S$. The more negative this term, the more likely the reaction is to be spontaneous. This term is more important at high temperatures.

Multiple Choice

1. (d). Absorbing heat increases the energy, while doing work decreases it.
2. (b). See Section 15-4.
3. (c). The ice must *absorb* this 334 joules, so $\Delta H = +334$ joules.
4. (c)
5. (c).

$H_2S(g) \rightarrow S(s) + H_2(g)$	$\Delta H^0 =$	$+20.6$ kJ/mol (reversed)
$H_2(g) \rightarrow 2H(g)$	$\Delta H^0 =$	436.0 kJ/mol
$S(s) \rightarrow S(g)$	$\Delta H^0 =$	278.8 kJ/mol
$H_2S(g) \rightarrow S(g) + 2H(g)$	$\Delta H^0 =$	735.4 kJ/mol

Breaking both H−S bonds requires 735.4 kJ/mol. Each requires 735.4/2 = 367.7 kJ/mol. Compare this to the value of 347 kJ/mol given in Table 15-2; remember that the tabulated values represent *average* bond energies in a number of compounds.

6. (a). Recall that $\Delta H = \Delta E + (\Delta n)RT$, as shown in Section 15-11. Remember that Δn is the change in number of moles of gaseous substances shown in the balanced reaction. In the case mentioned here, $\Delta n = 0$, so $\Delta H = \Delta E$.
7. (c). Now $\Delta n > 0$, so $\Delta H > \Delta E$.
8. (b). See Section 15-5.
9. (c). As seen in this chapter, exothermic processes *may* be spontaneous, but not necessarily. The factor of change in order/disorder (entropy change) also affects spontaneity.
10. (c). Combine the equations, as shown in Examples 15-7 and 15-8, so that they add up to the desired equation. You need to *reverse* the last two equations (while changing the signs of their ΔH^0 values) and also *double* the second and third equations (while doubling their ΔH^0 values). If you got answer (a), you forgot to double the appropriate equations and their ΔH^0 values. If you got answer (d), you forgot to reverse the appropriate equations and change the signs of ΔH^0. If you got answer (b), you forgot both the doubling and the reversing. If you got answer (e), you forgot to double the H_2O equation. If you think about it, you would realize that the *positive* answers could not be correct, because this is a combustion reaction.

You know that burning liberates heat, so the reaction must be exothermic and have a negative ΔH^0 value.

11. (a) The value of 2220 kJ is for the combustion of *1 mole* of propane. In the problem, we have 11.0 g, or $\dfrac{11.0 \text{ g}}{44.0 \text{ g/mol}} = 0.25$ mol.

12. (b). The gas is less ordered than the liquid.

13. (d). $q + w = \Delta E$. As for (a), see Appendix K. ΔH_f^0 can be positive, negative, or zero. As for (b), $\Delta H = \Delta E$ when no $P\Delta V$ work is done. As for (c), see the answer to True-False question 30. As for (e), the standard entropy of an element, or a compound, is zero at absolute zero, if it forms a perfect crystal.

14. (a). $\Delta G = \Delta H - T\Delta S$ at constant temperature and pressure.
$\Delta G = -38.468 \text{ kJ/mol} - (298 \text{ K})(0.0514 \text{ kJ/mol·K})$
$\Delta G = -38.468 \text{ kJ/mol} - 15.3 \text{ kJ/mol}$
$\Delta G = -53.8 \text{ kJ/mol}$

15. (a). $\Delta G < 0$

16. (a). We must put heat into a liquid to boil it—vaporization is an endothermic process.

17. (a). Molecules of a gas are in a greater state of dispersal than as a liquid at the same temperature.

18. (c). The boiling point of water at 1 atm is 100°C so the process is at equilibrium. This means that $\Delta G = 0$.

19. (b). This is class 4 of Table 15-7. See the discussion in Section 15-17.

20. (d). This is class 2 of Table 15-7.

21. (c). This is class 3 of Table 15-7.

16 Chemical Kinetics

Chapter Summary

In the preceding chapter, you studied the topic of chemical thermodynamics, which is concerned with the question of *whether* a reaction can go. In Chapter 16, you will study an equally important question—that of *how fast* a reaction goes. The subject of **chemical kinetics** deals with the rates of chemical reactions, the factors affecting these rates, and the information that a study of reaction rates can give about the detailed set of steps by which the reaction proceeds. Thus, a study of chemical kinetics will serve several purposes, from the practical one of allowing us to speed up or slow down a reaction to the fundamental one of learning about reactions on the molecular level.

In Section 16-1, you learn how to describe quantitatively the rate of a reaction. The rate of travel of an automobile could be described by the expression: $\dfrac{\text{change in location}}{\text{change in time}}$. The greater the change in location in a given length of time, the higher is the rate of travel. In just the same way, we describe the rate of a reaction by the expression: $\dfrac{\text{change in concentration of reactant or product}}{\text{change in time}}$.

Several points should be noted here:
(1) As you have already seen in Chapter 15, the symbol Δ means "change in;"
(2) the notation [X] means **molar** concentration of the substance X, whose formula appears in the brackets;
(3) the term for the change in each reactant or product is divided by its coefficient in the balanced equation; and
(4) this change term is multiplied by –1 for reactants, to account for their decrease in concentration. Study very carefully the discussion in Section 16-1 of the H_2 + ICl reaction. It will introduce you to several important ideas such as the **determination** of the rate from a plot of concentration vs. time, the concept of **initial rate**, and the realization that **rates change with time** (as reactants are consumed, the reaction slows).

You are then introduced to the factors that affect reaction rates: **nature of the reactants**, **concentrations of reactants**, **temperature**, and **presence of a catalyst.** Section 16-2 describes the influence of the nature of the reactants—how finely divided solid reactants are, whether reactants are in a well-mixed solution, and so on.

The presence of more substance means more can react in a given time, so it is not surprising that concentration of reactants plays a very important part in determining reaction rates. The central idea of Section 16-3 is the **rate-law expression**, which relates the rate of reaction to the concentrations of reactants. Keep in mind that the values of the **specific rate constant (k)**, and of the **orders** with respect to the reactants *must* be determined experimentally. They are not necessarily related to the coefficients in the balanced overall chemical equation. The several examples in this section will clarify the deduction of rate-law expressions from experimental data. Exercises 13 through 29 at the end of the textbook chapter will give you some necessary practice at this type of data analysis.

In Section 16-4, the rate-law expression is reformulated into the **integrated rate equation**. Here the emphasis is on the **time** required for the reaction to proceed by a specified amount. This is sometimes expressed in terms of the **half-life** of reactants. The integrated rate equation is different for reactions with different order rate laws. You will study only the most straightforward cases, those of zero-, first-, and second-order reactions. Again, careful attention to the examples will aid you in understanding the central idea of this mathematically complex section. Table 16-2 summarizes important concentration relationships from Sections 16-3 through 16-4. You will find the optional Enrichment section (Calculus Derivation of Integrated Rate Equations, page 633) useful as an illustration of the use of calculus in chemistry.

The integrated rate equation also provides us a method for finding reaction orders from graphical and other analyses of concentration-*vs.*-time data (Enrichment section: Using Integrated Rate Equations to Determine Reaction Order, page 634). Once again, the examples will help you to understand the details of this method.

The **collision theory** of reaction rates, Section 16-5, takes the viewpoint that for molecules to react they must collide (1) with sufficient energy to react—that is, to break bonds and form new ones—and (2) with the proper orientation for a reaction to occur. This idea is extended by the **transition-state theory**, Section 16-6, which says that reactants pass through a short-lived, high-energy state known as the **transition state** when they react. The reactants must increase their energy by an amount called the **activation energy (E_a)** to reach the transition state. Thus, the activation energy can be viewed as a barrier to the reaction. If the colliding molecules do not have this much energy, they cannot react; if they do have sufficient energy, the appropriate bonds can break, new bonds can then form, and the reaction (for those particular colliding molecules) can proceed. Some energy is then released as the new molecules are formed from the transition state. **Potential energy diagrams** such as those in Figure 16-10 help us to understand the relation between the activation energies for the forward and reverse reactions and the value of ΔE for the reaction.

One of the most important applications of the study of chemical kinetics is the information this study gives us about the detailed series of elementary steps by which reactions occur, called the **reaction mechanism**. Section 16-7 relates the reaction mechanism and the experimentally determined rate law, introducing the important ideas of **reaction intermediates** and the **rate-determining step.** This interpretation of kinetics makes use of both the collision theory and the transition-state theory presented earlier in the chapter.

Any reaction proceeds faster if the temperature rises. Section 16-8 relates this observation to the increase in the fraction of molecules with sufficient energy to react (recall **activation energy** and review the kinetic-molecular theory, Section 12-13). This idea is expressed mathematically in the **Arrhenius equation**. Of course, instead of increasing the number of molecules that can "leap the energy barrier" in order to speed up the reaction, we could lower the barrier, thereby lowering E_a. This is the function of **catalysts**, discussed in Section 16-9. The catalyst provides a lower energy pathway by which the reaction can occur, causing the reaction to speed up. It is important to realize that catalysts cannot cause thermodynamically impossible reactions to occur; they can only alter the rates of reactions that are spontaneous. The discussion in this section of several **homogeneous** and **heterogeneous** catalysts will illustrate how catalysis can occur.

The ideas of chemical kinetics from this chapter will be one basis for your study in Chapter 17 of chemical equilibrium.

Study Goals

Chapter Introduction

1. Distinguish between chemical kinetics and chemical thermodynamics. Know what each subject does and does not cover. *Work Exercise 4.*

Sections 16-3, 16-4, 16-6, and 16-7 through 16-9

2. Summarize the factors that affect reaction rates. *Work Exercise 1.*

Section 16-3

3. For a given reaction, express reaction rate in terms of changes in the concentrations of reactants and products per unit time. *Work Exercises 3 and 6 through 10.*

4. Describe how reaction rates are determined. Know what must be measured and how the data are analyzed. *Work Exercise 3.*

5. Identify all the terms in an expression such as: rate $= k[A]^x[B]^y$, and describe the significance of each. Be familiar with the terminology of order of a reaction. *Work Exercises 3, 6, and 13 through 16.*

6. Given experimental data for a particular reaction at a specified temperature, consisting of initial rates corresponding to different combinations of initial concentrations of reactants, determine the rate law for the reaction. *Work Exercises 17, 18, and 21 through 27.*

7. Given the rate-law expression for a specific reaction at a certain temperature, calculate the initial rate of reaction at the same temperature for any initial concentrations of reactants. Use changes in concentrations to predict changes in initial rates. *Work Exercises 19, 20, 27, 28, and 29.*

Section 16-4

8. Know the relationships for various orders of reaction that are summarized in Table 16-2.

9. Perform various calculations with the integrated rate equations for zero-, first-, and second-order reactions relating rate constants, half-lives, initial concentrations, and concentrations of reactants remaining at some later time. *Work Exercises 30 through 39 and 44.*

Enrichment, pages 634-637

10. Analyze concentration-*vs.*-time data to determine reaction order. *Work Exercises 41 through 43.*

Sections 16-5 and 16-6

11. Describe simple one-step reactions in terms of (a) collision theory and (b) transition state theory. *Work Exercises 49 and 58.*

Section 16-6

12. Describe activation energy and illustrate it graphically for exothermic and endothermic reactions. Be able to relate E_a for forward and reverse reactions to ΔE_{rxn}. *Work Exercise 49.*

Section 16-7

13. Understand how rate-law expressions are related to reaction mechanisms. *Work Exercises 5 and 65.*

14. Derive the rate law based on a proposed mechanism. Given the observed rate-law expression for a reaction, (a) recognize whether a proposed mechanism could be the correct one, or (b) postulate mechanisms consistent with the data, specifying the slow and fast steps. *Work Exercises 66 through 75.*

Section 16-8

15. Understand, in terms of the distribution of energies of the reactant molecules, how the reaction rate depends on temperature. *Work Exercises 49 and 52.*

16. Use the Arrhenius equation to:
    (1) find the rate constant from the collision frequency factor (A) and activation energy (E_a), and
    (2) relate rate constants at two different temperatures.
    Know how to represent and interpret this equation graphically.
    Work Exercises 40, 47, 48, and 53 through 64.

Section 16-9

17. Understand, in terms of potential energy diagrams, how a reaction rate is altered by the presence of a catalyst. Give examples of homogeneous and heterogeneous catalysts. *Work Exercises 50 and 51.*

General

18. Be able to recognize and solve a variety of types of problems based on the principles of this chapter. *Work **Mixed Exercises** 76 through 81.*

19. Answer conceptual questions based on this chapter. *Work **Conceptual Exercises** 82 through 94.*

20. Apply concepts and skills from earlier chapters to the ideas of this chapter. *Work **Building Your Knowledge** exercises 95 through 100.*

21. Exercises at the end of chapter direct you to sources outside the textbook for information to use in solving them. Work **Beyond the Textbook** *exercises 101 through 106.*

Some Important Terms in This Chapter

Some important new terms from Chapter 16 are listed below. Fill the blanks in the following paragraphs with terms from the list. Each term is only used once. Check the Key Terms list and the chapter reading. Study other new terms, and review terms from preceding chapters if necessary.

activation energy	heterogeneous catalyst	reaction mechanism
Arrhenius equation	homogeneous catalyst	specific rate constant
catalysis	integrated rate equation	transition state theory
collision theory	rate-determining step	
half-life	rate-law expression	

The relationship between the *rate* of a reaction and the *concentrations* of the reactants is described mathematically by the [1]_____, which is often called simply the *rate law*. It is important to realize that the rate-law expression is determined experimentally for each reaction by studying how its rate varies with the concentrations. The rate law generally has the format: rate = $k[A]^x[B]^y$ ..., where A, B, ... are the reactants, [ ] means "molar concentration of," and k is the [2]_____, which is usually referred to simply as the *rate constant*. The values of the exponents x, y, ..., and the rate constant k must be determined experimentally.

The relationship between the *concentration* and the *time* (since the reaction started) is given by the [3]_____. The integrated rate equation can also be used to calculate the [4]_____, $t_{1/2}$, of a reactant, which is the time it takes for half of the reactant to be converted into product.

The concept that for a reaction to occur, molecules, atoms, or ions must first collide is the basis of the theory known as the [5]_____ of reaction rates. Furthermore, to get to the products, the reactants must pass through a short-lived, high-energy transition state, according to the theory called the [6]_____. In order to reach the transition state, reactant molecules must have a minimum amount of kinetic energy, called the [7]_____.

The step-by-step pathway by which a reaction occurs is called the [8]_____ _____. The slowest step in the reaction mechanism determines the rate of the reaction because a reaction can never occur faster than its slowest step. Therefore, this slow step is called the [9]_____.

The mathematical relationship among the activation energy, the absolute temperature, and the specific rate constant of a reaction at a particular temperature, determined by Svante Arrhenius, is called the [10]_____.

Catalysts are substances that can be added to reacting systems to increase the rate of reaction. The use of a catalyst is called [11]_____. There are two kinds of catalysts: a [12]_____ exists in the same phase as the reactants, and a [13]_____ is present in a different phase from the reactants.

Preliminary Test

As in other chapters, this test will check your understanding of basic concepts and types of calculations. Be sure to practice *many* of the additional textbook Exercises, including those indicated with the Study Goals.

True-False

Mark each statement as true (T) or false (F).

_____ 1. All reactions taking place at the same temperature occur at the same rate.
_____ 2. All exothermic reactions taking place at the same temperature occur at the same rate.
_____ 3. All reactions taking place at the same temperature with the same value of ΔH occur at the same rate.

_____ 4. All reactions taking place at the same temperature with the same activation energy occur at the same rate.

_____ 5. Reactions with lower (more negative) values of ΔG are more spontaneous, so they usually proceed at a higher rate.

_____ 6. The transition state is a short-lived, high-energy state, intermediate between reactants and products in the reaction.

_____ 7. According to the collision theory, whenever molecules collide with one another, they react.

_____ 8. Once the reactants have absorbed the necessary activation energy to reach the transition state, they are trapped and cannot react further.

_____ 9. Once the reactants have absorbed the necessary activation energy to reach the transition state, they must then go ahead to form products.

_____ 10. The activation energy E_a is usually the same as ΔE for the reaction.

_____ 11. A reaction in which the activation energy of the forward reaction is greater than the activation energy of the reverse reaction is endothermic.

_____ 12. The exponents in the rate law must match the coefficients in the balanced chemical equation for the reaction.

_____ 13. The exponents in the rate law may match the coefficients in the balanced chemical equation for the reaction.

_____ 14. If the exponents in the rate law match the coefficients in the balanced chemical equation, then we know that the reaction takes place in one step.

_____ 15. The rate law for a reaction can be correctly predicted from the balanced chemical equation.

_____ 16. The value of k for a given reaction is constant at all conditions.

_____ 17. The value of k for a given reaction is constant at a given temperature.

_____ 18. If we know the mathematical form of the rate-law expression for a reaction, and we have also determined the value of the specific rate constant at some temperature, we can predict the rate at which the reaction would occur at that temperature for any specified reactant concentrations.

_____ 19. As a rule of thumb, every 10°C increase in temperature roughly doubles the rate of many reactions.

_____ 20. A catalyst works by increasing the average energy of reacting molecules.

_____ 21. A catalytic converter on an automobile is an example of a heterogeneous catalyst.

_____ 22. An effective catalyst increases the rate of the forward reaction but does not affect the rate of the reverse reaction.

_____ 23. A homogeneous catalyst is composed of a single compound, whereas a heterogeneous catalyst is a mixture of compounds.

_____ 24. A catalyst can make a thermodynamically nonspontaneous reaction spontaneous.

_____ 25. If the initial concentration of the reactant in a first-order reaction is changed, then the half-life changes.

_____ 26. If the initial concentration of the reactant in a second-order reaction is changed, then the half-life changes.

Short Answer

1. The symbol "[X]" means _____
   ___.

2. In the rate-law expression, the quantity represented by the symbol k is called the
   _____.

3. For a particular reaction, the rate law is determined to be: rate $= k[A]^2[B]$. The reaction is
   said to be _____ order in A, _____ order in B, and _____ order overall.

4. The rate of a reaction may be expressed either in terms of the _____ in reactant
   concentration with time or the _____ in product concentration with time.

5. In a reaction for which the rate-determining step is: $A + B \rightarrow$ products,
   doubling the concentration of A and doubling the concentration of B would cause the rate
   to _____ by a factor of _____.

6. The amount of energy that reactants must gain above their ground states in order to react is
   called the _____ of the forward reaction.

7. Four   factors   that   can   influence   the   rate   of   a   chemical   reaction   are
   _____,   _____,
   _____,   and   _____
   _____.

8. When heated, magnesium metal reacts with the oxygen in air to form magnesium oxide.
   Suppose we have a cube of magnesium metal, 1 cm on edge, and measure the initial rate at
   which it forms MgO. Then we cut the magnesium into cubes 1 mm on edge. We would
   expect that at the same temperature the initial rate of MgO formation would
   _____ by a factor of _____.

9. The principal reason that a reaction speeds up at elevated temperatures is an increase in the
   _____
   _.

10. A substance that alters the rate of a chemical reaction but is not itself consumed or
    produced in the reaction is called a _____.

11. The sequence of individual steps by which a reaction occurs is called the _____
    of the reaction.

12. The   slowest   step   in   a   reaction   sequence   is   called   the
    _____.

13. According to the collision theory, when a reaction requires the collision between two
    reactant molecules for the rate-determining step to proceed, that step is termed
    _____ and the reaction is called _____ order.

14. The time required for half of a reactant to disappear in a reaction is called the
    _____ of the reactant.

15. The rate equation $\dfrac{1}{[A]} - \dfrac{1}{[A]_0} = akt$ describes the dependence of concentration on time

    for a reaction that is _____ order; the rate law for this reaction would be
    _____.

16. For a reaction that is first order in A, the equation that relates concentration and time is
_____.

___

The next five questions, 17 through 21, refer to the reaction

$$2A + 2B \rightarrow C + 2D$$

in which all reactants and products are gases. The reaction is observed to follow the rate law: rate = $k[A][B]^2$. The reaction is also strongly exothermic. In each question, we make *only* the one change in the reaction conditions indicated. For each, predict the effect on (a) the initial rate of the forward reaction, (b) the value of the rate constant, k, and (c) the activation energy, E_a, for the forward reaction. The possible answers are: increase (I), decrease (D), or remain unchanged (U).

Change	Initial Rate of Forward Reaction	k	E_a
17. The concentration of A is doubled.	____	____	____
18. The concentration of D is doubled.	____	____	____
19. A catalyst is introduced.	____	____	____
20. The temperature is decreased.	____	____	____
21. The volume of the container is doubled.	____	____	____

___

22. For a particular reaction, the rate law is determined to be: rate = $k[A]^2[B]$. We first carry out the reaction with some starting concentrations of A and B. Then we repeat it, after doubling the concentrations of A and B. The rate will change by the factor:

$$\frac{\text{new rate}}{\text{old rate}} = \underline{\hspace{2cm}} \text{ (a number).}$$

23. For a particular reaction, the rate law is determined to be: rate = $k[A]^2[B]$. We first carry out the reaction with some starting concentrations of A and B. Then we repeat it, after halving the concentration of A and doubling the concentration of B. The rate will change by the factor:

$$\frac{\text{new rate}}{\text{old rate}} = \underline{\hspace{2cm}} \text{ (a number).}$$

24. The half-life of a certain first-order reaction is 2.5 hours. After 10 hours, the fraction of the initial amount of reactant remaining is _____.

Multiple Choice

_____ 1. Which one of the following reactions is expected to be the slowest at a given temperature?

(a) $Ag^+(aq) + Cl^-(aq) \rightarrow AgCl(s)$

(b) $H^+(aq) + OH^-(aq) \rightarrow H_2O(\ell)$

(c) $CH_4(g) + 2O_2(g) \rightarrow CO_2(g) + 2H_2O(g)$

(d) $Pb^{2+}(aq) + CrO_4^{2-}(aq) \rightarrow PbCrO_4(s)$

(e) $H^+(aq) + CN^-(aq) \rightarrow HCN(aq)$

_____ 2. One of the reactions used commercially to produce hydrogen gas is

$$H_2O(g) + CO(g) \rightarrow H_2(g) + CO_2(g)$$

Which of the following is a proper expression for the rate of this reaction?

(a) $-\dfrac{\Delta[CO_2]}{\Delta t}$ (b) $-\dfrac{\Delta[H_2]}{\Delta t}$ (c) k (d) $\dfrac{\Delta[CO]}{\Delta t}$ (e) $-\dfrac{\Delta[H_2O]}{\Delta t}$

The following potential energy diagram is to be used in answering Questions 3 through 6.

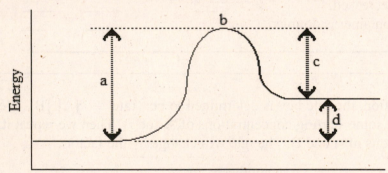

Reaction coordinate

_____ 3. The reaction "reactants → products" carried out at constant volume and temperature

(a) is exothermic.   (b) is endothermic.   (c) takes place without energy change.

(d) is impossible.   (e) may either be exothermic or endothermic.

_____ 4. The energy difference represented by line d is the

(a) energy content of products.   (b) activation energy for the forward reaction.

(c) energy content of reactants.   (d) activation energy for the reverse reaction.

(e) energy change of the reaction.

_____ 5. The region designated by b in the diagram is

(a) the energy of the mixture when half of the reactants have been converted to products.

(b) the energy of the transition state.

(c) the number of moles of transition state that must be formed.

(d) the energy of the forward reaction.

(e) the energy of the reverse reaction.

_____ 6. The energy difference represented by line *c* is the
(a) reaction energy of the forward reaction. (b)   activation energy of the forward reaction.
(c) reaction energy of the reverse reaction. (d)   activation energy of the reverse reaction.
(e) energy content of the reactants.

_____ 7. A reaction has an activation energy of 40 kJ and an overall energy change of reaction of −100 kJ. In each of the following potential energy diagrams, the horizontal axis is the reaction coordinate and the vertical axis is potential energy in kJ. Which potential energy diagram best describes this reaction?

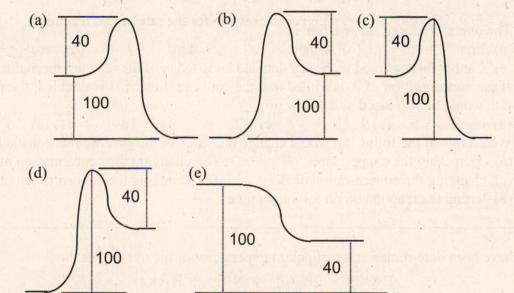

_____ 8. Four of the following changes can affect the forward rate of a chemical reaction. Which one cannot affect this rate?
(a) raising the temperature   (b) adding a catalyst   (c)   removing some of the reactants
(d) removing of some of the products   (e)   grinding a solid reactant into smaller pieces

_____ 9. For a given reaction the rate law is
(a) a constant of proportionality between reaction rate and the concentrations of reactants.
(b) the sum of the powers to which reactant concentrations appear.
(c) an equation in which reaction rate is equal to a mathematical expression involving concentrations of reactants.
(d) the additional energy that the reactants must obtain to be able to react.
(e) 70 miles per hour.

_____ 10. A catalyst
   (a) is used up in a chemical reaction.
   (b) changes the value of ΔG^0 of the reaction.
   (c) is always a solid.
   (d) does not take part in the reaction in any way.
   (e) changes the activation energy of the reaction.

_____ 11. The units of the rate constant for a second order reaction can be
   (a) $M^{-1} \cdot s^{-1}$ (b) M (c) $M \cdot s^{-1}$ (d) s^{-1} (e) $M^2 \cdot s^{-1}$

Questions 12 through 14, refer to the gaseous reaction: A + B → C, for which the reaction rate is experimentally observed to follow the relationship: rate $= k\,[A]^2[B]$.

_____ 12. The overall order of the reaction is
   (a) first.   (b) second.   (c) third.   (d) zero.
   (e) Cannot be answered without a detailed knowledge of the reaction mechanism

_____ 13. If the concentration of A is tripled and the concentration of B is doubled, the reaction rate would be increased by a factor of ____.
   (a) 6   (b) 9   (c) 12   (d) 18   (e) 36

_____ 14. Which one of the following would change the value of the specific rate constant k?
   (a) decreasing the temperature   (b) changing the concentration of A
   (c) changing the concentration of B   (d) changing the concentration of C
   (e) letting the reaction go on for a long time

Rate data have been determined at a particular temperature for the overall reaction

$$2NO(g) + 2H_2(g) \rightarrow N_2(g) + H_2O(g)$$

in which all reactants and products are gases. These data are shown in the table:

Trial Run	Initial [NO]	Initial [H$_2$]	Initial Rate ($M \cdot s^{-1}$)
1	0.10	0.20	0.0150
2	0.10	0.30	0.0225
3	0.20	0.20	0.0600

Questions 15 through 18 refer to this reaction and these data.

_____ 15. The rate law is
   (a) rate $= k\,[NO]^2[H_2]^2$ (b) rate $= k\,[NO][H_2]^2$
   (c) rate $= k\,[NO]^2[H_2]$ (d) rate $= k\,[NO][H_2]$
   (e) None of the preceding answers is correct.

_____ 16. The value of the specific rate constant at this temperature is

(a) $0.75\ M^{-1} \cdot s^{-1}$ (b) $7.5\ M^{-2} \cdot s^{-1}$ (c) $3.0 \times 10^{-3}\ M^{-2} \cdot s^{-1}$

(d) $3.0 \times 10^{-4}\ M^{-1} \cdot s^{-1}$ (e) $375\ M^{-2} \cdot s^{-1}$

_____ 17. What would be the initial rate with the molar concentration of NO = 0.30 M and the molar concentration of H_2 = 0.10M?

(a) $0.0675\ M \cdot s^{-1}$ (b) $0.225\ M \cdot s^{-1}$ (c) $0.0225\ M \cdot s^{-1}$

(d) $0.0400\ M \cdot s^{-1}$ (e) $0.1000\ M^{-1}$

_____ 18. Does the reaction take place in one step?
(a) Yes   (b) No
(c) Probably yes, but we need more information to be certain.
(d) Probably no, but we need more information to be certain.

_____ 19. An instructor dismisses class early. Which of the following factors is most important in determining the rate of students exiting from the room (students/min)?
(a) realizing that the instructor really meant it
(b) gathering up books and notes
(c) walking from the seats to the aisle
(d) walking down the aisle to the door
(e) crawling through a 1-ft by 2-ft hole in the locked door

_____ 20. Consider the following proposed mechanism.

       (1)   2A $\rightarrow$ C + I   (specific rate constant = k_1)

       (2)   I + B $\rightarrow$ C + D   (specific rate constant = k_2)

If this mechanism for the overall reaction were correct, and if k_1 is much less than k_2, then the observed rate law would be

(a) rate = $k_1 k_2 [A]^2 [I][B]$ (b) rate = $k_2 [I][B]$ (c) rate = $k_1 [A]^2$

(d) rate = $k_1 [A]^2 k_2 [C][D]$ (e) rate = $k_1 [A]$

_____ 21. A reaction mechanism will usually be
(a) the only possible explanation for the reaction.
(b) difficult to verify experimentally.
(c) proven experimentally to be the only possible mechanism.
(d) obvious from a consideration of the balanced chemical equation.
(e) obvious from a consideration of the reaction rate data.

_____ 22. According to the Arrhenius equation for the rate constant of a chemical reaction, the higher the activation energy, the ____ at a given temperature.
(a) slower a reaction is   (b) more exothermic a reaction is
(c) more endothermic a reaction is   (d) more atoms are involved in a reaction
(e) faster a reaction is

_____ 23. The first-order rate constant for the conversion of: A $\rightarrow$ B is 0.0551 min^{-1}.
What mass of A would be left after 1 hr if the reaction started with 80 g of A at t = 0?
(a) 2.89 g   (b) 4.42 g   (c) 10.0 g   (d) 52.2 g
(e) cannot be determined from the data given.

_____ 24. The decomposition of dimethylether at 504°C is first-order with a half-life of 1570 seconds. What fraction of an initial amount of dimethylether remains after 4710 seconds?

(a) $\frac{1}{3}$ (b) $\frac{1}{6}$ (c) $\frac{1}{8}$ (d) $\frac{1}{16}$ (e) $\frac{1}{32}$

Answers to Some Important Terms in This Chapter

1. rate-law expression
2. specific rate constant
3. integrated rate equation
4. half-life

5. collision theory
6. transition state theory
7. activation energy
8. reaction mechanism
9. rate-determining step

10. Arrhenius equation
11. catalysis
12. homogeneous catalyst
13. heterogeneous catalyst

Answers to Preliminary Test

Do not just look up answers. *Think about the reasons for the answers.*

True-False

1. False. Temperature is not the only factor that determines the rate of a reaction.
2. False. Temperature is not the only factor that determines the rate of a reaction.
3. False. Temperature is not the only factor that determines the rate of a reaction.
4. False. Temperature and activation energy are not the only factors that determine the rate of a reaction.
5. False. One of the most important ideas of this chapter is that the rate of the reaction depends on the availability of a low-energy pathway from reactants to products. Spontaneity of a reaction does not determine its rate. Think about Study Goal 1.
6. True
7. False. According to this theory, collision is necessary but not sufficient for reaction. Be sure that you understand what is required for a collision to be effective (Section 16-5).
8. False. They either react to give products or go back to form reactants again, giving up energy.
9. False. They either react to give products or go back to form reactants again, giving up energy.
10. False. See the potential energy diagrams in Figure 16-10. ΔE_{rxn} is the difference between the E_a of the forward reaction and the E_a of the reverse reaction.
11. True. Draw a potential energy diagram to help you think about this.
12. False. They must be determined experimentally.
13. True. They could, but they don't have to.
14. False. This important idea is discussed in Section 16-7. What this tells us is that it *may* take place in one step, but not necessarily. If the exponents do *not* match the coefficients in the

balanced equation, then we know that the reaction definitely *does not* take place in one step.

15. False. This important point can cause much confusion. The reaction may take place in a series of steps with some fast and some slow, so the overall rate law is generally impossible to predict simply from the balanced equation. Study Section 16-7.

16. False. It can be altered either by changing the temperature or by adjusting the activation energy with the introduction of a catalyst.

17. False. It can still be changed by the introduction of a catalyst.

18. True

19. True. Of course, the actual change depends on the activation energy—this amount of increase is just typical of many reactions with activation energies near 50 kJ/mol. A calculation of this type is done in Section 16-8.

20. False. It lowers (or, in general, alters) the activation energy, so that at a given temperature, more molecules have enough energy to react.

21. True. The catalyst is solid, and the reactants are gases.

22. False. The action of a catalyst is to change the energy of the activated complex (transition state). This changes the values of E_a for both the forward and the reverse reactions, so the rates of both reactions are altered.

23. False. The terms homogeneous and heterogeneous refer to the number of phases of the reaction mixture. When the catalyst is in a different physical state than the reaction mixture, it is called a heterogeneous catalyst.

24. False. Catalysts cannot cause thermodynamically impossible reactions to occur; they can only alter the rates of reactions that are spontaneous.

25. False. For a first-order reaction, the time required for half of the initial substance to react is the same regardless of amount of starting material.

26. True. For a second-order reaction, the half-life is different depending how much material we start with. Watch out for this in calculations or reasoning involving half-lives.

Short Answer

1. the molar concentration of substance X
2. specific rate constant. Sometimes it is just referred to as the rate constant.
3. second; first; third
4. decrease; increase
5. increase; 4. If this is the rate-determining step, we would observe that rate = $k[A][B]$.
6. activation energy
7. concentrations of reactants; temperature; nature of reactants; presence of a catalyst
8. increase; 10. The rate would depend on the surface area of magnesium exposed to air. In the 1 cm cube, the total area is 6 cm^2 (6 faces with 1 cm^2 each). After we cut it up, the total area is 60 cm^2. Can you figure this out?
9. fraction of collisions that have energy equal to or exceeding E_a. A secondary factor is the increased number of collisions, but this is less important than the increase in the fraction of high-energy molecules. Can you see this in the Arrhenius equation, Section 16-8?
10. catalyst
11. mechanism

12. rate-determining step
13. bimolecular; second
14. half-life. You may see this referred to in some books as the "half-time."
15. second; rate $= k[A]^2$. Study Section 16-4.
16. $\ln\left[\dfrac{[A]_0}{[A]}\right] = akt$
17. I, U, U. Changing concentration does *not* alter either k or E_a. What does?
18. U, U, U. Changing concentration of a product does not change the rate of the *forward* reaction.
19. I, I, D. Generally, when we talk about a catalyst, we assume that it is a catalyst that enhances the rate. If it were an inhibitory catalyst we would have specified it so.
20. D, D, U. Be sure that you know how a temperature change alters the rate.
21. D, U, U. Think about what doubling the volume does to concentrations.
22. 8. Doubling [A] would make the rate $2^2 = 4$ times as fast; doubling [B] would account for an additional factor of 2, so the total effect would be $4 \times 2 = 8$.
23. $\frac{1}{2}$. That is, it would go only $\frac{1}{2}$ as fast. Halving [A] would make the rate $(\frac{1}{2})^2 = \frac{1}{4}$ as fast; doubling [B] would account for an additional factor of 2, for a total change of $\frac{1}{2}$.
24. $\frac{1}{16}$. For this reaction ten hours is four half-life periods (10/2.5). This is a first-order reaction, so the half-life means that $\frac{1}{2}$ of *any* starting amount remains after 2.5 hours. Thus, after 2.5 hours, $\frac{1}{2}$ of the original remains. After the next 2.5 hours, $\frac{1}{2}$ of the amount at the beginning of *that* 2.5 hour period, or $\frac{1}{4}$ of the original, remains. Reasoning in this fashion for four half-lives, we obtain the answer. Be sure you know how to answer a similar question (a) if the time is not an integer number of half-lives, (b) if k instead of half-life is given, or (c) for a second-order reaction. Study the Examples in Section 16-4 carefully.

Multiple Choice

1. (c). This reaction requires that covalent bonds be broken. The others involve only aqueous ionic reactants, either forming covalent bonds or forming ionic solids. Reactions that involve the breaking of covalent bonds are usually slower. Can you tie this in with what you know about activation energies?
2. (e). The signs for (a), (b), and (d) are wrong. (c) is the rate constant, not the rate.
3. (b). The energy of products is higher than that of reactants.
4. (e). It is the difference between the energy content of the reactants and products.
5. (b)
6. (d)
7. (a)
8. (d). The concentration of the products does not affect the rate of forward reaction.
9. (c). (a) is the rate constant, (b) is the order of the reaction, (d) is the activation energy, and (e) is a speed limit.
10. (e). A catalyst is *not* used up in a chemical reaction (a), does *not* change the value of ΔG^0 (b), can be a liquid, a gas, or a solution (c), and *does* take part in the reaction, providing a new reaction pathway (d).
11. (a). Write down a second-order rate law and think about units.

12. (c)
13. (d). $(3)^2(2) = 18$
14. (a). The rate constant is only affected by changing the temperature.
15. (c). The rationale for this is like that in Examples 16-3 and 16-4 in the text. From trial 1 to trial 3, [NO] doubles, [H$_2$] remains the same, and the initial rate quadruples, so the exponent for [NO] is a 2. From trial 1 to trial 2, [NO] remains the same, [H$_2$] increases 50%, and the initial rate increases 50%, so the exponent for [H$_2$] is 1.
16. (b). Put the observed rate and the initial concentrations from any one of the three trial runs into the rate law and solve for k. Be careful of the units. Check to see that you get the same answer regardless of which set of data you use. This can be a check that you deduced the correct rate law as well as a check on your arithmetic.

    Trial 1: $0.0150 \ M \cdot s^{-1} = k (0.10 \ M)^2 (0.20 \ M)$

    Trial 2: $0.0225 \ M \cdot s^{-1} = k (0.10 \ M)^2 (0.30 \ M)$

    Trial 3: $0.0600 \ M \cdot s^{-1} = k (0.20 \ M)^2 (0.20 \ M)$

    Each produces $k = 7.5 \ M^{-2} \cdot s^{-1}$.
17. (a). Put these molar concentrations into the rate law, together with the value for k that was obtained in Question 16.

    rate $= (7.5 \ M^{-2} \cdot s^{-1})(0.30 \ M)^2 (0.10 \ M) = 0.0675 \ M \cdot s^{-1}$
18. (b). The exponents do not match the stoichiometric coefficients, so we know that the reaction *cannot* take place in one step.
19. (e). The slow step determines the rate of overall process.
20. (c). Because k_1 is much less than k_2, the first step is much slower than the second. Thus, the rate of the first step determines the rate of the entire reaction. Can you write the equation for the overall reaction? Note that I is an intermediate—the reaction is: $2A + B \rightarrow 2C + D$.
21. (b). Read the discussion in Section 16-7.
22. (a)
23. (a). The calculation is similar to that for N_2O_5 in Example 16-6. Proceed as follows:

    $\ln\left(\dfrac{[A]_0}{[A]}\right) = akt$, where $a = 1$, k is $0.0551 \ min^{-1}$, and $t = 60 \ min$

    Then $\ln\left(\dfrac{[A]_0}{[A]}\right) = 3.31$; taking antilogarithms, $\dfrac{[A]_0}{[A]} = 27.3$.

    We cannot calculate either molar concentration of A because we do not know its formula weight or the volume of the container. However, the volume does not change, so the volumes will cancel in the ratio, as will molecular weight:

    $$\frac{[A]_0}{[A]} = \frac{\dfrac{mass_0}{MW_A \cdot volume}}{\dfrac{mass}{MW_A \cdot volume}} = \frac{mass_0}{mass}$$

    Thus, $\dfrac{mass_0}{mass} = 27.3$ or mass $= \dfrac{mass_0}{27.3}$. Then mass $= \dfrac{80.0}{27.3} = 2.93 \ g$.

24. (c). 4710 seconds is $\dfrac{4710}{1570}$ = 3 half-lives. This is a first-order reaction, so each half-life period consumes $\frac{1}{2}$ of the amount present at the beginning of *that* period.

17 Chemical Equilibrium

Chapter Summary

Most chemical reactions do not go to completion; rather they attain a state of **chemical equilibrium** that consists of some mixture of reactants and products. Some important questions regarding this equilibrium state are these: How can we describe the equilibrium mixture in terms of concentrations? What factors determine the equilibrium, and how can we use these factors to enhance the occurrence of some desirable reaction? How can we predict what amounts of products will have been formed and what amounts of reactants will remain when equilibrium has been reached? In Chapter 17, we bring together many ideas from our earlier studies to get a general answer to these and other questions about chemical equilibrium. We will first describe from the standpoint of chemical kinetics (Chapter 16) how chemical equilibrium is attained. Then you will use this description to carry out calculations of equilibrium concentrations of reactants and products and to understand the factors that influence equilibrium. Finally, you will see the relation of the equilibrium discussion to chemical thermodynamics. This general discussion of chemical equilibrium ties together many important ideas that you have been developing over the past several chapters. It will also be the basis for your study of several specific types of reactions in the next three chapters.

Some basic ideas of equilibrium are presented in Section 17-1. Central to this discussion are three points you must keep in mind:

(1) The main conceptual point is that equilibrium is a dynamic situation (sound familiar?) in which two processes, the forward reaction and the reverse reaction, are proceeding at equal rates.

(2) You know how rates depend on concentrations (Chapter 16), so you can see that the condition rate$_{forward}$ = rate$_{reverse}$ can be satisfied for many different combinations of concentrations.

(3) The *changes* in concentrations of all reactants and products due to reaction must always be in the ratio of the stoichiometric coefficients, even though the concentrations themselves may not be.

Please learn the general layout of analyzing equilibrium problems presented in this section. For each substance,

initial concentration + change in concentration due to reaction = equilibrium concentration

This approach will also apply to equilibrium-concentration problems in the next several chapters.

Section 17-2 defines one of the most important quantities in your study of chemistry, the **equilibrium constant**, *K*. The algebraic expressions for this constant, unlike those for the rate law expressions, can *always* be written down directly from the balanced chemical equation. Keep in mind that the concentrations referred to in this expression for the equilibrium constant are the concentrations of reactants and products *when the reaction has reached equilibrium*. Of course,

247

you could put any other values of the concentrations into such an expression and carry out the arithmetic, but this gives the equilibrium constant value *only* if the concentrations described are the concentrations at equilibrium. (See the discussion of reaction quotient, later in this chapter). You also see in this section how the *numerical value* of the equilibrium constant for any given reaction can be calculated from a measurement of the concentrations at equilibrium. Some further points should be noticed about equilibrium constants: (1) The symbol for equilibrium constant expressed in terms of molar concentrations is K_c. (2) A value of K_c for a particular reaction is valid only at a given temperature. We shall see later in the chapter how this value changes with temperature. In Section 17-3 you see how the form and value of K_c vary with the exact form of the balanced equation for a given reaction. You could use any of these forms to solve any particular problem and would always get the same answer for concentration, so long as you are careful to know to which representation of the chemical equation the value of K_c corresponds.

The **reaction quotient, Q,** presented in Section 17-4, is a very useful aid to analyzing whether a reacting system is at equilibrium. You see that it has the same *algebraic form* in terms of concentrations as does K_c, but for Q the concentrations are *not necessarily* equilibrium concentrations. They might be, but only if those concentrations represent equilibrium. This means that you can use the reaction quotient with any specified combination of concentrations as a criterion (1) of whether or not the reaction is at equilibrium and (2) if not, how it must adjust to get there. The diagram in the middle of page 675 summarizes these interpretations.

In Section 17-5, you get to the heart of the matter—the **calculation of concentrations from equilibrium constants.** In this most important kind of calculation, you typically know what the concentrations of all species are at the beginning of the reaction, as well as the value of K_c, and you are asked to calculate the concentrations of some or all of the substances when equilibrium has been reached. Be sure that you see how this differs from the stoichiometric calculations you learned in Chapter 3. In the calculations in this section, the reactants may or may not be present initially in stoichiometric amounts. In stoichiometry problems, you were finding out the maximum amount of product that could be formed *if* the reaction went to completion; now you are recognizing that few reactions actually go to completion.

The strategy of analyzing these problems as shown in Section 17-5 will be very useful. You are sometimes required to solve a quadratic equation in problems of this kind. Pay close attention to the comments in the text regarding which of the two mathematically correct roots of the quadratic equation is the one to be used for the problem at hand. Do not look for a rule to tell you whether to take the + or the − sign in applying the quadratic formula, as this will depend on how you set up the problem in the first place. There is only one good rule—*think* about the physical significance of the answers. The criterion that will always work is that no concentration can be negative—you could not have less than none of a substance! To put it another way, it is not possible for more of a substance to react than was present to begin with. (Of course, a *change* in concentration can be negative, meaning that the substance is consumed in the reaction.) Whereas chemical equilibrium problems all use essentially the same reasoning, they can appear in many different disguises and with varied wordings. You must work many of these problems to become comfortable with them. The advice early in Chapter 1 of this guide to be systematic as you interpret and work through problems is never more important than in learning to do equilibrium calculations!

The various factors that may affect chemical equilibria are assessed descriptively in Section 17-6. Think about a reaction at balance (equilibrium), and then throw it out of balance and use LeChatelier's Principle to tell how it could regain equilibrium. Four "stresses" are considered:

(1) **Changes in concentration.** The reaction quotient is useful in analyzing the response to this stress. At equilibrium, Q is equal to K. Changing one or more of the concentrations, by adding or removing a reactant or product, alters the value of Q (but *not K*!). You then reason whether Q must increase or decrease to get back to the equilibrium value of K. Because product concentrations appear in the numerator and reactant concentrations in the denominator, this tells you which way the reaction must proceed to re-establish equilibrium. It is much safer (and surely more satisfying) to learn to analyze the situation yourself than simply to memorize the directions of shift summarized in the text.

(2) **Changes in volume and pressure.** These changes can be interpreted simply in terms of the resulting changes in concentrations and then analyzed as just outlined. Solids and liquids are nearly incompressible, so pressure changes are important only if one or more reactant or product is a gas. Notice again that these changes do not change the value of K, though the value of Q may be altered.

(3) **Changes in temperature.** The direction of such a change can be reasoned out by thinking of heat as a product of an exothermic reaction, or as a reactant in an endothermic reaction, and then applying LeChatelier's Principle. But it is important to see that a change in temperature, like *no other change,* alters the numerical value of K. The change in K caused by changing T will be calculated later in the chapter.

(4) **Introduction of a catalyst.** You saw in Chapter 16 that introduction of a catalyst can change reaction rates dramatically. However, you must remember that it *does not* affect the position of equilibrium. You should understand how this can be true in terms of the action of a catalyst in lowering the activation energies of both the forward and the reverse reactions. Because not all reactions reach equilibrium quickly, we might desire to introduce a catalyst to help it reach equilibrium in a useful time.

A thoughtful study of the discussion of the Haber process (Section 17-7) for the industrial synthesis of ammonia will help you understand these aspects of chemical equilibrium.

These qualitative considerations of Section 17-6 regarding stresses applied to a system at equilibrium are studied quantitatively in Section 17-8. Study the examples carefully, not just so you can learn to do these kinds of problems, but for the understanding that you can gain of the nature of chemical equilibrium and LeChatelier's Principle. In all equilibrium problems, it is helpful to use the relationship

equilibrium concentration = initial concentration + change in concentration due to reaction

This approach will also be very helpful to you in learning to work problems that involve disturbances from equilibrium. Just think of imposing a change on the equilibrium concentrations to produce the "new initial concentrations." These are then changed by the response of the reaction to the stress as discussed above, to reach the "new equilibrium concentrations." When you have finished working a problem of this type, reason whether the answer is consistent with the changes expected from the qualitative approach you learned in Section 17-6.

The next three sections deal with other ways to represent the equilibrium constant that are sometimes more useful. In Sections 17-9 and 17-10, we see that the equilibrium condition can also be described in terms of the partial pressures of gases, an approach that may be more useful for reactions that take place in the gas phase. The relation between the **pressure equilibrium constant**, K_P, and the **concentration equilibrium constant**, K_c, is discussed in Section 17-10. Remember that Δn has the same meaning here that it had in relating ΔE to ΔH in Chapter 15, the change in number of moles of gaseous substances *in the balanced equation*. Section 17-11 shows how the equilibrium constant is formulated and interpreted in the case of **heterogeneous reactions**, that is, when at least one phase is a pure solid or a pure liquid.

In your study of chemical thermodynamics, you saw that ΔG could be used as a criterion for spontaneity of a reaction. A negative value of ΔG means that the reaction is product-favored (spontaneous in the forward direction) and a positive value means that it is reactant-favored (spontaneous in the reverse direction). This implies that when ΔG is neither negative nor positive, but zero, the reaction is spontaneous in neither direction, i.e., that it is at equilibrium. As you can see in Section 17-12, this means that you can relate the **value of ΔG_{rxn}^0**, which corresponds to unit concentrations of reactants and products, to the **value of the equilibrium constant, K**. Review two methods for the calculation of ΔG_{rxn}^0 from tabulated values (Section 15-16). The relation between ΔG_{rxn}^0 and K has two practical uses: (1) If a value of a desired equilibrium constant cannot be found in tables, then more readily available tables of ΔG_f^0 (or ΔH_f^0 and S^0) values can be used to obtain ΔG_{rxn}^0. Then the numerical value of K can be found, from which the equilibrium concentrations can be calculated. (2) Experimental determination of equilibrium constant values can sometimes be used to evaluate ΔG_f^0 and other thermodynamic quantities for substances that are not otherwise available.

The final section of the chapter, Section 17-13, returns to a quantitative consideration of the change of K with temperature, first described in general terms in Section 17-6. The key relationship of this section is the **van't Hoff equation.** This equation has two practical uses:

(1) If you know the value of K at one temperature, T_1, and also know the standard molar enthalpy of reaction, ΔH_{rxn}^0 (calculated from tabulated values of ΔH_f^0), you can calculate the value of K at another desired temperature, T_2. This is demonstrated in Example 17-20.

(2) From experimental measurements of K at different temperatures, you can calculate ΔH_{rxn}^0.

This latter application is useful for reactions that involve substances for which ΔH_f^0 values are not tabulated or for reactions that may be difficult to measure with a calorimeter. This might be due, for instance, to scarcity of material, very small amounts of heat given off, or unfavorably slow reaction rates.

It is important to carry out the calculations of these last two sections using the proper form of K. For your calculations, remember that when the reaction takes place in the gas phase, the K referred to in Sections 17-12 and 17-13 is the *pressure* equilibrium constant, K_P; when the reaction is in liquid solution, K refers to the *concentration* equilibrium constant, K_c.

At first encounter, the topic of chemical equilibrium is often difficult for students. But it is one of the most important parts of a study of chemistry, and understanding it well is worth your patient, systematic, and diligent study. In the next chapters, you will see the application of these concepts of chemical equilibrium to particular types of processes, such as reactions of acids and bases, hydrolysis reactions, and the dissolution and precipitation of slightly soluble ionic substances.

Study Goals

Section 17-1
1. Describe the concept of chemical equilibrium as a dynamic equilibrium. Be able to use real or hypothetical reactions in your explanation. Relate chemical equilibrium to forward and reverse reaction rates. *Work Exercises 1, 3, 4, 6 through 8, and 10.*

Sections 17-2, 17-3, and 17-9 through 17-11
2. Write the equilibrium constant expression for any balanced chemical equation, for either a homogeneous or a heterogeneous reaction. *Work Exercises 2, 9, 14 through 21, and 30.*

Section 17-2
3. Be able to reason qualitatively (i.e., without calculations) from information about the magnitude of K. *Work Exercises 5, 11 through 13, 22, 23, 35, and 36.*

Sections 17-2, 17-9, and 17-10
4. Given the equilibrium concentrations or the equilibrium partial pressures, evaluate the equilibrium constant, K_c or K_P, for a reaction at a particular temperature. *Work Exercises 24 through 26, 28, and 31 through 34.*

Section 17-4
5. Distinguish between the reaction quotient, Q, and the equilibrium constant, K. Be able to use Q to assess whether or not a system is at equilibrium, and if not, how it must proceed to approach equilibrium. *Work Exercises 29 and 37 through 43.*

Section 17-5
6. Given initial concentrations of all species in a reaction and the value of K_c for the reaction, calculate equilibrium concentrations of all species. *Work Exercises 27 and 44 through 52.*

Sections 17-5 and 17-6
7. Use LeChatelier's Principle to determine the direction in which a system at equilibrium will shift (if it does) when changes of the following kinds are made: (a) change in pressure, (b) change in volume, (c) change in temperature, (d) change in amounts of reactants or products present, (e) addition of a catalyst. *Work Exercises 53 through 63.*

Section 17-8
8. Given concentrations or partial pressures of all species in a system at equilibrium, to which a stress is then applied, determine the concentrations or partial pressures of all species after equilibrium is re-established. *Work Exercises 64 through 68.*

Section 17-10
9. Carry out calculations as in Study Goals 4 and 6, but involving partial pressures and K_P. Be able to convert between K_c and K_P for gas phase reactions. *Work Exercises 69 through 79.*

Section 17-12
10. Given the equilibrium constant for a reaction at a particular temperature, calculate the standard free-energy change, ΔG^0_{rxn}, at that temperature, and vice versa. *Work Exercises 81 through 86 and 88.*

Section 17-13

11. Use the van't Hoff equation to relate the standard enthalpy change (ΔH^0) and the equilibrium constant values at two different temperatures. *Work Exercises 80 and 87.*

General

12. Be able to recognize and solve a variety of types of problems based on the principles of this chapter. *Work* Mixed Exercises *92 through 99.*

13. Answer conceptual questions based on this chapter. *Work* **Conceptual Exercises** *100 through 108.*

14. Apply concepts and skills from earlier chapters to the ideas of this chapter. *Work* **Building Your Knowledge** *exercises 109 through 113.*

15. Exercises at the end of chapter direct you to sources outside the textbook for information to use in solving them. Work **Beyond the Textbook** *exercises 114 through 117.*

Some Important Terms in This Chapter

Some important new terms from Chapter 17 are listed below. Fill the blanks in the following paragraphs with terms from the list. Each term is only used once. Check the Key Terms list and the chapter reading. Study other new terms, and review terms from preceding chapters if necessary.

chemical equilibrium	heterogeneous equilibrium	reaction quotient
dynamic equilibrium	homogeneous equilibrium	reversible reaction
equilibrium constant	LeChatelier's Principle	van't Hoff equation

A reaction that does not go to completion *and* that can occur in either direction is called a [1]_____. A reversible reaction can be represented, in general, by the equation: $a\text{A} + b\text{B} \rightleftharpoons c\text{C} + d\text{D}$, where A and B are considered the reactants, and C and D are considered the products. The double arrow ($\rightleftharpoons$) indicates that both the forward and reverse reactions occur simultaneously. [2]_____ exists when the two opposing reactions occur simultaneously at the same rate. A chemical equilibrium is an example of [3]_____ because individual molecules are continually reacting, even though the overall composition of the reaction mixture does not change. The extent to which the reversible reaction occurs is expressed quantitatively by the [4]_____, K_c or simply K, a kind of ratio of the amounts of the products to the amounts of the reactants. A small value for the equilibrium constant indicates little product produced and a lot of reactant left, while a large value suggests a lot of product produced and little reactant left. If the equilibrium constant for a reaction is known, whether the system is at equilibrium, and, if not, which direction the reaction must proceed for equilibrium to be established can be determined by calculating Q, the [5]_____. The reaction quotient has the same *form* as the equilibrium constant expression, but the concentrations of the reactants and products are not necessarily the equilibrium concentrations, unless the value calculated for Q is equal to K. When a system is at equilibrium, any change in the conditions that moves it out of equilibrium is a stress to the system. According to the guiding concept of chemical equilibrium, [6]__________________________________, the system shifts in the direction that reduces the stress, to move toward a new state of equilibrium.

A chemical equilibrium can be one of two types: a [7]______________________________________ involves species all in a single phase (all gases, all solutions, etc.), but a [8]____ involves species in more than one phase.

The relationship between ΔH^0 for a reaction and its equilibrium constants at two different temperatures is expressed quantitatively by the [9]_____________________________.

Quotefall

Fit the letters in each vertical column into the squares directly below them to form words across. The letters do not necessarily go into the squares in the same order in which they are listed. Each letter given is used once. A black square indicates a space. Words starting at the end of a line may continue in the next line. Punctuation in the statement is included in the boxes. When all the letters have been placed in their correct squares, you will be able to read the complete quotation across the diagram from left to right, line by line. Hint: It may be helpful at times to fill in small words, like "the," to use up some letters to help in determining other longer words. Some letters have been "seeded" to help you get started. The solution is on page 243.

Puzzle #1:

M	N	G	I	I	G	Q	L	T	H	A	B	O	N	U	R	E	C	C	I	D	L	S	E	S	R	B	E	E	S	S	Y	A	D	H
A	E	C	H	S	T	N	I	I	L	E	T	S	I	S	M	U	I	O	N	S	I	H	T	U	P	I	N	I	T	H	Y	S	C	I
T	E	A	T	N	E	E	U	T	H	R	C	R	Y	E	D	T	M	S	S	P	T	E	A	P	I	F	R	D	N	G	E		T	E
L	R	C	T	A	O			I	E	I		S	R	P	T	I	N	E	S	H	I	F	T		S	L	Y	A		B				S
R		E	S					T	S					D	I		I									T				S				
		C									'		■		R								:	■			■		■					
	■			■					L							■			■							B			■			■		
						■				■				D							■	(								G	■		■	
			S		)	,	■				■							■							■			■				■		
R							■	T				■					C			■				■							.	■	■	■

Preliminary Test

This test will check your understanding of basic concepts and types of calculations. Most students require more practice than they anticipate on the calculations and reasoning of chemical equilibrium. Be sure to practice *many* of the additional text Exercises, including those indicated with the Study Goals.

True-False

Mark each statement as true (T) or false (F).

_____ 1. In any equilibrium mixture, the amounts of substance present must be in ratios given by the stoichiometric coefficients in the balanced chemical equation.

_____ 2. Generally the larger the numerical value of K_c, the more nearly to completion the reaction goes before reaching equilibrium.

_____ 3. A living organism is in a state of chemical equilibrium.

_____ 4. In writing the expression for K, we omit species present as pure solids or pure liquids, but include all substances present in solution or as gases.

_____ 5. When we need to solve a quadratic equation in a chemical equilibrium problem, we always use the positive sign before the square root expression in the quadratic formula.

_____ 6. Because the reaction quotient has the same algebraic form in the concentrations as does the equilibrium constant, Q must always equal K_c.

_____ 7. For the reaction $A + B \rightleftharpoons C$, for which $\Delta G^0_{rxn} = -44.5$ kJ, there will be only C left after a long enough period of time if stoichiometric amounts of A and B are allowed to react.

The next eleven questions, 8 through 18, refer to the endothermic equilibrium reaction:

$$CoCO_3(s) \rightleftharpoons CoO(s) + CO_2(g)$$

_____ 8. At equilibrium conditions, the forward and the reverse reactions have stopped.

_____ 9. The same equilibrium condition would result if we started with *only* pure $CoCO_3(s)$ in a closed container or if we started with *only* pure $CoO(s)$ in a closed container.

_____ 10. Equal concentrations of $CoO(s)$ and $CO_2(g)$ result from the decomposition of a given amount of $CoCO_3(s)$.

_____ 11. Equal numbers of moles of $CoO(s)$ and $CO_2(g)$ result from the decomposition of a given amount of $CoCO_3(s)$.

_____ 12. Decreasing the volume of the closed system at constant temperature results in more $CoCO_3(s)$ being formed.

_____ 13. Introducing 1.00 atm additional pressure of $CO_2(g)$ to the system at equilibrium in a closed container results in more $CoCO_3(s)$ being formed.

_____ 14. Introducing 1.00 atm additional pressure of $N_2(g)$ to the system at equilibrium in a closed container results in more $CoCO_3(s)$ being formed.

_____ 15. Increasing the temperature of the reaction mixture, while holding the pressure constant by allowing the container to expand or contract, results in more $CoCO_3(s)$ being formed.

_____ 16. The equilibrium constant expression is: $K_c = [CO_2]$.

_____ 17. The reaction is a heterogeneous reaction.

_____ 18. If the temperature increases, then the value of the equilibrium constant increases.

Short Answer

1. LeChatelier's Principle may be stated as follows: _____
   _____
   _____ .

2. If an equation for a reaction is multiplied by a number n, the original value of K_c is altered by
   _____
   .

3. Suppose we know the numerical value of K_c for a reaction (call this numerical value x). If we reverse the direction in which the equation is written, the numerical value of the equilibrium constant for this reversed equation will be _____.

4. What is the equilibrium constant expression, K_c, for the reaction below?

   $$NO_2(g) + SO_2(g) \rightleftharpoons NO(g) + SO_3(g)$$

   _____

5. What is the pressure equilibrium constant expression, K_P, for the reaction in Question 4?
   _____

6. What is the equilibrium constant expression, K_c, for the reaction below?

   $$4FeS_2(s) + 11O_2(g) \rightleftharpoons 2Fe_2O_3(s) + 8SO_2(g)$$

   _____

7. What is the pressure equilibrium constant expression, K_P, for the reaction in Question 6?
   _____

8. In qualitative analysis for metal ions, sulfide ions are often produced by the reaction

   $$H_2S(aq) + 2H_2O(\ell) \rightleftharpoons 2H_3O^+(aq) + S^{2-}(aq)$$

   The addition of _____ would decrease the S^{2-} concentration in the equilibrium mixture.

9. The slightly soluble substance barium fluoride dissolves only slightly in water, according to the equation below. What is the equilibrium constant expression for this reaction?

   $$BaF_2(s) \rightleftharpoons Ba^{2+}(aq) + 2F^-(aq)$$ _____

In the Mond process for refining nickel, the gaseous substance nickel tetracarbonyl, $Ni(CO)_4$, is heated to cause it to decompose into solid nickel and gaseous carbon monoxide, CO. Questions 10 through 13 refer to this reaction.

10. The expression for the equilibrium constant, K_c, for this reaction is
    _____ .

11. What is the pressure equilibrium constant expression, K_P, for this reaction?
    _____

12. The value for Δn for this reaction is _____.

13. The numerical value of K_P for this reaction at room temperature is _____ (larger than, smaller than, the same as) the numerical value for K_c.

Questions 14 through 18 refer to the reaction

$$2NO(g) + 2H_2(g) \rightleftharpoons N_2(g) + 2H_2O(g)$$

This reaction is strongly exothermic. In each question, we allow the reaction to attain equilibrium; then we make *only* the one change in the conditions indicated and allow the reaction to re-establish equilibrium. For each, predict the effect on (a) the amount of N_2 present compared with the original equilibrium conditions and (b) the numerical value of K. The possible answers are: increase (I), decrease (D), or remain unchanged (U).

Change	Amount of N_2	Value of K
14. More NO is added to the reaction mixture.	_____	_____
15. A catalyst is introduced.	_____	_____
16. Some $H_2O(g)$ is removed from the reaction mixture by introducing a drying agent.	_____	_____
17. The temperature is decreased.	_____	_____
18. The volume of the reaction container is doubled.	_____	_____

19. The thermodynamic quantity that is most closely related to the value of the equilibrium constant is _____.
20. The thermodynamic quantity that is most closely related to the change in value of the equilibrium constant is _____.

Multiple Choice

_____ 1. When a reaction reaches equilibrium,
   (a) the rate of the forward reaction and the rate of the reverse reaction are both zero.
   (b) all reaction stops.
   (c) the rate of the forward reaction is zero.
   (d) the rate of the reverse reaction is zero.
   (e) the forward and reverse reaction rates are equal.

_____ 2. In which of the following cases will the least time be required to arrive at equilibrium?
   (a) K_c is a very small number.
   (b) K_c is a very large number.
   (c) K_c is approximately 1.
   (d) Cannot tell without knowing the value of K_c.
   (e) Cannot tell because the time to arrive at equilibrium does not depend on K_c.

_____ 3. Which one of the following statements is *not* true?
   (a) A system that is disturbed from an equilibrium condition reacts in a manner to restore equilibrium.
   (b) Equilibrium in molecular systems is static, all molecular interactions having stopped.
   (c) The nature of an equilibrium mixture is the same regardless of the direction from which it was reached.
   (d) A system moves spontaneously toward a state of equilibrium.
   (e) The equilibrium constant usually depends on temperature.

_____ 4. In the reaction

$$A + B \rightleftharpoons C + D$$

for which k_f and k_r are the forward and reverse specific rate constants, respectively, which of the following is correct at equilibrium?
   (a) $k_f = k_r$ (b) $k_f > k_r$ (c) $k_f < k_r$ (d) $\text{rate}_f = \text{rate}_r$ (e) $K_c = \dfrac{[A][B]}{[C][D]}$

_____ 5. For the reaction

$$2A + B \rightleftharpoons C + 2D \qquad \text{at } 35°C$$

the value of k_f is 3.0×10^{-3} $M^{-1}\cdot s^{-1}$ and the value of k_r is 1.5×10^{-2} $M^{-1}\cdot s^{-1}$. Calculate the value of K_c for this reaction.
   (a) 2.0   (b) 0.5   (c) 0.2   (d) 5.0   (e) 4.5×10^{-5}

_____ 6. Solid ammonium carbamate, $NH_4CO_2NH_2$, dissociates completely into ammonia and carbon dioxide. When pure ammonium carbamate is put into an evacuated container and allowed to come to equilibrium with the gaseous products, the total pressure at 45°C is found to be 0.6 atm. What is the value of K_P for this reaction at 45°C?
   (a) 6.0×10^{-2} (b) 4.0×10^{-3} (c) 8.0×10^{-2}
   (d) 3.2×10^{-2} (e) 2.16×10^{-1}

_____ 7. The equilibrium constant for the gaseous reaction

$$HCHO \rightleftharpoons H_2 + CO$$

has the numerical value 0.50 at 600°C. A mixture of HCHO, H_2, and CO is introduced into a flask at 600°C. After a short time, analysis of a small sample of the reaction mixture shows the molar concentrations to be [HCHO] = 1.5, [H_2] = 0.5, and [CO] = 1.0.
Which of the following statements about this reaction mixture is true?
   (a) The reaction mixture is at equilibrium.
   (b) The reaction mixture is not at equilibrium, but no further reaction will occur.
   (c) The reaction mixture is not at equilibrium, but will move toward equilibrium by forming more HCHO.
   (d) The reaction mixture is not at equilibrium, with will move toward equilibrium by using up more HCHO.

_____ 8. The numerical value of the concentration equilibrium constant for the gaseous reaction

$$2SO_2 + O_2 \rightleftharpoons 2SO_3$$

is 0.5 at some temperature. When a reaction mixture is brought to equilibrium at that temperature, $[O_2]$ is found to be 0.5 M and $[SO_3]$ is found to 5.0 M.

What is the equilibrium concentration of SO_2 in this mixture, in mol/L?

(a) 0.5   (b) 1.0   (c) 2.0   (d) 5.0   (e) 10.

_____ 9. The equilibrium constant for the gaseous reaction

$$CO + H_2O \rightleftharpoons CO_2 + H_2$$

is 4.0 at a certain temperature. A reaction is carried out at this temperature starting with 2.0 mol/L of CO and 2.0 mol/L of H_2O. What will be the equilibrium concentration of H_2, in mol/L?

(a) 2.0   (b) 0.75   (c) 1.33   (d) 0.67   (e) 1.5

_____ 10. The introduction of a catalyst into a reaction mixture at equilibrium would

(a) cause a change in the amount of heat absorbed or evolved.

(b) change the value of $\rightleftharpoons$ for the reaction.

(c) change the value of the equilibrium constant.

(d) change the relative amounts of reactants and products present at equilibrium.

(e) cause none of the changes listed above.

_____ 11. Suppose the reaction mixture

$$N_2(g) + 3H_2(g) \rightleftharpoons 2NH_3(g)$$

is at equilibrium at a given temperature and pressure. The pressure is then increased at constant temperature by compressing the reaction mixture. The mixture is then allowed to re-establish equilibrium. At the new equilibrium

(a) there is more ammonia present than there was originally.

(b) there is less ammonia present than there was originally.

(c) there is the same amount of ammonia present as there was originally.

(d) the nitrogen is used up completely.

(e) the amount of ammonia present may be either greater or smaller than it was originally, depending on the value of K.

_____ 12. Suppose the reaction mixture

$$H_2(g) + Cl_2(g) \rightleftharpoons 2HCl(g)$$

is at equilibrium at a given temperature and pressure. The pressure is then increased at constant temperature by compressing the reaction mixture. The mixture is then allowed to re-establish equilibrium. At the new equilibrium

(a) there is more hydrogen chloride present than there was originally.

(b) there is less hydrogen chloride present than there was originally.

(c) there is the same amount of hydrogen chloride present as there was originally.

(d) the hydrogen and chlorine are completely used up.

(e) the amount of hydrogen chloride present may be either greater or smaller than it was originally, depending on the value of K.

_____ 13. An acetic acid solution is allowed to come to equilibrium:
$$CH_3COOH(aq) + H_2O(aq) \rightleftharpoons H_3O^+(aq) + CH_3COO^-(aq)$$
If some silver ion, Ag^+, is then added to the solution, solid silver acetate, $AgCH_3COO$, is formed. The resulting amount of undissociated acetic acid, CH_3COOH, in solution would be
(a) unchanged from the amount in the original solution.
(b) higher than in the original solution.
(c) lower than in the original solution.
(d) zero.

_____ 14. The reaction
$$CaCO_3(s) + 2H_3O^+(aq) \rightleftharpoons Ca^{2+}(aq) + 2H_2O(\ell) + CO_2(g)$$
may be forced to completion (that is, forced to consume all of the $CaCO_3$ present) by which one of the following methods?
(a) Removing some H_3O^+ from the reaction mixture by neutralizing it with base.
(b) Adding more Ca^{2+} to the solution.
(c) Removing CO_2 gas as it is formed.
(d) Adding CO_2 gas to the reaction mixture.
(e) None of the preceding would have any effect on the amount of $CaCO_3$ consumed.

_____ 15. For the heterogeneous reaction
$$2MnO_2(s) \rightleftharpoons 2MnO(s) + O_2(g)$$
the equilibrium constant expression for K_c is which one of the following?

(a) $K_c = \dfrac{[MnO]^2[O_2]}{[MnO_2]^2}$

(b) $K_c = [O_2]$

(c) $K_c = \dfrac{1}{[O_2]}$

(d) $K_c = k[MnO_2]^2$

(e) $K_c = \dfrac{[MnO_2]^2}{[MnO]^2[O_2]}$

_____ 16. Given the equilibrium reaction

$$MgSO_4(s) \rightleftharpoons MgO(s) + SO_3(g)$$

which one of the following statements is true?
(a) If the container were opened to the atmosphere, eventually only $MgSO_4(s)$ would remain.
(b) Decreasing the volume of the closed container at constant temperature results in more $MgO(s)$ being formed.
(c) The same equilibrium condition would result if we started with *only* pure $MgSO_4(s)$ in a closed container as if we started with *only* pure $MgO(s)$ in a closed container.
(d) At equilibrium the forward and the reverse reactions have stopped.
(e) If pure $MgSO_4(s)$ were placed in a closed container filled with $N_2(g)$ at 1 atm, the total pressure would increase due to $SO_3(g)$ being formed.

_____

The following three questions, 17 through 19, refer to the exothermic reaction:

$$2SO_2(g) + O_2(g) \rightleftharpoons 2SO_3(g)$$

_____ 17. Suppose we let the reaction come to equilibrium. Then we decrease the temperature of the reaction mixture. What will be the effect on the net amount of $SO_3(g)$ present?
(a) It increases.　　(b) It decreases.　　(c) It does not change.
(d) The question cannot be answered without knowing the value of K_c or K_P.
(e) The question cannot be answered without knowing the value of ΔH^0.

_____ 18. Suppose we let the reaction come to equilibrium. Then we decrease the total pressure by increasing the volume of the container. What will be the effect on the net amount of $SO_3(g)$ present?
(a) It increases.　　(b) It decreases.　　(c) It does not change.
(d) The question cannot be answered without knowing the value of K_c or K_P.
(e) The question cannot be answered without knowing the value of ΔH^0.

_____ 19. Suppose we let the reaction come to equilibrium. Then we decrease the temperature and decrease the total pressure simultaneously. What will be the effect on the net amount of $SO_3(g)$ present?
(a) It increases.　　(b) It decreases.　　(c) It does not change.
(d) It may increase, decrease, or remain the same, depending on the sizes of the temperature and pressure changes.
(e) It may increase, decrease, or remain the same, depending on the numerical value of the equilibrium constant.

_____

_____ 20. At 25°C, the value of K_c for a particular reaction is 30. At 75°C, the value of K_c for this same reaction is 40. Which one of the following statements about this reaction is true?
(a) The reaction is at equilibrium.
(b) LeChatelier's Principle does not apply to this reaction.
(c) The reaction is endothermic.
(d) The reaction is exothermic.
(e) Increasing the pressure would decrease the value of K_c.

Answers to Some Important Terms in This Chapter

1. reversible reaction
2. chemical equilibrium
3. dynamic equilibrium

4. equilibrium constant
5. reaction quotient
6. LeChatelier's Principle

7. homogeneous equilibrium
8. heterogeneous equilibrium
9. van't Hoff equation

Answer to Quotefall Puzzle

Puzzle #1:
LeChatelier's Principle: If a system at equilibrium is disturbed by changing its conditions (applying a stress), the system shifts in the direction that reduces the stress.

Answers to Preliminary Test

Do not just look up answers. *Think about the reasons for the answers.*

True-False

1. False. The *changes* in concentrations of all reactants and products due to reaction must always be in the ratio of the stoichiometric coefficients, but the amounts themselves may not be.
2. True. K_c is a kind of ratio of the amounts of the products to the amounts of the reactants. A small value for the equilibrium constant indicates little product produced and a lot of reactant left, while a large value suggests a lot of product produced and little reactant left.
3. False. Obviously not all reactions in a living organism are at equilibrium at any given time.
4. True. Since the concentrations of pure solids or pure liquids are dependent upon the density of the substance and do not change during the course of the reaction, their activity is taken as one.
5. False. The only sure way to know is to find out which root leads to physically reasonable results. Usually this means that the incorrect root would predict negative concentrations for one or more substances in the reaction.
6. False. This is true only at equilibrium. Study Section 17-4 and the uses of the reaction quotient outlined in Section 17-6.

7. False. The negative value for ΔG_{rxn}^0 only tells us that the reaction is spontaneous at the standard state conditions. As A and B react, their concentrations decrease, and ΔG (not ΔG_{rxn}^0) approaches 0 as the reaction approaches equilibrium.

8. False. Chemical equilibrium is a *dynamic* equilibrium—both the forward and reverse reactions continue, but at the same rate, canceling their effects.

9. False. Starting with only pure CoO(s), equilibrium could never be reached. Some CO_2(g) is necessary to allow the reverse reaction to proceed at all.

10. False. We do not refer to "concentrations" of pure solids. The activity of a pure liquid or solid is one. Equal numbers of *formula units* of CoO and CO_2 are produced.

11. True

12. True. Decreasing the volume increases the pressure of CO_2(g), which increases the reverse reaction.

13. True. The reverse reaction is increased.

14. False. Because N_2(g) is neither a reactant nor a product, its presence does nothing to disturb the equilibrium.

15. False. Since the reaction is endothermic, increasing the temperature will shift the equilibrium to the right. More $CoCO_3$(s) will decompose.

16. True. Remember that solids do not appear in the equilibrium constant expression.

17. True. Solid and gas are different phases.

18. True. Because the reaction is endothermic (consumes heat), the forward reaction would be favored at higher temperatures. This would form more products and consume reactants, so the value of K would increase.

Short Answer

1. If a system at equilibrium is disturbed by changing its conditions (applying a stress), the system shifts in the direction that reduces the stress. (Of course, it is much better for you to know this and be able to state it in your own words than just to memorize a set statement.)

2. raising it to the *n*th power. You do not need to remember this rule—if you understand how to write an equilibrium constant from the chemical equation, you can easily figure it out.

3. $\frac{1}{x}$. Same comments as for Question 2.

4. $K_c = \dfrac{[NO][SO_3]}{[NO_2][SO_2]}$

5. $K_P = \dfrac{(P_{NO})(P_{SO_3})}{(P_{NO_2})(P_{SO_2})}$

6. $K_c = \dfrac{[SO_2]^8}{[O_2]^{11}}$

7. $K_P = \dfrac{(P_{SO_2})^8}{(P_{O_2})^{11}}$

8. H_3O^+. This is an example of a weak acid equilibrium. This will be covered in more detail in Chapter 18, but you can answer this with your general knowledge about equilibrium from Chapter 17.

9. $[Ba^{2+}][F^-]^2$. Equilibria involving slightly soluble salts will be covered in more detail in Chapter 20. This is clearly just a case of heterogeneous equilibrium, which you have learned about in Chapter 17.

10. $K_c = \dfrac{[CO]^4}{[Ni(CO)_4]}$. From the description given for the reaction, you should be able to write the chemical equation as: $Ni(CO)_4(g) \rightleftharpoons Ni(s) + 4CO(g)$

11. $K_P = \dfrac{(P_{CO})^4}{(P_{Ni(CO)_4})}$

12. 3. Remember that this is the number of moles of gas products minus the number of moles of gas reactants in the balanced equation.

13. larger than. Remember that $K_P = K_c(RT)^{\Delta n}$ (Section 17-10). From the preceding question, Δn is positive.

14. I, U

15. U, U

16. I, U

17. I, I. Only a temperature change can alter the value of K.

18. D, U. The only way to change the value of K is to change the temperature, so K does not change. Doubling the volume of the container has the effect of multiplying all gas concentrations (or all gas pressures) by a factor of $\frac{1}{2}$. Because there are more moles of gas on the left than on the right, the denominator of Q decreases more than the numerator. Thus, Q is now greater than K. The reaction shifts to re-establish equilibrium by proceeding to the left, making more moles of gas, and decreasing the amount of N_2 present.

19. ΔG^0_{rxn}. $\Delta G^0_{rxn} = -RT \ln K$

20. ΔH^0. $\ln\left(\dfrac{K_{T_2}}{K_{T_1}}\right) = \dfrac{\Delta H^0}{R}\left(\dfrac{1}{T_1} - \dfrac{1}{T_2}\right)$

Multiple Choice

1. (e). Chemical equilibrium is a *dynamic* equilibrium—both the forward and reverse reactions continue, but at the same rate, canceling their effects.

2. (e)

3. (b). See the answer to question 1.

4. (d). As equilibrium is reached, the rates change, but the values of k_f and k_r do not. Do not get the idea that $k_f = k_r$ at equilibrium; rather, the concentrations are such that (d) is true. Answer (e) must be inverted to be correct.

5. (c). From Section 17-2, $K_c = \dfrac{k_f}{k_r} = \dfrac{3.0\times10^{-3}\ M^{-1}\cdot s^{-1}}{1.5\times10^{-2}\ M^{-1}\cdot s^{-1}} = 0.2$

6. (d). You should be able to figure out that the reaction is

263

$$NH_4CO_2NH_2(s) \rightleftharpoons 2NH_3(g) + CO_2(g)$$

If the total pressure is 0.6 atm, the partial pressure of NH_3 is 0.4 atm and that of CO_2 is 0.2 atm. You should be able to find these from the stoichiometry of the reaction. Then put them into the expression $K_P = (P_{NH_3})^2(P_{CO_2}) = (0.4)^2(0.2) = 0.032$.

7. (d). Putting the given values of concentrations into the expression for Q, you find that $Q = 0.333$ $[\frac{(0.5)(1.0)}{1.5}]$. This is less than K, so the reaction must proceed to make more products, raising the value of the numerator and lowering the denominator. This uses up HCHO.

8. (e). Put the known values of concentrations into the expression for K_c and then solve for $[SO_2]$.

$K_c = \dfrac{[SO_3]^2}{[SO_2]^2[O_2]}$, so $0.5 = \dfrac{(5.0)^2}{[SO_2]^2(0.5)}$; $[SO_2]^2 = \dfrac{(5.0)^2}{(0.5)(0.5)}$; $[SO_2]^2 = 100$; $[SO_2] = 10$

9. (c). This calculation is like Examples 17-5 through 17-7.
Let $x = M$ of CO that react. Then $x = M$ of H_2O that react, and $x = M$ of CO_2 and H_2 formed.

	CO	+	H_2O	$\rightleftharpoons$	CO_2	+	H_2
initial	2.0 M		2.0 M		0 M		0 M
change due to rxn	$-x$ M		$-x$ M		$+x$ M		$+x$ M
equilibrium	$(2.0-x)$ M		$(2.0-x)$ M		x M		x M

$K_c = \dfrac{[CO_2][H_2]}{[CO][H_2O]} = 4.0$ so $\dfrac{(x)(x)}{(2.0-x)(2.0-x)} = 4.0$ or $\dfrac{x^2}{(2.0-x)^2} = 4.0$

Taking the square root of both sides, $\dfrac{x}{2.0-x} = 2.0$ or $x = 1.33$ $M = [H_2]$ at equilibrium.

10. (e). Introduction of a catalyst does not affect an equilibrium, except to get the mixture to equilibrium sooner.

11. (a). There are more moles of gas on the left $(3 + 1 = 4)$ than on the right (2) of the balanced equation. Thus, an increase in pressure would disturb the equilibrium. This stress could be relieved by the reaction proceeding more to the right.

12. (c). In this case, there are equal numbers of moles of gas on the two sides of the balanced equation $(1 + 1 = 2)$. Thus, an increase in pressure would not disturb the equilibrium.

13. (c). This combines ideas that will be covered quantitatively in Chapters 18 and 19 (weak acid equilibrium) and in Chapter 20 (slightly soluble salt equilibrium). It can be answered easily from the general principles of equilibrium presented in Chapter 12.

14. (c). Removing CO_2 gas shifts the equilibrium to the right (increases the forward reaction).

15. (b). O_2, a product, is in the numerator. The MnO and MnO_2 are both solids, so they are not included in the equilibrium constant expression.

16. (e). As $MgSO_4$ decomposes, SO_3 is produced. The N_2 is irrelevant.

17. (a). Since the reaction is exothermic, decreasing the temperature would shift the equilibrium to the right (decreases the forward reaction less than the reverse reaction), as if heat were a product being removed.

18. (b). There are more moles of gas on the left than on the right of the balanced equation. Thus, a decrease in pressure would disturb the equilibrium. This stress could be relieved by the reaction proceeding more to the left, using SO_3.

19. (d). A decrease in temperature would increase the amount of SO_3 (question 17), and a decrease in pressure would decrease the amount of SO_3 (question 18). How much is each one changed?

20. (c). You should be able to figure this out by the ideas of Section 17-6. You should also be able to reason it out, without calculation, from the van't Hoff equation, Section 17-13.

18 Ionic Equilibria I: Acids and Bases

Chapter Summary

Many important reactions take place in water solutions—in the laboratory, in living organisms, and in industrial chemistry. You can apply what you learned about chemical equilibrium in Chapter 17 to solutions of strong and weak electrolytes (Section 6-1), with special emphasis on weak acids and bases (Section 6-1 parts 2 and 4, Chapter 10). Some review of the appropriate sections of the chapters just mentioned would be helpful before you begin your detailed study of Chapter 18.

The chapter begins with a review of the terminology of **strong electrolytes** (Section 18-1). Until now, you have seen this topic only in a descriptive way. Here you use it to calculate the concentrations of each ionic species present in solution when a strong electrolyte (a strong acid, a strong base, or a soluble ionic salt) dissolves in water. The key idea here is that you consider each reaction to go *virtually to completion*, so you can calculate concentrations based solely on interpreting the stoichiometric coefficients in the balanced chemical equation. Do not apply the ideas of chemical equilibrium to solutions of strong electrolytes. However, you will see in this chapter that when one of the ions is the conjugate acid of a weak base or the conjugate base of a weak acid, you still must apply the concepts of equilibrium.

Application of chemical equilibrium is also important when you consider the **autoionization of water** to form H_3O^+ and OH^- (Section 18-2). This reaction occurs only slightly, so we describe it in terms of an equilibrium constant, $K_w = [H_3O^+][OH^-] = 1.0 \times 10^{-14}$ at 25°C. A very important point is that this equilibrium constant expression *must be satisfied in all aqueous solutions,* no matter what the source of H_3O^+ or OH^- and no matter what other ions may be present. Thus, in any aqueous solution, $[H_3O^+]$ and $[OH^-]$ are related; their product must be equal to K_w. You must become very familiar with the terminology at the end of Section 18-2, by which we refer to a solution as acidic, basic, or neutral, depending on whether $[H_3O^+]$ is greater than, less than, or equal to $[OH^-]$, respectively. Such solutions can contain Brønsted-Lowry or Lewis acids and bases, but it is an overall characteristic of the solution described here.

Because of the inconvenience of the great range of numbers necessary to describe $[H_3O^+]$ or $[OH^-]$, chemists use the **pH** and **pOH scales** (Section 18-3). On these scales,

(1) a ten-fold change in $[H_3O^+]$ or $[OH^-]$ is described by a change of one pH or pOH unit,

(2) increasing $[H_3O^+]$ corresponds to decreasing pH, and

(3) increasing $[OH^-]$ corresponds to decreasing pOH.

Use of the terminology of pH is prevalent in science and engineering, so study this section carefully.

266

There are only a few common strong acids and bases, but many solutions involve **weak acids and weak bases.** These substances are involved in dissociation or ionization reactions that reach equilibrium without going to completion. Thus, we describe the ionization reactions of these substances in terms of equilibrium constants (Section 18-4). This is the most important section of this chapter—be sure you understand it well. These equilibrium constants are **ionization constants**, and represented by K_a for weak acids and K_b for weak bases. Of course, you are really just applying the same kind of equilibrium constant calculations that you learned in Chapter 17. The *very small* values typical of K_a and K_b often help us to solve problems more easily. For instance, you can often neglect very small values when added to or subtracted from much larger numbers. See the Problem-Solving Tip entitled "Simplifying Quadratic Equations" in Section 18-4; be sure that you understand how and when you can make this approximation. In addition to learning to work problems to solve for concentrations, you should

(1) have a feel for relative strengths of acids or bases from looking at K_a or K_b values,

(2) be familiar with the **percent ionization** concept, and

(3) see how the weak acid-strong conjugate base relationships from Chapter 10 apply here.

You will use pH notation throughout these calculations.

You can also use equilibrium constants and expressions to describe **polyprotic weak acids** (Section 18-5), those that ionize stepwise (with varying strengths), each step yielding H_3O^+. The first ionization constant is K_{a1}, the second is K_{a2}, and the third (if it exists) is K_{a3}. Notice that successive steps are always less complete, so that $K_{a1} > K_{a2}$, and so on. There are no *common* examples of weak bases that ionize in a stepwise fashion, although a few inorganic bases and many weak organic bases do so.

In the remainder of this chapter, you continue your study of chemical equilibrium. You now examine equilibria in **hydrolysis reactions** (Section 18-6), those in which water or one of its ions is a reactant. Several sections deal with hydrolysis reactions involving salts of Brønsted-Lowry acids or bases, each of which may be weak or strong. As background to this material, you must be familiar with the Brønsted-Lowry terminology, especially with regard to **conjugate acid-base pairs** and their relative strengths, Section 10-4. At the end of the chapter, you will consider reactions in which water interacts with metal ions.

So far in this chapter, you have dealt with solutions in which an acid or base, either strong or weak, has been added. In Sections 18-7 through 18-10, you are concerned with solutions in which only a salt has been added, but no free acid or base. In pure water $[H_3O^+] = [OH^-]$. In aqueous solutions, reactions can occur in which this $[H_3O^+]/[OH^-]$ balance is upset; thus, the resulting solutions could be acidic (excess H_3O^+), basic (excess OH^-), or neutral. Recall that you can view salts can as being composed of the cation of a base and the anion of an acid (Section 10-8). You can classify each acid or base as either strong or weak, so you will observe four classes of salts; see the summaries of the hydrolysis reactions for these four classes of salts in the table on the next page.

Be sure you can calculate the solution concentrations that result from hydrolysis of salts of the strong base/weak acid and weak base/strong acid types. These results can be expressed in terms of either H_3O^+, OH^-, pH, pOH, or percent hydrolysis. The relation $K_aK_b = K_w$ is very useful in equilibrium constant calculations involving conjugate acid-base pairs.

Another type of hydrolysis reaction occurs in solutions containing **small, highly charged metal ions** (Section 18-11). Water molecules coordinate with these ions; the resulting strong interaction of the metal ion with the oxygen atom of water weakens the O−H bond. This causes the hydrated complex ion to lose H^+ to a solvent water molecule; the hydrated ion thus acts as a Brønsted-Lowry acid. Thus, the solution contains an excess of H_3O^+ over OH^-, and is acidic. As with other reactions that occur only to a limited extent, the principles of chemical equilibrium apply. Because the reaction produces excess H_3O^+, the equilibrium constant used for this type of hydrolysis reaction has the symbol K_a, although it is a **hydrolysis constant.** Do not get so involved with the calculation of concentrations that you forget to see which metal ions hydrolyze readily. Observe how the ease of hydrolysis of metal ions correlates with position in the periodic table, with charge density, and with oxidation state of the metal.

Salt of			Nature of	
Base	Acid	Section	Solution	Comments
strong	strong	18-7	neutral	The conjugate base of the strong acid is too weak to react with H_3O^+; the conjugate acid of the strong base is too weak to react with OH^-. The $[H_3O^+]/[OH^-]$ balance remains undisturbed, so the solution remains neutral.
strong	weak	18-8	basic	The anion of the weak acid is a strong base so it reacts with H_3O^+ (hydrolyzes), leaving excess OH^-. The water equilibrium shifts, making more H_3O^+ and OH^- in equal amounts, so OH^- is still in excess, and the solution is basic. The equilibrium constant for this hydrolysis reaction is $K_b = K_w/K_a$, where K_a is the ionization constant for the parent weak acid.
weak	strong	18-9	acidic	The conjugate acid of the weak base is strong so it reacts with OH^- (hydrolyzes), leaving excess H_3O^+. The water equilibrium shifts, making more H_3O^+ and OH^- in equal amounts, so H_3O^+ is still in excess and the solution is acidic. The equilibrium constant for this hydrolysis reaction is $K_a = K_w/K_b$, where K_b is the ionization constant for the parent weak acid.
weak	weak	18-10	acidic, basic, or neutral	Both ions hydrolyze; the result depends on the relative strengths of the weak acid and weak base.

Study Goals

There is large amount of apparently diverse material in this chapter. The common themes are acid-base behavior and equilibrium considerations. The following Study Goals will help you to organize the material. Be systematic in your study. Write your own summary of each section, with these goals in mind. Also, be systematic in answering Exercises from the text.

Review Sections 6-1, 10-4, 10-7, and 10-8

1. Review the fundamental ideas of acids and bases from earlier chapters. Memorize the lists of the common strong acids (Table 6-1) and the common strong bases (Table 6-3). Classify common acids as strong or weak; classify common bases as strong, insoluble, or weak. *Work Exercises 1 through 3.*

Review Section 10-4

2. Understand the Brønsted-Lowry acid-base theory. Identify Brønsted-Lowry acids and Brønsted-Lowry bases in reactions. Understand and apply the Brønsted-Lowry terminology of conjugate acid-base pairs and their relative strengths.

Section 18-1

3. Calculate the concentrations of constituent ions in solutions of strong electrolytes (whether they are strong acids, strong bases, or soluble salts). *Work Exercises 4 through 9.*

Section 18-2

4. Describe the autoionization of water, including the chemical equation and the significance and uses of its equilibrium constant, K_W. *Work Exercises 10 through 16 and 33 through 35.*

Section 18-3

5. Describe the pH and pOH scales. Given any one of the following for any aqueous solution, calculate the other three: hydronium ion concentration, hydroxide ion concentration, pH, or pOH. *Work Exercises 17 through 37.*

Section 18-4

6. Write the equation for the ionization of any weak acid or weak base. Write and interpret the expression for the corresponding equilibrium constant. *Work Exercises 27 through 31 and 36.*

7. Perform equilibrium calculations for solutions of weak acids and weak bases. Relate initial concentrations, value of ionization constants, and equilibrium concentrations of all species present. Given any one of these quantities, calculate the others. *Work Exercises 38 through 57.*

Section 18-5

8. Apply Study Goal 7 to solutions of polyprotic weak acids. *Work Exercises 58 through 63.*

Section 18-6 (Review Section 10-7)

9. Describe solvolysis and hydrolysis reactions. Review the relationships between the strengths of conjugate acids and bases. *Work Exercises 64 through 66.*

Sections 18-7 through 18-10

10. Classify salts into one of the following categories: (a) salts of strong bases and strong acids, (b) salts of strong bases and weak acids, (c) salts of weak bases and strong acids, or (d) salts of weak bases and weak acids. *Work Exercise 67.*

11. Write the chemical equation and the equilibrium constant expression describing any possible hydrolysis of salts from the categories in Study Goal 10. Tell whether each resulting solution will be acidic, basic, or neutral. Evaluate the hydrolysis constant in terms of the parent weak acid or base. *Work Exercises 68 through 77, 81 through 85, and 87 through 89.*

12. Given the initial concentration of a salt and the ionization constant for the weak acid or base to which it is related, calculate the concentrations of all species in the solution and the solution's pH, pOH, and percent hydrolysis. *Work Exercises 78 through 80 and 86.*

Section 18-11

13. Write equations for and describe the hydrolysis of small, highly charged metal ions. *Work Exercises 90 and 91.*

14. Given the hydrolysis constant of a small, highly charged metal ion, calculate the concentrations of all species in a solution of a given concentration of a salt of the cation. *Work Exercises 92 and 93.*

General

15. Recognize and solve problems of various types from this chapter. *Work* **Mixed Exercises** *94 through 104.*

16. Answer conceptual questions based on this chapter. *Work* **Conceptual Exercises** *105 through 113.*

17. Apply concepts and skills from earlier chapters to the ideas of this chapter. *Work* **Building Your Knowledge** *exercises 114 through 118.*

18. Exercises at the end of chapter direct you to sources outside the textbook for information to use in solving them. Work **Beyond the Textbook** *exercises 119 through 124.*

Some Important Terms in This Chapter

You have previously encountered many of the important terms in Chapter 18, especially in Chapters 6 and 10. Fill in the blanks in the following paragraphs with terms from the list. Use each term only once. Check the Key Terms list and the chapter reading. Study other new terms, and review as needed.

autoionization	hydrolysis constant	pH
Brønsted-Lowry acid	ion product for water	pOH
Brønsted-Lowry base	ionization	polyprotic acid
conjugate acid-base pair	ionization constant	solvolysis
hydrolysis	monoprotic acid	

A molecule or ion that *donates* a proton in a chemical reaction is classified as a [1]_____, and a proton *acceptor* is classified as a [2]_____. Each Brønsted-Lowry acid can be paired up with the base it becomes when it donates a proton to form a team called a [3]_____ _____. A characteristic that distinguishes the Brønsted-Lowry acid-base concept from that of Arrhenius is that at least one member of the conjugate acid-base pair must be an ion. [4]_____ is the formation of ions from a molecule. Even water can form ions by reacting with itself to transfer a proton from one molecule to another, a process known as [5]_____. Quantitatively, the concentrations of the ions from water can be expressed by the equilibrium constant expression: $K_w = [H_3O^+][OH^-]$, where the equilibrium constant, K_w, is called the [6]_____, because it is the product of the ion concentrations in water. Because these concentrations are very small, chemists find it easier to express them on a logarithmic scale. The concentration of H_3O^+ (often thought of as simply H^+) is expressed by the [7]_____, which is calculated as $-\log [H_3O^+]$. Similarly, the concentration of OH^- is expressed by the [8]_____, which is calculated as $-\log [OH^-]$.

The concepts of chemical equilibrium help us to understand the ionization of weak acids and bases. An acid that ionizes only *one* hydrogen per molecule is a [9]_____ _____. An acid that ionizes *more than one* hydrogen per molecules is a [10]_____. The equilibrium constant for the ionization of a weak acid or base is the [11]_____.

The salt of a weak acid or base contains the conjugate base or acid, respectively, which reacts with the solvent in which it dissolves, a process called [12]_____. If the solvent is water, the reaction is specifically [13]_____, and, therefore, the ionization constant for the equilibrium reaction is called the [14]_____, regardless of whether it is labeled K_b or K_a.

Preliminary Test

This test will check your understanding of basic concepts and types of calculations. Be sure to practice *many* of the additional textbook Exercises, including those indicated with the Study Goals.

Short Answer

1. Strong electrolytes _____ electricity better than weak electrolytes in aqueous solution, because they are more completely _____.
2. The seven common strong acids are
    _____,-
    _____,
    _____,-
    _____,
    _____,-_____,
    and _____.
    (Give name and formula for each.)
3. The eight strong bases are
    _____,-
    _____,
    _____,-
    _____,
    _____,-
    _____,
    _____, and
    _____. (Give name and formula for each.)
4. The five weak monoprotic acids listed in Table 18-4 of the text, arranged according to increasing acid strength, are _____.

5. Table 18-4 also lists five conjugate bases (excluding water). Arranged according to increasing base strength, these bases are _____.

6. The common weak bases discussed in this chapter are the compound _____ and several of its organic derivatives, called _____.

7. The symbol for the ionization constant of a weak base is _____.

8. The symbol for the ionization constant of a weak acid is _____.

9. The weaker the acid, the _____ is the percent ionization of the acid in solution.

10. The definition of pH is _____.

11. In a solution of pH 9.0, $[H_3O^+]$ = _____.

12. In a solution of pH 9.3, $[H_3O^+]$ = _____.

13. In any aqueous solution, as $[H_3O^+]$ increases, $[OH^-]$ _____.

14. In any aqueous solution at 25°C, _____ = 1.0×10^{-14} and _____ = 14.00 .

15. In 0.10 M HCl, $[H_3O^+]$ = _____ M.

16. In 0.10 M HCl, $[OH^-]$ = _____ M.

17. The pH of a solution in which $[H_3O^+]$ = 1.0×10^{-5} is _____.

18. The pH of a solution in which $[H_3O^+]$ = 2.0×10^{-5} is _____.

19. In a solution with pH = 3.0, pOH = _____.

20. In a solution with pH = 3.0, $[H_3O^+]$ = _____ M and $[OH^-]$ = _____ M.

21. A solution with pH = 3.0, would be described as _____ (acidic, basic, or neutral?).

22. By definition, an acidic solution is one in which $[H_3O^+]$ _____ $[OH^-]$, a basic solution is one in which $[H_3O^+]$ _____ $[OH^-]$, and a neutral solution is one in which $[H_3O^+]$ _____ $[OH^-]$.

23. If the concentration of a strong monoprotic acid is x molar, then the concentration of hydronium ions, $[H_3O^+]$, will be equal to _____ M.

24. In a solution of RbOH that is x molar, $[Rb^+]$ = _____ M and $[OH^-]$ = _____ M.

25. In a solution of $Ba(OH)_2$ that is x molar, $[Ba^{2+}]$ = _____ M and $[OH^-]$ = _____ M.

26. In a 0.10 M solution of any strong monoprotic acid, pH = 1.00; in a 0.10 molar solution of any weak monoprotic acid, pH is _____ (greater than, less than, or equal to?) 1.00 .

27. In a solution containing only a weak monoprotic acid HA, $[H_3O^+]$ is _____ (greater than, less than, or equal to?) $[A^-]$; if the solution is not very dilute, the concentration of nonionized HA is approximately equal to the _____ of the solution.

28. H_3PO_4 is a _____ (stronger, weaker?) acid than $H_2PO_4^-$, and
    $H_2PO_4^-$ is a _____ (stronger, weaker?) acid than HPO_4^{2-}.

29. For all diprotic weak acids, K_{a2} _____ K_{a1}.

30. The reaction of a substance with a solvent in which it dissolves is _____; when the solvent is water, it is _____.

31. The stronger an acid, the _____ (stronger, weaker?) is its conjugate base.
32. Two types of salts produce neutral solutions when dissolved in water. These are: (1) salts of _____ acids and strong bases and (2) salts of _____ acids and _____ bases, but only in the latter case if _____ _____ are the same.
33. One reaction whose equilibrium is always disturbed when a hydrolyzable salt dissolves in water is: _____.
34. The hydrolysis reaction that occurs when ammonium chloride dissolves in water is: _____.
35. The hydrolysis reaction that occurs when potassium fluoride dissolves in water is: _____.
36. HX is a weak acid. When a salt such as NaX dissolves in water, the net hydrolysis reaction can be represented as: _____.
37. When the salt of a strong base and a strong acid dissolves in water, the resulting solution is _____. An example is _____.
38. When the salt of a strong base and a weak acid dissolves in water, the resulting solution is _____. An example is _____.
39. When the salt of a weak base and a strong acid dissolves in water, the resulting solution is _____. An example is _____.
40. An example of a salt of a weak base and a weak acid is _____.
41. Whether the solution of a salt of a weak acid and a weak base is acidic, basic, or neutral depends on the value of _____ and on the value of _____.
42. The value of K_a for benzoic acid, C_6H_5COOH, is 6.3×10^{-5}. When a salt of benzoic acid, such as sodium benzoate, dissolves in water, the reaction that occurs is:
    _____.
    This leads to a solution that is _____ (acid, basic, or neutral?).
    The value of the equilibrium constant for this reaction is _____.
43. The value of K_b for methylamine, CH_3NH_2, is 5.0×10^{-4}.
    When methylammonium chloride, CH_3NH_3Cl, dissolves in water, the reaction that occurs is:
    _____.
    The resulting solution is _____ (acid, basic, or neutral?).
    The value of the equilibrium constant for this reaction is _____.
44. When a salt such as methylammonium benzoate dissolves in water, the two hydrolysis reactions that occur are:
    _____ and
    _____.
    The resulting solution would be _____ because _____.
45. The _____ of the ionization constant for an acid and the hydrolysis constant for its conjugate base is equal to K_w.
46. When a salt of a small, highly charged metal ion such as Al^{3+} dissolves in water, the resulting solution is _____ (acid, basic, or neutral?).

47. Comparing isoelectronic cations in the same period of the periodic table, the extent of hydrolysis increases as the _____ increases.

_____

In questions 48 through 70, classify each salt as a salt of a *strong*, *weak*, or *insoluble* base and a *strong* or *weak* acid. Tell whether an aqueous solution of each salt would be *acidic*, *basic*, or *neutral*. It should usually not be necessary to refer to tables of ionization constants, but you may find them helpful in a few cases. Write the equation for any hydrolysis reaction.

| | Salt of | | Nature of | |
Salt	*Base*	*Acid*	Solution	**Hydrolysis Reaction**
48. KCH_3COO	_____	_____	_____	_____
49. RbI	_____	_____	_____	_____
50. $CaBr_2$	_____	_____	_____	_____
51. NH_4Cl	_____	_____	_____	_____
52. $(CH_3)_2NH_2Cl$	_____	_____	_____	_____
53. K_3AsO_4	_____	_____	_____	_____
54. $Ba(NO_2)_2$	_____	_____	_____	_____
55. $Ba(NO_3)_2$	_____	_____	_____	_____
56. NaF	_____	_____	_____	_____
57. $NaCl$	_____	_____	_____	_____
58. $NaBr$	_____	_____	_____	_____
59. $NaClO_4$	_____	_____	_____	_____
60. Na_2CO_3	_____	_____	_____	_____
61. NH_4ClO_4	_____	_____	_____	_____
62. KCN	_____	_____	_____	_____
63. NH_4Br	_____	_____	_____	_____
64. $AlCl_3$	_____	_____	_____	_____
65. $Fe(NO_3)_3$	_____	_____	_____	_____
66. $BiCl_3$	_____	_____	_____	_____
67. K_2S	_____	_____	_____	_____
68. $BeCl_2$	_____	_____	_____	_____

69. NH$_4$CN _____ _____ _____ _____

70. NH$_4$OCN _____ _____ _____ _____

Multiple Choice

_____ 1. Which one of the following is a weak electrolyte in aqueous solution?
(a) HClO$_4$ (b) NaOH   (c) NaCl   (d) HNO$_2$ (e)
Ca(NO$_3$)$_2$

_____ 2. Ammonia dissolves in water by the reaction

$$NH_3(g) + H_2O(\ell) \rightleftharpoons NH_4^+(aq) + OH^-(aq)$$

We would expect the resulting solution to be
(a) acidic.   (b) neutral.   (c) basic.

_____ 3. We describe a solution in which the pH is 12.5 as
(a) very basic.   (b) slightly basic.   (c) neutral.   (d) slightly acidic. (e)   very
acidic.

_____ 4. We add enough base to a solution to cause the pH to increase from 7.5 to 8.5. This
means that
(a) [OH$^-$] increases by a factor of 10.   (b) [H$_3$O$^+$] increases by a factor of 10.
(c) [OH$^-$] increases by 1 M.   (d) [H$_3$O$^+$] increases by 1 M.
(e) [OH$^-$] increases by a factor of 8.5/7.5.

_____ 5. Calculate the pH of a 0.04 M HNO$_3$ solution.
(a) 2.4   (b) 12.5   (c) 1.4   (d) 1.6   (e) 4.0×10$^-$$_2$

_____ 6. What would be the [OH$^-$] concentration in a 0.04 M HNO$_3$ solution?
(a) 2.5×10^{-10} M (b) 12.6 M (c) 1.4 M (d) 2.5×10^{-13} M (e) 0 M

_____ 7. For a given acid HA, the value of K_a
(a) will change with pH.   (b) cannot be less than 10^{-7}.
(c) cannot be greater than 10^{-7}.   (d) does not change with temperature.
(e) None of the preceding answers is correct.

Hypobromous acid, HOBr, is a weak acid with an ionization constant of 2.5×10^{-9}. Questions 8 through 10 refer to a 0.010 M HOBr solution.

_____ 8. Calculate the concentration of [H$_3$O$^+$] in this solution.
(a) 5.0×10^{-6} M (b) 5.0×10^{-5} M (c) 2.5×10^{-7} M (d) 2.5×10^{-11} M (e) 5.0×10^{-7} M

_____ 9. What is the pH of this solution?
(a) 3.5   (b) 4.7   (c) 5.3   (d) 5.7   (e) 6.3

_____ 10. What is the concentration of [OH⁻] in this solution?
(a) $2.0 \times 10^{-9} \, M$ (b) $2.0 \times 10^{-8} \, M$ (c) $1.0 \times 10^{-7} \, M$ (d) $5.0 \times 10^{-6} \, M$ (e) $5.0 \times 10^{-5} \, M$

_____ 11. In a sample of pure water, only one of the following statements is *always* true at all conditions of temperature and pressure. Which one is *always* true?
(a) $[H_3O^+] = 1.0 \times 10^{-7} \, M$ (b) $[OH^-] = 1.0 \times 10^{-7} \, M$ (c) pH = 7.0
(d) pOH = 7.0   (e) $[H_3O^+] = [OH^-]$

_____ 12. Which one of the following expressions correctly describes the relationship between pH and pOH in any aqueous solution at 25°C?
(a) (pH)(pOH) = 10^{-14} (b) pH/pOH = 10^{-14} (c) pOH/pH = 10^{-14}
(d) pH + pOH = 14   (e) pH − pOH = 14

_____ 13. Which one of the following expressions correctly describes the relationship between $[H_3O^+]$ and $[OH^-]$ in any aqueous solution at 25°C?
(a) $[H_3O^+][OH^-] = 10^{-14}$ (b) $[H_3O^+]/[OH^-] = 10^{-14}$
(c) $[OH^-]/[H_3O^+] = 10^{-14}$ (d) $[H_3O^+] + [OH^-] = 14$
(e) $[H_3O^+] − [OH^-] = 14$

_____ 14. In water solution, HCN only very slightly ionizes, according to the equation:

$$HCN + H_2O \rightleftharpoons H_3O^+ + CN^-$$

This observation shows that CN⁻ is a _____ than H_2O.
(a) stronger acid   (b) stronger base   (c) weaker acid   (d) weaker base

_____ 15. When solid NaCN dissolves in water,
(a) the pH remains at 7.
(b) the pH becomes greater than 7 because of hydrolysis of Na^+.
(c) the pH becomes less than 7 because of hydrolysis of Na^+.
(d) the pH becomes greater than 7 because of hydrolysis of CN^-.
(e) the pH becomes less than 7 because of hydrolysis of CN^-.

_____ 16. When the salt NH_4CH_3COO dissolves in water, the resulting solution is
(a) acidic.   (b) basic.   (c) neutral.

_____ 17. When the salt NH_4NO_2 dissolves in water, the resulting solution is
(a) acidic.   (b) basic.   (c) neutral.

_____ 18. Each of the following salts completely ionizes when dissolved in water. Which one of these salts will produce a basic solution when dissolved in water?
(a) sodium chloride   (b) sodium acetate   (c)   ammonium chloride
(d) calcium nitrate   (e) rubidium perchlorate

_____ 19. Which of the following salts will produce an acidic solution when dissolved in water?
(a) sodium chloride   (b) sodium acetate   (c)   ammonium chloride
(d) calcium nitrate   (e) rubidium perchlorate

____ 20. The value of K_a for acetic acid is 1.8×10^{-5}. Find the value of K_b for

$$CH_3COO^- + H_2O \rightleftharpoons CH_3COOH + OH^-$$

    (a) 1.8×10^{-5} (b) 1.8×10^{-10} (c) 1.0×10^{-14} (d) 5.6×10^{-10} (e) 5.6×10^{-5}

> Be sure to work many of the text exercises involving weak acid equilibria and salt hydrolysis.

Answers to Some Important Terms in This Chapter

1. Brønsted-Lowry acid
2. Brønsted-Lowry base
3. conjugate acid-base pair
4. ionization
5. autoionization
6. ion product for water
7. pH
8. pOH
9. monoprotic acid
10. polyprotic acid
11. ionization constant
12. solvolysis
13. hydrolysis
14. hydrolysis constant

Answers to Preliminary Test

Do not just look up answers. *Think about the reasons for the answers.*

Short Answer

1. conduct; ionized (dissociated)
2. hydrochloric, HCl; hydrobromic, HBr; hydroiodic, HI; nitric, HNO_3; perchloric, $HClO_4$; chloric, $HClO_3$; sulfuric, H_2SO_4. It may help to remember that three of these are binary acids of halogens. Remember that HF is a weak acid.
3. lithium hydroxide, LiOH; sodium hydroxide, NaOH; potassium hydroxide, KOH; rubidium hydroxide, RbOH; cesium hydroxide, CsOH; calcium hydroxide, $Ca(OH)_2$; strontium hydroxide, $Sr(OH)_2$; barium hydroxide, $Ba(OH)_2$.
4. $HCN < HOCl < CH_3COOH < HNO_2 < HF$. Smaller values of K_a mean weaker acids.
5. $F^- < NO_2^- < CH_3COO^- < OCl^- < CN^-$. Remember that the stronger the acid, the weaker its conjugate base.
6. ammonia; amines
7. K_b
8. K_a
9. smaller
10. $-\log [H_3O^+]$
11. 10^{-9}
12. 5.0×10^{-10}
13. decreases

14. $[H_3O^+][OH^-]$; pH + pOH
15. 0.10. Remember that HCl is a strong acid and is completely ionized.
16. 1.0×10^{-13}. $[H_3O^+] = 0.10$ (from preceding question) and $[OH^-] = K_w/[H_3O^+]$.
17. 5.00
18. 4.70
19. 11.0
20. 10^{-3}; 10^{-11}
21. acidic
22. $>$; $<$; $=$
23. x. Strong acids are completely dissociated, so x moles of a strong monoprotic acid would yield x moles of H_3O^+ on ionization.
24. x; x. RbOH is a strong base.
25. x; $2x$. Ba(OH)$_2$ is a strong base that gives 2 moles OH$^-$ for each mole dissolved.
26. greater than. In a 0.10 M weak acid solution, $[H_3O^+]$ would be less than 0.10, so pH would be greater than 1.00
27. equal to; molarity
28. stronger; stronger
29. $<$
30. solvolysis; hydrolysis
31. weaker
32. (1) strong; (2) weak; weak; the strengths of the parent acid and base
33. $2H_2O \rightleftharpoons H_3O^+ + OH^-$
34. $NH_4^+ + H_2O \rightleftharpoons NH_3 + H_3O^+$. NH_4^+ is the conjugate acid of the weak base NH_3.
35. $F^- + H_2O \rightleftharpoons HF + OH^-$. F^- is the conjugate base of the weak acid HF.
36. $X^- + H_2O \rightleftharpoons HX + OH^-$.
37. neutral. The example discussed in the chapter is NaCl. There are many other correct answers, such as $NaNO_3$, KBr, $RbClO_4$, and CaI_2.
38. basic. Examples discussed in the chapter include $NaCH_3COO$ and NaCN. Many other examples are possible.
39. acidic. Among the examples in this chapter are NH_4NO_3 and NH_4Cl.
40. The textbook lists several examples in the chapter: NH_4CH_3COO, NH_4CN, and NH_4F.
41. K_a for the acid from which the salt is derived; K_b for the base from which the salt is derived.
42. $C_6H_5COO^- + H_2O \rightleftharpoons C_6H_5COOH + OH^-$; basic; 1.6×10^{-10}.
    Remember that the hydrolysis constant, K_b, for the conjugate base of a weak acid with ionization constant K_a is given by $K_aK_b = K_w$.
43. $CH_3NH_3^+ + H_2O \rightleftharpoons CH_3NH_2 + H_3O^+$; acidic; 2.0×10^{-11}.
    The hydrolysis constant, K_a, for the conjugate acid of a weak base with ionization constant K_b is given by the relationship $K_aK_b = K_w$.
44. $C_6H_5COO^- + H_2O \rightleftharpoons C_6H_5COOH + OH^-$; $CH_3NH_3^+ + H_2O \rightleftharpoons CH_3NH_2 + H_3O^+$; basic; $K_{base} > K_{acid}$.

Be sure that you understand how to reason this out (Section 18-10). Compare the K values for the hydrolysis reactions that you determined for Questions 42 and 43. From these you can see that the hydrolysis of benzoate occurs (slightly) more readily than the hydrolysis of methylammonium ion. This makes the solution slightly basic.

45. product

46. acidic. Section 18-11 discusses the hydrolysis reactions of such ions. Note that this type of reaction can only cause the solution to become acidic, never basic.

47. charge.

48. strong; weak; basic; $CH_3COO^- + H_2O \rightleftharpoons CH_3COOH + OH^-$

49. strong; strong; neutral; none

50. strong; strong; neutral; none

51. weak; strong; acidic; $NH_4^+ + H_2O \rightleftharpoons NH_3 + H_3O^+$

52. weak; strong; acidic; $(CH_3)_2NH_2^+ + H_2O \rightleftharpoons (CH_3)_2NH + H_3O^+$

53. strong; weak; basic; $AsO_4^{3-} + H_2O \rightleftharpoons HAsO_4^{2-} + OH^-$.

   Probably $HAsO_4^{2-}$ hydrolyzes further, but the reaction shown is the predominant one.

54. strong; weak; basic; $NO_2^- + H_2O \rightleftharpoons HNO_2 + OH^-$

55. strong; strong; neutral; none

56. strong; weak; basic; $F^- + H_2O \rightleftharpoons HF + OH^-$

57. strong; strong; neutral; none

58. strong; strong; neutral; none

59. strong; strong; neutral; none

60. strong; weak; basic; $CO_3^{2-} + H_2O \rightleftharpoons HCO_3^- + OH^-$.

   Hydrolysis of HCO_3^- also occurs to a limited extent.

61. weak; strong; acidic; $NH_4^+ + H_2O \rightleftharpoons NH_3 + H_3O^+$

62. strong; weak; basic; $CN^- + H_2O \rightleftharpoons HCN + OH^-$

63. weak; strong; acidic; $NH_4^+ + H_2O \rightleftharpoons NH_3 + H_3O^+$

64. insoluble; strong; acidic; $Al^{3+} + 2H_2O \rightleftharpoons Al(OH)^{2+} + H_3O^+$

   or, more explicitly, $[Al(OH_2)_6]^{3+} + H_2O \rightleftharpoons [Al(OH)(OH_2)_5]^{2+} + H_3O^+$

65. insoluble; strong; acidic; $Fe^{3+} + 2H_2O \rightleftharpoons Fe(OH)^{2+} + H_3O^+$

   or, more explicitly, $[Fe(OH_2)_6]^{3+} + H_2O \rightleftharpoons [Fe(OH)(OH_2)_5]^{2+} + H_3O^+$

66. insoluble; strong; acidic; $Bi^{3+} + 2H_2O \rightleftharpoons Bi(OH)^{2+} + H_3O^+$

   or, more explicitly, $[Bi(OH_2)_6]^{3+} + H_2O \rightleftharpoons [Bi(OH)(OH_2)_5]^{2+} + H_3O^+$

67. strong; weak; basic; $S^{2-} + H_2O \rightleftharpoons HS^- + OH^-$.

   The HS^- ion also hydrolyzes to some extent, but the reaction shown is the predominant one.

68. insoluble; strong; acidic; $Be^{2+} + 2H_2O \rightleftharpoons Be(OH)^+ + H_3O^+$

   or, more explicitly, $[Be(OH_2)_4]^{2+} + H_2O \rightleftharpoons [Be(OH)(OH_2)_3]^+ + H_3O^+$

69. weak; weak; basic; $NH_4^+ + H_2O \rightleftharpoons NH_3 + H_3O^+$ *and* $CN^- + H_2O \rightleftharpoons HCN + OH^-$.

   K_b for NH_3 is greater than K_a for HCN, so K_a for NH_4^+ is less than K_b for CN^-. Thus, the second hydrolysis reaction occurs more than the first, making the solution *basic*. See the discussion in Section 18-10.

70. weak; weak; acidic; $NH_4^+ + H_2O \rightleftharpoons NH_3 + H_3O^+$ *and* $OCN^- + H_2O \rightleftharpoons HOCN + OH^-$.

K_b for NH_3 is less than K_a for HOCN, so K_a for NH_4^+ is greater than K_b for CN^-. Thus, the first hydrolysis reaction occurs more than the first, making the solution *acidic*. See the discussion in Section 18-10.

Multiple Choice

1. (d). Remember the lists of strong acids and strong bases. Also, review the solubility rules. HNO_2 is a weak acid, so it only slightly dissociates into ions in water solution.

2. (c). An excess of OH^- ions is produced.

3. (a). Any solution with a pH greater than 7 is basic. The physical extreme for pH is about 15. A pH of 12.5 is more basic than household ammonia.

4. (a). Any increase of one pH unit indicates a *decrease* in $[H_3O^+]$ to $1/10^{th}$, and an *increase* in $[OH^-]$ of 10 times.

5. (c). Because HNO_3 is a strong acid, $[H_3O^+] = 0.04$; pH $= -\log (0.04) = 1.4$

6. (d). $[H_3O^+][OH^-] = 1\times10^{-14}$; $[OH^-] = \dfrac{1\times10^{-14}}{0.04} = 2.5\times10^{-13}$

7. (e). K_a is an equilibrium constant. It could be any number, depending on the strength of the acid, and it is not affected by any change except temperature.

8. (a). This problem must be solved by the methods in Section 18-4. See Example 18-10.
Let $x =$ mol/L of HOBr that ionizes.

	HOBr	+	H_2O	$\rightleftharpoons$	H_3O^+	+	OBr^-
initial	0.010 M				$\approx 0\ M$		$0\ M$
change due to rxn	$-x\ M$				$+x\ M$		$+x\ M$
equilibrium	$(0.010 - x)\ M$				$x\ M$		$x\ M$

$K_a = \dfrac{[H_3O^+][OBr^-]}{[HOBr]} = \dfrac{(x)(x)}{0.010 - x} = 2.5\times10^{-9}$

Assuming that $x \ll 0.010$, $\dfrac{x^2}{0.010} \approx 2.5\times10^{-9}$ or $x^2 \approx 2.5\times10^{-11}$ or $x = 5.0\times10^{-6}\ M$

(Notice that x is 0.05 % of 0.010, so the assumption is valid.)

9. (c). pH $= -\log [H_3O^+] = -\log (5.0\times10^{-6}) = 5.3$

10. (a). $[H_3O^+][OH^-] = 1\times10^{-14}$; $[OH^-] = \dfrac{1\times10^{-14}}{5.0\times10^{-6}} = 2.0\times10^{-9}$

11. (e). The others all depend on $K_w = 1.0\times10^{-14}$, and this is only true at 25°C. Remember that K_w, like all equilibrium constants, changes with temperature. See Table 18-2.

12. (d). See question 13.

13. (a). See question 12.

14. (b). HCN only very slightly ionizes because the equilibrium is far to the left (reactants). The reverse reaction predominates, indicating that H_3O^+ is a stronger acid than HCN and CN^- is a stronger base than H_2O.

15. (d). The hydrolysis reaction $CN^- + H_2O \rightleftharpoons HCN + OH^-$ occurs. This raises $[OH^-]$, lowers $[H_3O^+]$, and thus raises pH. This reasoning makes it unnecessary to carry out the calculation of pH to find the answer.

16. (c). Remember this is the only common weak base-weak acid salt for which the acid and base strengths are essentially equal. See Section 18-10.

17. (a). The K_a for HNO_2 (4.5×10^{-4}) is greater than the K_b for NH_3 (1.8×10^{-5}). Thus, NH_4^+ would hydrolyze more than NO_2^-, and the resulting solution would be acidic. See Section 18-10.

18. (b). Acetate is the conjugate base of acetic acid, a weak acid. Sodium chloride (a), calcium nitrate (d), and rubidium perchlorate (e) are salts of a strong acid and a strong base, so their solutions will be neutral. Ammonium chloride is the salt of a weak base and a strong acid, so its solution will be acidic.

19. (c). Ammonium is the conjugate acid of ammonia, a weak base. Sodium chloride (a), calcium nitrate (d), and rubidium perchlorate (e) are salts of a strong acid and a strong base, so their solutions will be neutral. Sodium acetate is the salt of a strong base and a weak acid, so its solution will be basic.

20. (d). This is the hydrolysis reaction of the base CH_3COO^-. For any conjugate acid-base pair, $K_a K_b = K_w = 1.0 \times 10^{-14}$.

19 Ionic Equilibria II: Buffers and Titration Curves

Chapter Summary

LeChatelier's Principle (Chapter 17) tells you that the addition from an outside source of one of the products to a reaction at equilibrium results in "shifting the equilibrium to the left," i.e., tending to suppress the reaction. When the reaction is the ionization of a weak acid (or base) and the product added is the conjugate base (or acid) from an appropriate soluble salt, this shift is referred to as the **common ion effect** (Section 19-1). The resulting solution, which is able to absorb modest amounts of acid or base and thus to resist changes in pH, is called a **buffer solution.**

The buffering action (Section 19-2) of such a solution may be understood as follows. Consider a solution of an acid, either weak or strong, to which you then add some additional acid from an external source. If the original solution contained only a strong acid, its conjugate base would be far too weak to react with the added acid, so the pH would change markedly. If the original solution contained a weak acid, its conjugate base would be strong, but there would be only a tiny bit of that strong base present to react with the added acid, so the pH still would change markedly. But suppose the original solution had been made by including *both* the weak acid HA *and* a separate source of its conjugate base A^- (Section 19-2). There would then be high concentrations of both the undissociated acid and its conjugate base. This solution would be able to absorb added acid by reacting with the base A^-. On the other hand, if you were to introduce a little OH^-, the weak acid HA could ionize additionally to provide more H_3O^+. Thus, such a solution is far more resistant to large pH changes than most other solutions. A solution made by dissolving a weak base and a separate source of its conjugate acid shows similar buffering action. Buffer solutions of desired pH are made either by mixing appropriate solutions or by dissolving the appropriate salt in an acidic or basic solution (Sections 18-8 and 18-9).

The calculations of Sections 19-1 through 19-3 will help you to understand buffer solutions more clearly. The key relationship for use in buffer calculations is the **Henderson-Hasselbalch equation** (19-1) for acid/salt buffers (and the analogous equation for base/salt buffers later in the section). In application of this equation and related ones, the concentrations used are the total concentrations that were added of weak acid, weak base, or salt. Do not be content with plugging values into equations. Be sure you understand how buffer solutions work and the chemical reactions and equilibrium calculations associated with them. Then you will better understand the nature of acid-base equilibria in aqueous solution.

Acid-base indicators are discussed in Section 19-4. As you see, these useful substances are either weak acids or weak bases that undergo a marked color change when they convert to their conjugate bases or conjugate acids, respectively.

As you slowly add a solution of a strong base to a solution of a strong acid **(titration)** you gradually neutralize the acid. Before a stoichiometric amount of base has been added, the

282

solution is still acidic; at the **equivalence point,** the acid and base have just neutralized each other; and beyond this point the solution is basic. Section 19-5 shows how this information may be quantitatively represented in a graph called a **titration curve.** You see in this section how an indicator may provide a visual signal when the **end point** of a titration has been reached. Of course, an analogous discussion would apply to the titration of a strong base by a strong acid solution. The calculations in this section provide useful practice in stoichiometry.

Now that you understand hydrolysis reactions, you can consider the titration curves that describe the **neutralization of weak acids with strong bases** (Section 19-6). You see that before the equivalence point, a buffer is formed so you can calculate pH as you did in Section 19-1. At the equivalence point, the solution is that of a strong base-weak acid salt, so the calculations can be done as in Section 18-8. Similar considerations apply to **titration of a weak base with a strong acid.**

Each of these two types of titration curves can be considered in four parts. Be sure you see how pH calculations in Sections 19-5 and 19-6 fall into the four categories listed in the Problem-Solving Tip on page 771. The pH changes in weak acid/weak base titration curves are far less extreme than when one component is strong (Section 19-7).

The various acidic or basic solutions of Chapters 18 and 19 are summarized in Section 19-8. Study Table 19-7 carefully.

Study Goals

Section 19-1

1. Describe the common ion effect in terms of LeChatelier's Principle. Use this effect to understand and explain buffering action. *Work Exercises 7 through 10, 14, 19 through 22.*

Sections 19-2 and 19-3

2. Know how buffer solutions are prepared. Calculate $[H_3O^+]$, $[OH^-]$, pH, and pOH in buffer solutions of the acid/salt type and the base/salt type. *Work Exercises 11 through 13, 15 through 18, and 32 through 39.*

Section 19-2

3. Calculate changes in the $[H_3O^+]$, $[OH^-]$, pH, and pOH when specified amounts of acid or base are added to the buffer solution. *Work Exercises 23 through 31.*

Section 19-4

4. Describe why acid-base indicators change color over a definite pH range and why this range differs for different indicators. *Work Exercises 40 through 47.*

Section 19-5

5. Calculate the $[H_3O^+]$ and pH values of solutions resulting from the mixture of any volumes of specified concentrations of a strong acid and a strong base. (Alternatively, the acid or the base may be added as a pure substance.) Use the results of these calculations to plot titration curves for titrations of strong acids and strong bases. *Work Exercises 48 through 52.*

Section 19-6

6. Calculate the $[H_3O^+]$ and pH values of solutions resulting from the mixing of any volumes of specified concentrations of acids and bases when one of these may be weak. Plot

titration curves for weak acid/strong base or weak base/strong acid titrations. *Work Exercises 53 through 63.*

General

7. Recognize and answer a variety of questions based on the concepts in this chapter. *Work* **Mixed Exercises** *64 through 68.*

8. Answer conceptual questions based on this chapter. *Work* **Conceptual Exercises** *69 through 79.*

9. Apply concepts and skills from earlier chapters to the ideas of this chapter. *Work* **Building Your Knowledge** *exercises 80 through 85.*

10. Exercises at the end of chapter direct you to sources outside the textbook for information to use in solving them. *Work* **Beyond the Textbook** *exercises 86 through 90.*

Some Important Terms in This Chapter

Some important new terms from Chapter 19 are listed below. Fill the blanks in the following paragraphs with terms from the list. Each term is only used once. Check the Key Terms list and the chapter reading. Study other new terms, and review terms from preceding chapters if necessary.

acid/salt buffer	end point	titration
base/salt buffer	equivalence point	titration curve
buffer solution	Henderson-Hasselbalch equation	
common ion effect	indicator	

A solution that resists changes in pH when strong acids or strong bases are added to it is called a [1]_____. The two most common types of buffer solutions are the [2]_____, a solution of a weak acid and a soluble ionic salt of its conjugate base, and the [3]_____, a solution of a weak base and a soluble ionic salt of its conjugate acid. In either of these solutions the ionization of the weak acid or base is suppressed by the addition of its conjugate base or acid from a salt; this behavior is termed the [4]_____ because the conjugate base or acid is an ion the weak acid or base and the salt have in common. By calculation, we can predict the pH (or pOH) of a buffer solution, using the [5]_____.

For an acid/salt buffer, this equation has the form $pH = pK_a + \log \dfrac{[\text{conjugate base}]}{[\text{acid}]}$,

but for a base/salt buffer, it is $pOH = pK_b + \log \dfrac{[\text{conjugate acid}]}{[\text{base}]}$.

Some organic compounds exhibit different colors in solutions of different pH. Any of these can be used as an [6]_____ to indicate the point in a titration at which reaction between an acid and a base is complete. An indicator is used in the quantitative measurement of the amounts of acid and base that neutralize each other, a process called [7]_____. The indicator changes color at the [8]_____, when the titration should be stopped. If the proper indicator has been chosen this end point will be essentially the same as the

[9]_____, the point in the titration at which chemically equivalent amounts of acid and base have reacted. A plot of the pH values at various volumes of acid or base solution added in a titration produces a [10]_____.

Preliminary Test

This test will check your understanding of basic concepts and types of calculations. Be sure to practice *many* of the additional textbook exercises, including those indicated with the Study Goals.

Short Answer

Answer with a word, a phrase, a formula, or a number with units as necessary.

1. In a solution that is 0.10 *M* in acetic acid and 0.10 *M* in sodium acetate, the $[H_3O^+]$ is _____ (greater than, less than, or equal to?) the $[H_3O^+]$ in a solution of 0.10 *M* acetic acid.

2. There are two common kinds of buffer solutions; one is made by dissolving _____ _____ and _____, and the other is made by dissolving _____ and _____ _____.

3. The utility of a buffer solution is that it _____ when moderate amounts of acid or base are added to the solution.

4. The Henderson-Hasselbalch equation for pH in a weak acid buffer is: _____.

5. An acid-base indicator changes color when the _____ changes sufficiently.

6. For most indicators, one color is distinct if the ratio of $[In^-]/[HIn]$ is greater than about _____ and the other color is seen distinctly if this ratio is less than about _____; thus most indicators change color over a range of about _____ pH units.

7. After the start of titration of a weak acid with a strong base, but before the equivalence point is reached, the solution acts as a _____.

8. A procedure in which a solution containing one reactant at a known concentration is slowly added to a solution of unknown concentration of another reactant until the reaction between them is complete is called _____.

9. The point in a titration at which chemically equivalent amounts of acid and base have reacted is called the _____; the point at which the indicator changes color is known as the _____.

10. We can describe the titration curve for the reaction in which a strong acid solution is titrated by a strong base as follows: first there is a region in which the pH is relatively constant at a _____ value, followed by a region where the pH _____ _____, and then a region where the pH is relatively constant at a _____ value.

Multiple Choice

_____ 1. A buffer solution contains 0.56 M acetic acid and 1.00 M sodium acetate. K_a for acetic acid is 1.8×10^{-5}. The $[H_3O^+]$ in this solution would be closest to which of the following?
(a) $10^{-3} M$ (b) $10^{-4} M$ (c) $10^{-5} M$ (d) $10^{-6} M$ (e) $10^{-7} M$

_____ 2. The acid form of an indicator is yellow and its anion is blue. The K_a value for this indicator is 10^{-5}. What will be the approximate pH range over which this indicator changes color?
(a) 3–5   (b) 4–6   (c) 5–7   (d) 8–10   (e) 9–11

_____ 3. What will be the color of the indicator in Question 2 in a solution of pH 3?
(a) red   (b) orange   (c) yellow   (d) green   (e) blue

_____ 4. Which of the following indicators would be most appropriate to use for the detection of the equivalence point of the titration of the weak base ammonia with a strong acid such as HCl?
(a) methyl red   (b) neutral red   (c) phenolphthalein
(d) Any of the preceding would be equally useful.
(e) None of the indicators mentioned would be suitable.

_____ 5. At the equivalence point in the titration of 0.10 M acetic acid with 0.20 M sodium hydroxide, which of the following would be true?
(a) pH = 7.
(b) For every 1 mL of acid used, 2 mL of base will have been used.
(c) For every 2 mL of acid used, 1 mL of base will have been used.
(d) For every 1 mL of acid used, 1 mL of base will have been used.
(e) For every 1 g of acid used, 2 g of base will have been used.

_____ 6. When we titrate a strong acid with a strong base, the pH at the equivalence point is
(a) 7.   (b) less than 7.   (c) greater than 7.
(d) Cannot be predicted unless we know which acid and base are used.

_____ 7. When we titrate a weak base with a strong acid, the pH at the equivalence point is
(a) 7.   (b) less than 7.   (c) greater than 7.
(d) Any of the preceding answers may be correct, depending on the K_b value of the base.

_____ 8. When we titrate a weak acid with a strong base, the pH at the equivalence point is
(a) 7.   (b) less than 7.   (c) greater than 7.
(d) Any of the preceding answers may be correct, depending on the K_a value of the acid.

_____ 9. When we titrate a weak base with a weak acid, the pH at the equivalence point is
(a) 7.   (b) less than 7.   (c) greater than 7.
(d) Any of the preceding answers may be correct, depending on the K_a value of the acid and the K_b value of the base.

Be sure to work many of the textbook exercises involving buffers and titrations.

Answers to Some Important Terms in This Chapter

1. buffer solution
2. acid/salt buffer
3. base/salt buffer
4. common ion effect
5. Henderson-Hasselbalch equation
6. indicator
7. titration
8. end point
9. equivalence point
10. titration curve

Answers to Preliminary Test

Do not just look up answers. *Think about the reasons for the answers.*

Short Answer

1. less than. No calculations are needed; think about the common ion effect. Of course you could do these two calculations if you wish. (See Sections 18-4 and 19-1.)
2. a weak acid; a soluble salt of this weak acid; a weak base; a soluble salt of this weak base.
3. resists changes in pH
4. $\text{pH} = pK_a + \log\frac{[\text{salt}]}{[\text{acid}]}$
5. pH (or $[H_3O^+]$)
6. 10; 0.1; 2.0
7. buffer
8. titration
9. equivalence point; end point
10. low; increases rapidly over a very small volume of base added; high

Multiple Choice

1. (c). The calculation is done as shown in Section 19-3.

$$K_a = \frac{[H_3O][CH_3COO^-]}{[CH_3COOH]} = \frac{(x)(1.00+x)}{(0.56-x)} = 1.8\times10^{-5}$$

Assuming that $x \ll 0.56$, $\frac{(x)(1.00)}{0.56} = 1.8\times10^{-5}$ or $x = 1.0\times10^{-5}\ M$ (which is $\ll 0.56$)

2. (b). An indicator changes color when $[\text{In}^-]/[\text{HIn}] = 1$ (Section 19-4). This condition is satisfied when $[H_3O^+] = K_a$. For this indicator, this is at $[H_3O^+] = 10^{-5}\ M$, or pH = 5. The color change becomes discernible from about 1 pH unit below this to about 1 pH unit above it, or from pH 4 to pH 6.

3. (c). At pH = 3, there is considerable excess H_3O^+. In this excess H_3O^+, the indicator equilibrium is shifted to less dissociation, or predominantly to the form HIn. This form is yellow.

4. (a). From Figure 19-5 we see that the equivalence point occurs at about pH 5. Of the indicators listed, methyl red has a color change at a pH nearest this. See Table 19-4.

5. (c). If you thought the answer was (a), study Section 19-6. If you picked one of the other incorrect answers, think more carefully about the stoichiometry and about the relation between number of moles and volume of solution.
6. (a). See Section 19-5.
7. (b). See Figure 19-5 and the associated discussion in Section 19-6.
8. (c). See Figure 19-4 and the associated discussion in Section 19-6.
9. (d). See Figure 19-6 and Section 18-10 regarding the salt produced at the equivalence point.

20 Ionic Equilibria III: The Solubility Product Principle

Chapter Summary

This chapter continues your study of equilibria in aqueous solutions. Here you consider equilibria involving *slightly soluble ionic compounds*. This is, in many ways, the simplest of the types of ionic equilibria. It is the basis for many useful laboratory techniques, such as separation by fractional precipitation and the dissolving of precipitates. Recall from your study of acid-base equilibria that some acids are so strong (completely ionized) that equilibrium considerations do not apply. In similar fashion, the equilibrium descriptions of this chapter do not apply to the dissolving of soluble substances. (But you will need to know how to find the concentrations of ions in such solutions.) Before you begin your study of this material, be sure you *know* the solubility guidelines, Section 6-1.5 and Table 6-4. Be aware, however, that the substances classified in those rules as "insoluble" are actually very slightly soluble. *These are the types of compounds considered in Chapter 20.*

As you shall see, the reactions studied here can be viewed in either of two ways: as the dissolving of slightly soluble substances to form ions in solution (**dissolution**) or as the ions in solution coming together to form ionic solids (**precipitation**). From either point of view, the equation for the reaction may be written as **solid ⇌ dissolved ions.** We can define an equilibrium constant for this equilibrium process (Section 20-1). As you saw in our discussion in Chapter 18 of heterogeneous equilibria, the solid does not appear in the expression for the equilibrium constant. This equilibrium constant is given the symbol K_{sp} and is referred to as the **solubility product constant,** or sometimes simply the **solubility product.** The remainder of the chapter contains many applications of the **solubility product principle.**

In Section 20-2, you see how the value of K_{sp} for any slightly soluble substance is determined from an experimental measurement of its solubility. In carrying out these calculations, remember the following points:

(1) The measured solubility, which is often stated in terms of grams of substance dissolving to make a specified volume of **saturated** solution, must be converted to a *molar* basis.

(2) You must use the stoichiometry of the dissolution process to find out the concentration, in moles/liter, of each ion present.

(3) Solubilities of solids in liquid are usually quite dependent on temperature, so values of K_{sp} can vary considerably as temperature changes.

Several uses of K_{sp} are detailed in Section 20-3. Pay close attention to the reasoning in the examples in this section. All subsequent sections of this chapter will apply the calculations you study in Section 20-3. The major uses of the solubility product principle are these:

(1) *To calculate the solubility*, in terms of moles or grams of a substance that will dissolve to make a specified amount of saturated solution. This calculation can also let us find the concentration of each ionic species in the saturated solution.

(2) *To calculate the concentrations of ions that can exist together in solution.* A very useful extension of this is deciding whether or not a precipitate will form at specified concentrations. The reaction quotient (Section 17-4), Q_{sp}, is very useful in arriving at this decision.

(3) *To calculate the concentration of an ion necessary to initiate precipitation* of a substance from a solution that contains other ions.

(4) *To calculate the concentrations of ions remaining in solution* after precipitation has occurred. This last type of calculation is useful in predicting the efficiency of precipitation methods of recovering valuable metals and removing objectionable ions from solution.

One application of the calculations from Section 20-3 is in designing methods to separate the ions of a mixture in solution. This technique, called **fractional precipitation** (Section 20-4) makes use of K_{sp} differences among salts having an ion in common. Examples 20-8 and 20-9 illustrate this with the silver halides. Notice which types of calculations from Section 20-3 you are using here.

An important aspect of chemical equilibrium is that in a mixture, *all species must be at equilibrium with respect to all relevant reactions.* For example, this principle was used throughout the discussions of Chapters 18 and 19, where the equilibrium involving the dissociation of water and the resulting $[H_3O^+]/[OH^-]$ ratio was affected by other reactions taking place in the same solution. This principle is emphasized again in Section 20-5 for simultaneous equilibria that involve slightly soluble compounds. Such calculations usually involve one species that is involved in more than one reaction. That species, therefore, affects the equilibria of several processes. In Examples 20-10 through 20-12, OH^- is involved in the equilibria for both the dissolution or precipitation of $Mg(OH)_2$ and the ionization of NH_3 as a weak base in aqueous solution. Thus, both equilibrium constant expressions must be satisfied. Of course, OH^- is also involved in a third reaction—that with H_3O^+ to form water (and the reverse of this reaction)—so that reaction must also be at equilibrium. This equilibrium condition may also be used, for example, to calculate the $[H_3O^+]$ or pH of the solution.

Section 20-6, which discusses methods of **dissolving precipitates,** is merely an application of the ideas presented in Section 20-5. Three methods are presented:

(1) converting an ion into a weak electrolyte,

(2) converting an ion to another species (usually by oxidation or reduction), and

(3) forming complex ions.

In each method, the essential strategy is to lower the concentration of one of the products (ions) of the dissolution process. In the first and third methods, other equilibria are used to help remove one of the ions from solution; in the second, a less reversible reaction is used, for instance the oxidation of S^{2-} to elemental sulfur. You should see that each of these methods is an application of LeChatelier's Principle. The latter part of Section 20-6 also includes the application of equilibrium principles to still another type of reaction, the **formation/dissociation of complex ions**. We usually express the equilibrium constant for such reactions as K_d, referring to the reaction written as a dissociation.

This chapter brings you to the end of four chapters that deal with chemical equilibrium and its applications. You have studied many different types of reactions, each with its own characteristic terminology and notation. However, you should recognize that it is the same set of principles that you have been applying:

(1) the general formulation of the concentration or pressure relationships in the equilibrium constant expression and

(2) LeChatelier's Principle.

Examine your understanding of these chapters by summarizing the different types of reactions or processes you have studied and by comparing the different terminologies you have used to discuss these two principles for the various reaction types.

Study Goals

Review Section 6-1, part 5 and Table 6-4

1. Know the solubility rules and use them to be able to determine which ionic solids are soluble and which are slightly soluble ("insoluble"). *Review Exercises 88 through 98 of Chapter 6.*

Section 20-1

2. Write the solubility product expression for any slightly soluble compound that ionizes when it dissolves in water. *Work Exercises 1 through 7.*

Section 20-2

3. Calculate the solubility product constant for a compound, given its molar solubility or the amount of the substance that dissolves to give a specified volume of saturated solution. *Work Exercises 8 through 13.*

Section 20-3

4. Given its solubility product, calculate the solubility of a compound. Given solubility products for a series of slightly soluble compounds, determine which ones are the most and least soluble by comparing molar solubilities. *Work Exercises 14 through 18 and 21 through 25.*

5. Describe and apply the common ion effect as it pertains to solid-dissolved ion equilibria. *Work Exercises 19, 20, and 26.*

6. Given appropriate solubility products, determine whether precipitation will occur when:
   (1) two solutions of given concentrations of specified solutes are mixed, or
   (2) a specified amount of a solute is added to a solution of specified concentration of another solute.
   Work Exercises 27 through 30.

7. Given appropriate solubility products, determine how much or what concentration of a given solute must be added to a solution of known concentration of another solute:
   (1) to cause precipitation of a compound to begin and
   (2) to remove all but a desired amount, concentration, or percent of a particular dissolved ion.
   Work Exercises 31 through 33.

Section 20-4

8. Describe the process of fractional precipitation. Given appropriate solubility products, perform related calculations. *Work Exercises 34 through 41.*

Section 20-5

9. Given appropriate equilibrium constants, perform calculations involving determinations like those listed above for systems that simultaneously involve solid-dissolved ion

equilibria *and* one or more kinds of other previously studied equilibria. *Work Exercises 42 through 53*.

Section 20-6

10. Describe and give specific examples of three general methods of dissolving precipitates. Discuss each in terms of LeChatelier's Principle. Explain why the processes work. *Work Exercises 54 through 63*.

General

11. Recognize and solve various kinds of problems using the concepts in this chapter. *Work* **Mixed Exercises** *64 through 69*.

12. Answer conceptual questions based on this chapter. *Work* **Conceptual Exercises** *70 through 75*.

13. Apply concepts and skills from earlier chapters to the ideas of this chapter. *Work* **Building Your Knowledge** *exercises 76 through 82*.

14. Use sources outside the textbook to work **Beyond the Textbook** *exercises 83 through 86*.

Some Important Terms in This Chapter

Some important new terms from Chapter 20 are listed below. Fill the blanks in the following paragraphs with terms from the list. Each term is only used once. Check the Key Terms list and the chapter reading. Study other new terms, and review terms from preceding chapters (especially Chapters 17 through 19) if necessary.

complex ion	molar solubility	solubility product constant
dissociation constant	precipitate	solubility product principle
fractional precipitation	Q_{sp}	

In equilibria that involve slightly soluble compounds in water (often called "insoluble" compounds), the equilibrium constant, K_{sp}, is called the [1]_____ _____, or simply the solubility product. Its determination is based upon a concept known as the [2]_____: The solubility product expression for a slightly soluble compound is generally the product of the concentrations of its constituent ions, each raised to the power that represents the number of that type of ion produced by each formula unit of the compound. This quantity is constant at constant temperature for a saturated solution of the compound.

The solubility of a compound can be expressed as the number of moles that dissolve to give one liter of saturated solution, a value known as the [3]_____. The concentrations of the ions in the resulting solution can be used to calculate the solubility product.

Conversely, the solubility product can be use to determine whether a [4]_____, a solid, will be formed by mixing in solution the constituent ions of a slightly soluble compound. What makes this situation different from simply dissolving the compound to form a saturated solution is that the ions are not necessarily mixed in the stoichiometric ratio produced by dissociation of the compound. Therefore, chemists generally calculate the reaction quotient, [5]_____ and compare it with K_{sp}. If Q_{sp} is *less* than K_{sp}, then no precipitate will form and the solution is unsaturated. If they are equal, then the solution is saturated. If Q_{sp} is *greater* than K_{sp}, then precipitation will occur until the solution is saturated.

Frequently, several anions form compounds with the same cation or several cations form compounds with the same anion. Differences in the solubilities of the compounds can be used to separate the ions, removing some ions from solution while leaving others, a process known as [6]_____.

A simple cation can form coordinate covalent bonds with electron-donating groups (like NH_3, CN^-, OH^-, F^-, Cl^-, Br^-, or I^-) to produce a larger [7]_____. The extent of the dissociation of this complex ion can be expressed quantitatively by its [8]_____, K_d.

Preliminary Test

This test will check your understanding of basic concepts and type of calculations. It is very important to get much additional practice by working problems, so be sure to do *many* of the textbook Exercises. This is especially true for fractional precipitation and dissolution of precipitates.

Short Answer

1. The number of moles of a substance that will dissolve to give one liter of a saturated solution is called the _____ of the substance.
2. The symbol for the solubility product constant is _____.
3. The   expression   for   the   solubility   product   of $CuCO_3$ is

    _____.
4. The   expression   for   the   solubility   product   of $Mn(OH)_2$ is

    _____.
5. The   expression   for   the   solubility   product   of $Mg_3(PO_4)_2$ is

    _____.
6. When the ions of a slightly soluble ionic substance, from separate sources, are mixed, the salt will begin to precipitate if Q_{sp} _____ K_{sp}, and will continue to precipitate until Q_{sp} _____ K_{sp}.
7. Insoluble metal hydroxides are _____ (more or less?) readily soluble in acidic solution than in basic solution.
8. Lead chloride, $PbCl_2$, is much more soluble in warm water than in cold water. This tells us that the dissolution reaction of lead chloride is _____ (exothermic or endothermic?).
9. The dissociation constant for a complex ion in solution is represented by the symbol _____.
10. Three methods for dissolving a precipitate of a slightly soluble ionic substance are:

    _____,

_____ ,

and_____ .

11. Given the following values for equilibrium constants:

$$K_{sp}(FeS) = 4.9 \times 10^{-18} \quad \text{and} \quad K_d([Fe(CN)_6]^{4-}) = 1.3 \times 10^{-37},$$

complete the following.

The equation for the dissolution of FeS(s) in a solution that contains cyanide ions, CN^-, is:

_____

___

The value of K for this reaction is _____, which tells us that it would be _____ to dissolve FeS by addition of a soluble compound such as NaCN that dissociates completely into Na^+ and CN^-.

12. Given the following values of equilibrium constants:

$$K_{sp}(CdS) = 3.6 \times 10^{-29} \quad \text{and} \quad K_d([Cd(CN)_4]^{2-}) = 7.8 \times 10^{-18}$$

complete the following.

The equation for the dissolution of CdS(s) in a solution that contains cyanide ions, CN^-, is

_____

___

The value of K for this reaction is _____, which tells us that it would be _____ to dissolve CdS by addition of a soluble compound such as NaCN that dissociates completely into Na^+ and CN^-.

13. In order to dissolve a precipitate we must adjust concentrations of the constituent ions in the solution until Q_{sp} is _____ K_{sp} (less than, greater than, or equal to?); this is usually done by _____ from the solution.

14. To prevent precipitation of $Cu(OH)_2$ from a solution that is 0.10 M in $Cu(NO_3)_2$, the $[OH^-]$ must be kept lower than _____. $K_{sp}(Cu(OH)_2) = 1.6 \times 10^{-19}$

15. Another way to express the result of Question 14 is to say that the pH must be kept _____.

Multiple Choice

_____ 1. Which one of the following ionic chlorides is *least* soluble in water?
(a) NH_4Cl (b) $AgCl$ (c) $BaCl_2$ (d) KCl (e) $NaCl$

_____ 2. In a solution obtained by dissolving silver sulfate, Ag_2SO_4, in water, the concentration of silver ion is observed to be 2×10^{-5} M. What would be the molar concentration of sulfate ion in this solution?
(a) 1×10^{-5} M (b) 2×10^{-5} M (c) 3×10^{-5} M (d) 4×10^{-5} M
(e) Cannot be answered without knowing the value of K_{sp} for Ag_2SO_4.

_____ 3. The molar solubility of $PbSO_4$ is 1.14×10^{-4} mol/L. What is the value of K_{sp} for $PbSO_4$?

(a) 1.16×10^{-14} (b) 2.68×10^{-4} (c) 3.45×10^{-2}

(d) 3.75×10^{-7} (e) 1.30×10^{-8}

_____ 4. The solubility of PbF_2 in water is 0.0021 mol/L. K_{sp} for PbF_2 is closest to which one of the following values?

(a) 2.1×10^{-3} (b) 2.0×10^{-4} (c) 4.4×10^{-6}

(d) 9.2×10^{-9} (e) 3.7×10^{-8}

_____ 5. AgCl would be _least_ soluble in

(a) pure water.   (b) $0.1\,M\,CaCl_2$.   (c) $0.1\,M\,HCl$.   (d) $0.1\,M\,HNO_3$.

(e) AgCl would be equally soluble in all of these.

_____ 6. The value of K_{sp} for $SrSO_4$ is 2.8×10^{-7}. What is the solubility of $SrSO_4$ in mol/L?

(a) 7.6×10^{-7} M   (b) 1.4×10^{-7} M   (c) 5.3×10^{-4} M

(d) 5.7×10^{-3} M   (e) 1.3×10^{-8} M

_____ 7. The value of K_{sp} for CuI is 5.1×10^{-12}. What is the solubility of CuI expressed in grams per liter?

(a) 1.1×10^{-8} g/L   (b) 5.1×10^{-12} g/L   (c) 4.3×10^{-4} g/L

(d) 2.3×10^{-6} g/L   (e) 4.3×10^{-6} g/L

_____ 8. Iron(II) hydroxide is a slightly soluble base. In which of the following would $Fe(OH)_2$ be most soluble? It is not necessary to know K_{sp} for $Fe(OH)_2$.

(a) $0.1\,M\,HCl$ (b) $0.1\,M\,KOH$ (c) $0.1\,M\,Ca(OH)_2$.   (d) $0.1\,M\,FeCl_2$.

(e) $Fe(OH)_2$ would be equally soluble in all of these.

_____ 9. Values of K_{sp} for some slightly soluble sulfides are as follows:

Sulfide	K_{sp}
CdS	3.6×10^{-29}
CuS	8.7×10^{-36}
PbS	8.4×10^{-28}
MnS	5.1×10^{-15}

Which of the following ions must exist at lowest concentration in a solution in which the sulfide ion concentration had been fixed at some fairly large constant value?

(a) Cd^{2+} (b) Cu^{2+} (c) Pb^{2+} (d) Mn^{2+}

(e) Cannot be answered without knowing the sulfide ion concentration.

_____ 10. If NaCl is added to a 0.010 M solution of $AgNO_3$ in water, at what $[Cl^-]$ does precipitation of AgCl begin to occur? K_{sp} for AgCl is 1.8×10^{-10}.

(a) 1.0×10^{-10} M   (b) 1.8×10^{-6} M   (c) 1.8×10^{-8} M

(d) 1.8×10^{-12} M   (e) 1.8×10^{-10} M

Suppose we have a solution that is 0.10 M in Ca^{2+} and also 0.10 M in Pb^{2+}. We wish to separate these ions by fractional precipitation of one of them as the carbonate. We do this by adding a source of carbonate ion, such as solid Na_2CO_3. Questions 11 through 13 refer to this solution. Use the K_{sp} values 4.8×10^{-9} for $CaCO_3$ and 1.5×10^{-13} for $PbCO_3$. Neglect any change in volume due to addition of the sodium carbonate.

_____ 11. Which will begin to precipitate from solution first, $CaCO_3$ or $PbCO_3$, and what will be the CO_3^{2-} when this begins (whether or not we can see it)?
   (a) $CaCO_3$, 4.8×10^{-8} M (b) $CaCO_3$, 4.8×10^{-9} M (c) $PbCO_3$, 1.5×10^{-12} M
   (d) $PbCO_3$, 1.5×10^{-13} M (e) $PbCO_3$, 7.2×10^{-21} M

_____ 12. In the solution of Question 11, suppose we keep adding Na_2CO_3. Eventually the second salt will begin to precipitate. When this second salt does begin to precipitate, what will be the concentration of the first metal ion left in solution?
   (a) 0.10 M (b) 1.5×10^{-12} M (c) 3.2×10^{-5} M
   (d) 3.1×10^{-6} M (e) 1.3×10^{-5} M

_____ 13. When the condition in Question 12 is reached, that is, when the second salt begins to precipitate, what percent of the first metal ion will have been removed from the solution?
   (a) 4.8%   (b) 50%   (c) 95.7%   (d) 99.997%   (e) all of it

_____ 14. We have a solution that is 0.0010 M Pb^{2+}, 0.0050 M Ag^+, and 0.0020 M K^+. If the concentration of sulfide ion, S^{2-}, is slowly increased in this solution, which metal sulfide will precipitate first? $K_{sp}(PbS) = 8.4 \times 10^{-28}$, $K_{sp}(Ag_2S) = 1.0 \times 10^{-49}$
   (a) PbS   (b) Ag_2S (c) K_2S
   (d) All would begin to precipitate at the same time.
   (e) None would precipitate at any reasonable sulfide concentration.

_____ 15. K_{sp} for $Fe(IO_3)_3$ is 10^{-14}. We mix two solutions, one containing Fe^{3+} and one containing IO_3^-. At the instant of mixing, $[Fe^{3+}] = 10^{-4}$ M and $[IO_3^-] = 10^{-5}$ M. Which of the following statements is true?
   (a) A precipitate forms because $Q_{sp} > K_{sp}$
   (b) A precipitate forms because $Q_{sp} < K_{sp}$.
   (c) No precipitate forms because $Q_{sp} > K_{sp}$.
   (d) No precipitate forms because $Q_{sp} < K_{sp}$.
   (e) None of the preceding statements is true.

_____ 16. Magnesium hydroxide is a slightly soluble hydroxide with $K_{sp} = 1.5 \times 10^{-11}$. What is the pH of a saturated solution of magnesium hydroxide?
   (a) 3.51   (b) 10.51   (c) 5.39   (d) 8.60   (e) 7.00

Answers to Some Important Terms in This Chapter

1. solubility product constant
2. solubility product principle
3. molar solubility
4. precipitate
5. Q_{sp}
6. fractional precipitation
7. complex ion
8. dissociation constant

Answers to Preliminary Test

Short Answer

1. molar solubility
2. K_{sp}
3. $K_{sp} = [Cu^{2+}][CO_3^{2-}]$
4. $K_{sp} = [Mn^{2+}][OH^-]^2$
5. $K_{sp} = [Mg^{2+}]^3[PO_4^{3-}]^2$
6. $>;=$
7. more. The small amount of OH^- produced by the equilibrium with solid metal hydroxide is removed by reaction with the acid, and the equilibrium shifts to the right.
8. endothermic. Recall the applications of LeChatelier's Principle as you learned them in Sections 17-6 through 17-8. It is very hard to measure the value of ΔH^0 for a reaction that proceeds as slightly as does the dissolution of lead chloride in water. Can you design an experiment based on measuring the solubility of lead chloride at several temperatures to determine ΔH^0 for this reaction? Think about Section 17-13.
9. K_d
10. converting an ion to a weak electrolyte; converting an ion to another species by oxidation or reduction; forming a complex ion involving one of the constituent ions.
11. $FeS(s) + 6CN^-(aq) \rightleftharpoons [Fe(CN)_6]^{4-} + S^{2-}(aq); 3.8\times10^{19}$; very easy.

    First write out the equations, reversing the one for the dissociation of $[Fe(CN)_6]^{4-}$:

    $$FeS(s) \rightleftharpoons Fe^{2+}(aq) + S^{2-}(aq) \qquad K = 4.9\times10^{-18}$$

    $$Fe^{2+}(aq) + 6CN^-(aq) \rightleftharpoons [Fe(CN)_6]^{4-}(aq) \qquad K = \frac{1}{1.3\times10^{-37}} = 7.7\times10^{36}$$

    (Remember to invert K when you reverse an equation.) Then recall that adding the equations requires multiplying their K values:

    $$FeS(s) + 6CN^-(aq) \rightleftharpoons [Fe(CN)_6]^{4-}(aq) + S^{2-}(aq) \qquad K = 3.8\times10^{19}$$

    This very large value of K tells us that the dissolution goes essentially to completion.
12. $CdS(s) + 4CN^-(aq) \rightleftharpoons Cd(CN)_4^{4-} + S^{2-}(aq); 4.6\times10^{-12}$; difficult.

    Reason as outlined in the answer to Question 11. The equilibrium constant is very small.
13. less than; removing one of the constituent ions
14. 1.26×10^{-10} M. The condition for precipitation to begin (barely) is

$$Q_{sp} = (0.10)[OH^-]^2 = 1.6 \times 10^{-19}$$

Solve this for $[OH^-] = 1.26 \times 10^{-10}$. If $[OH^-]$ is kept less than this, then Q_{sp} would be less than K_{sp}, and the precipitate would not form.

15. less than 4.10. In order to keep $[OH^-]$ less than 1.26×10^{-10} we would have to keep $[H_3O^+]$ greater than 7.94×10^{-5}. This corresponds to pH less than 4.10.

Multiple Choice

1. (b). Remember the solubility rules from Section 6-1, part 5 and Table 6-4. AgCl is one of the compounds seen in Chapter 20 as an example of a slightly soluble salt.

2. (a). For every mole of Ag^+ that goes into solution, one-half mole of SO_4^{2-} goes in. Look at the formula of the salt.

3. (e). Each mole of $PbSO_4$ that dissolves gives 1 mole of Pb^{2+} and 1 mole of SO_4^{2-} in solution. So in the saturated solution the concentrations of the ions would be $[Pb^{2+}] = 1.14 \times 10^{-4}\ M$ and $[SO_4^{2-}] = 1.14 \times 10^{-4}\ M$. Then $K_{sp} = [Pb^{2+}][SO_4^{2-}] = 1.30 \times 10^{-8}$.

4. (e). Did you remember that 2 moles of F^- go into solution for every mole of Pb^{2+}?
$[Pb^{2+}] = 0.0021\ M$, $[F^-] = 0.0042$, $K_{sp} = [Pb^{2+}][F^-]^2 = (0.0021)(0.0042)^2 = 3.7 \times 10^{-8}$

5. (b). This would have the highest $[Cl^-]$ to begin with, so less AgCl could dissolve before the value of K_{sp} would be reached by Q_{sp}. Note that actual calculations based on the K_{sp} of AgCl are not necessary.

6. (c). Solve by the methods of Section 20-3, in the way illustrated in Example 20-3.
$$SrSO_4(s) \rightleftharpoons Sr^{2+}(aq) + SO_4^{2-}(aq)$$
Let $x = [Sr^{2+}] = [SO_4^{2-}]$, so $K_{sp} = [Sr^{2+}][SO_4^{2-}] = (x)(x) = 2.8 \times 10^{-7}$, and $x = 5.3 \times 10^{-4}\ M$

7. (c). Solve as in Question 6, but then remember to convert from mol/L to g/L.
$$CuI(s) \rightleftharpoons Cu^+(aq) + I^-(aq)$$
Let $x = [Cu^+] = [I^-]$, so $K_{sp} = [Cu^+][I^-] = (x)(x) = 5.1 \times 10^{-12}$, and $x = 2.3 \times 10^{-6}\ M$
$(2.3 \times 10^{-6}\ M)(190.4\ g/mol) = 4.3 \times 10^{-4}\ g/L$

8. (a). The H_3O^+ present in this solution would help to remove OH^- as it is formed by dissolution of $Fe(OH)_2$. According to LeChatelier's Principle, this would help to shift the equilibrium for the dissolution process to the right.

9. (b). No calculation is needed here, because all K_{sp} expressions would have the same algebraic form. If some of the salts had different formulas, for example some XY and some X_2Y, then you would need to calculate concentrations. See text Exercises 40 and 41.

10. (c). $AgCl(s) \rightleftharpoons Ag^+(aq) + Cl^-(aq)$ so $K_{sp} = [Ag^+][Cl^-] = 1.8 \times 10^{-10}$
$[Ag^+] = 0.010\ M$, so $[Cl^-] = \dfrac{1.8 \times 10^{-10}}{0.010} = 1.8 \times 10^{-8}$

11. (c). Calculate the concentration of CO_3^{2-} necessary to begin to precipitate each, using the method shown in Section 20-3 and Examples 20-5 and 20-7. This calculation leads to the conclusion that $CaCO_3$ will begin to precipitate when $[CO_3^{2-}] = 4.8 \times 10^{-8}$ M and $PbCO_3$ will begin when $[CO_3^{2-}] = 1.5 \times 10^{-12}$ M. Because the latter is the smaller concentration, it will be reached first. $PbCO_3$ will begin to precipitate first.

12. (d). When $[CO_3^{2-}]$ reaches 4.8×10^{-8} M and $CaCO_3$ begins to precipitate, $[Pb^{2+}]$ must be down to $\dfrac{1.5 \times 10^{-13}}{4.8 \times 10^{-8}} = 3.1 \times 10^{-6}$ M.

13. (d). From the results of Question 12, we know that only 3.1×10^{-6} M of the original 0.10 M Pb^{2+} remains. Thus, the percentage of lead ions remaining is $\dfrac{3.1 \times 10^{-6}}{0.10} \times 100 = 0.003\%$. In other terms, $(100 - 0.003)\% = 99.997\%$ has been removed. We might call this "getting the lead out."

14. (b). Use the methods of Section 20-3 to calculate the concentration necessary to begin precipitation of PbS and of Ag_2S. You should remember from the solubility rules that K_2S, like all common salts of the alkali metals, is a soluble salt. To precipitate PbS, $[S^{2-}]$ needed is 8.4×10^{-25} M; for Ag_2S to precipitate, $[S^{2-}]$ needed is 4.0×10^{-45} M. It is clear that Ag_2S would precipitate first.

$$PbS(s) \rightleftharpoons Pb^{2+}(aq) + S^{2-}(aq) \qquad [Pb^{2+}] = \frac{8.4 \times 10^{-28}}{0.0010}$$

$$Ag_2S(s) \rightleftharpoons 2Ag^+(aq) + S^{2-}(aq) \qquad [Ag^+] = \frac{1.0 \times 10^{-49}}{(0.0050)^2}$$

15. (d). If you got answer (a), you probably forgot to raise $[IO_3^-]$ to the third power.
$$Fe(IO_3)_3(s) \rightleftharpoons Fe^{3+}(aq) + 3IO_3^-(aq)$$
$$Q_{sp} = [Fe^{3+}][IO_3^-]^3 = (10^{-4})(10^{-5})^3 = 10^{-19} < 10^{-14} = K_{sp}(Fe(IO_3)_3)$$

16. (b). Remember that each mole of $Mg(OH)_2$ that dissolves gives two moles of OH^-.
$$Mg(OH)_2(s) \rightleftharpoons Mg^{2+}(aq) + 2OH^-(aq)$$
You can use K_{sp} and the methods of Section 20-3 to find the molar solubility of $Mg(OH)_2$ to be:

Let x = molar solubility of $Mg(OH)_2$. $[Mg^{2+}] = x$, $[OH^-] = 2x$

$K_{sp} = [Mg^{2+}][OH^-]^2 = (x)(2x)^2 = 1.5 \times 10^{-11}$ $\qquad$ so $x = 1.6 \times 10^{-4}$ M

Then $[OH^-] = 3.2 \times 10^{-4}$ M.

In any aqueous solution, $[H_3O^+][OH^-] = 1.0 \times 10^{-14}$, so $[H_3O^+] = 3.1 \times 10^{-11}$.

This corresponds to pH $= 10.51$.

21 Electrochemistry

Chapter Summary

As you learned in Chapter 6, oxidation-reduction (redox) reactions always involve transfer (actual or apparent) of electrons. **Electrochemistry** is the study of this type of reaction, with emphasis on the electron transfer. In **electrochemical cells,** the electron transfer takes place through an external circuit rather than directly between reacting species. In Chapter 21, you will learn about the conceptual aspects of electrochemistry and its relation to the spontaneity of reactions. You will also learn some of electrochemistry's practical applications.

The following questions summarize the central ideas of electrochemistry: How can you tell whether a particular electron transfer process is spontaneous? If it is not, how can you make it occur by supplying electrical energy from an outside source? If it is spontaneous, how can you use this reaction as a source of electrical energy? These questions suggest that your electrochemical studies can conveniently be approached in two parts. The first part (Sections 21-3 through 21-7) concerns those reactions in which an external source of electrical energy is used to cause a nonspontaneous reaction to occur. Electroplating and electrolytic refining of substances are applications of this aspect of electrochemistry. The second part of your study of electrochemistry (Sections 21-8 through 21-25) involves those reactions in which the spontaneity of the reaction is used to make it serve as a source of electrical energy, with applications such as batteries and fuel cells.

Sections 21-1 and 21-2 present some of the basic terminology of electrochemistry—**metallic conduction, electrolytic conduction, electrode, anode, cathode,** and so on. Such terms are used throughout this chapter. A review of Sections 5-7 (**oxidation states**) and 6-5 (**oxidation-reduction reactions**) would be a useful prelude to your study of Chapter 21. Concentrate especially on how terms such as oxidation, reduction, oxidizing agent, and reducing agent describe electron transfer.

Throughout this chapter, you will be describing oxidation-reduction reactions in terms of **half-reactions.** In Chapter 11, you saw that the idea of separating the oxidation process from the reduction process was an aid in balancing redox reactions. At that point, you examined reactions with direct electron transfer between reactant species, so that the "separation" was mental, largely a bookkeeping device. In most of the electrochemical studies of this chapter, the experiment are arranged so that these half-reactions take place in physically separate parts of the **cell**. The electrons are transferred through an external circuit, such as a wire.

The first major topic of Chapter 21 is **electrolysis,** in which redox reactions that are *not* spontaneous are caused to occur by supplying electrical energy from an external source. As you study the three sections that discuss the electrolysis of molten sodium chloride (21-3), aqueous sodium chloride (21-4), and aqueous sodium sulfate (21-5), be sure that you understand how such terms as oxidizing agent, substance oxidized, anode, cathode, oxidation half-reaction, reduction half-reaction, and so on, are used. You must be able to apply these terms to an electrolytic cell, given the observations associated with the operation of the cell.

As you study these three electrolytic processes, two types of questions may occur to you: (1) How much electrical charge must be supplied to cause a specified amount of reaction to occur? (2) Why do different reactions take place in the different electrolytic cells? That is, in the electrolysis of aqueous sodium chloride, why is water reduced in preference to sodium ions (which can, however, be reduced in molten NaCl)? In aqueous NaCl, why do chloride ions oxidize instead of water molecules? This second type of question ("Which of several possible oxidation or reduction process is favored?") will be addressed later in the chapter, beginning in Section 21-15. The first question is the subject of Section 21-6.

The main idea of **Faraday's Law of Electrolysis** is that a specific amount of electrical charge delivered to the cell corresponds to a specific number of electrons. There is a relationship between the amount of electrical charge and the extent of redox reaction that it causes. In particular, one **faraday** of charge, which is 96,485 coulombs, corresponds to one **mole** of electrons. Calculations related to Faraday's Law, including the relation of amount of charge to time and strength of current, are illustrated in Examples 21-1 and 21-2. Applications of the principles of electrolysis are discussed in the next section (21-7). Electrolysis may be used in such processes as electrolytic refining and electroplating.

You now turn to the opposite aspect of electrochemistry, that of using spontaneous reactions to produce electrical energy, in **voltaic** or **galvanic cells**. This is accomplished by separating halves of the redox reaction, forcing the electron transfer to take place through the external circuit. Section 21-8 describes the construction of simple voltaic cells for measurement of voltage (electrical potential). The next two sections illustrate the use of such experimental arrangements to study redox reactions. For now you are working only with cells in which all reactants are at standard conditions—solutions are 1 M in each solute and gases are at 1 atm partial pressures. In Section 21-9, you study the zinc-copper cell. Think about the *possible* results of redox processes in such a cell: metallic copper could be oxidized ($Cu \rightarrow Cu^{2+} + 2e^-$) or copper(II) ions could be reduced ($Cu^{2+} + 2e^- \rightarrow Cu$); this could be accompanied by zinc ions being reduced ($Zn^{2+} + 2e^- \rightarrow Zn$) or metallic zinc being oxidized ($Zn \rightarrow Zn^{2+} + 2e^-$), respectively. In actual experiments with this cell, you would observe that Cu^{2+} is reduced and Zn is oxidized and that the initial cell voltage (i.e., while both solutions are still 1 M) is 1.10 V. By comparison, in the copper-silver cell, Section 21-10, Cu is oxidized and Ag^+ is reduced, with an initial cell voltage of 0.46 V. Because of such experiments as these, you will observe that the magnitude of the cell potential is a measure of the spontaneity of the redox reaction, a concept that will extended later in the chapter. For now, use such results to rank the relative strengths of the oxidizing agents and the reducing agents in the cells.

First, you must measure and rank oxidizing and reducing strengths on a quantitative basis. This is accomplished with the determination and tabulation of **standard electrode potentials**. These are discussed and applied in the next eight sections of the chapter. Your first requirement is a reference value, to which you can refer all other tendencies to be oxidized or reduced. For this, the **standard hydrogen electrode (SHE**, Section 21-11) has been arbitrarily chosen to have an "electrode potential" of exactly 0.0000... V. You can then construct a cell with any other electrode versus the SHE and measure the cell potential. This strategy is applied in Section 21-12 for the Zn/Zn^{2+} electrode and in Section 21-13 for the Cu/Cu^{2+} electrode, each measured against the SHE. In these experiments, you are measuring separately, but against the same reference, the quantitative strengths of the oxidizing and reducing agents, by measuring the cell voltages. In each case you have some other electrode vs. the SHE (which has exactly 0 V electrode potential), so the cell voltage is called the **standard electrode potential** (E^0) for that other electrode or half-cell.

In this way, the **electromotive series** or **activity series** is built up (Section 21-14). Remember that by convention these potentials for half-reactions are always tabulated as *reductions* (Table 21-2). The more positive the E^0 value for a particular half-reaction, the easier that reduction is to accomplish. As a way of helping you remember this convention, just look at the table and find some reaction that you know to be easy (or hard) and notice its sign. For instance, in Table 21-2 the reduction of elemental fluorine to fluoride ion is shown as having E^0 = +2.87 V. If you recall that fluorine has a very high affinity for electrons (Section 5-4) or that it is the most electronegative element (Section 5-6), it will help you remember that this reduction is a very easy one. This should enable you to recall that more positive values of E^0 correspond to greater ease of the process. Alternatively, recall that it is easy to oxidize alkali metals to their cations (these metals have low ionization energies, Section 5-3), and therefore *hard* to reduce the cations to the free metals. Observing that these metals have very negative E^0 values will remind you of the convention—the lower (more negative) the E^0 value, the *harder* it is to carry out that reduction. These electrode potentials are tabulated *only* at standard conditions of 1 M for solutions and 1 atm for gases. You will encounter corrections for different concentrations or partial pressures later in the chapter.

The next four sections of the chapter deal with uses of the knowledge summarized in the electromotive series. One important use, detailed in Section 21-15, is to **determine whether a particular redox reaction would be spontaneous.** To do this, write E^0 for the possible half-reactions (from the electromotive series). Then add E^0 for reduction and E^0 for oxidation to determine E^0_{cell} for the overall redox process. The more positive the E^0_{cell} value, the more favorable the process would be at these conditions. Notice that this is not necessarily the reaction that takes place in an electrochemical cell—the same conclusions apply to the same reaction in any single solution at standard conditions. As you study Example 21-3 in this section and in your working of similar exercises at the end of the chapter, be very careful of the conventions. If possible, write the half-reaction with the more positive E^0 value as a reduction and reverse the other half-reaction, changing the sign of its E^0 value. The resulting E^0_{cell} will then be positive, indicating that the reaction is spontaneous in the direction written. (If a reaction has a negative E^0_{cell} then the reaction is spontaneous in the reverse direction.) It is important to remember, though, that we do *not* change the magnitude of E^0 when we reverse the half-reaction *or* when we multiply it by some factor. This is because E^0 is a measure of the *relative tendency* of the reaction to occur. This is different from the case in thermodynamic calculations where, for example, ΔH^0 is a measure of the *amount* of heat produced or consumed when the reaction proceeds by a certain amount.

In Section 21-16, you see how these techniques can also be applied to oxidation and reduction processes that take place in solution at **inert electrodes.** There are no new concepts here, but remember that you are still dealing only with 1 M solutions. You can see in this section, however, that what has developed is consistent with all of the observations cited earlier (Sections 21-3 through 21-5) about which reactions occurred preferentially in the various electrolytic cells. The very practical topic of corrosion and corrosion protection is actually an application of electrochemical principles, as emphasized in the discussion of Sections 21-17 and 21-18.

Up to this point in the chapter, all electrode potentials and cell potentials have been standard ones (E^0 and E^0_{cell}), corresponding to solution concentrations of 1 M and gas partial pressures of 1 atm. In Section 21-19, you see that you can take these standard conditions as references and then make corrections for the effects of other concentrations. This is accomplished by the **Nernst equation**. Be sure that you know the exact meaning of the various quantities in this important equation, especially n (the number of electrons transferred in the half-reaction) and Q (the reaction quotient, Section 17-4). As you will see in the examples in this section, the approach used is to first apply the Nernst equation to correct each electrode potential to experimental concentrations. Then combine these E values (no longer E^0) to get E_{cell} (not E^0_{cell}). Then E_{cell} is an indication of whether the reaction, at the *specified concentrations*, can proceed. Positive values of E_{cell} indicate that the reaction is spontaneous. Another important use of the Nernst equation is to use measured cell voltages to determine concentrations of ions in solution (Section 21-20).

You have now seen three different ways to express the spontaneity of a reaction: (1) the standard Gibbs free energy change, ΔG^0, in Chapter 15, (2) equilibrium constant, K, in Chapter 17, and (3) E^0_{cell} in Chapter 21. You need to know the criteria for spontaneity of the forward reaction in terms of all three of these quantities. It should be clear that these are all related quantities; you have already seen (Section 17-12) the relation between ΔG^0 and K. In Section 21-21, E^0_{cell} is related to the other two indicators of spontaneity. The ideas of this section are important at two levels. First, three important concepts from different areas of chemistry are now connected, which gives an important insight into the fundamental unity of the science and the validity of the descriptions. Second, on a more practical level, it provides several different ways to derive numerical quantities or results that you may desire. For some reactions, it may be easier to measure E^0_{cell} than to determine K from experimentally measured concentrations or to determine ΔG^0 from calorimetric measurements or from tables that might not contain all necessary values. For other reactions, one of the latter two determinations may be easier. The relationships in this section then provide you with the means to convert your knowledge of the value of any one of these three quantities into the other two. (Of course, cell potentials apply only to redox reactions.) Examples 21-9 and 21-10 emphasize calculations of this nature.

The final sections of Chapter 21 are concerned with the use of voltaic cells as practical sources of electrical energy. Such cells are referred to as either: (1) **primary voltaic cells**, in which the electrochemical reaction cannot be reversed once the reactants have been mixed, or (2) **secondary voltaic cells**, also called **reversible cells**, in which the electrochemical reaction can be reversed by the action of a direct current from an external source—"recharging." The applications discussed in the text are the **dry cell**, an example of a primary voltaic cell (Section 21-22), the automobile **lead storage battery** (Section 21-23), and the rechargeable nickel-cadmium or **NiCad** cell, an example of a secondary voltaic cell (Section 21-24). When the reactants are continuously supplied to the voltaic cell, it is referred to as a **fuel cell**. The hydrogen-oxygen fuel cell (Section 21-25) is an example.

Study Goals

Sections 21-1 through 21-3 and 21-8

1. Distinguish between: (a) metallic conduction and electrolytic (ionic) conduction; (b) oxidation and reduction; (c) electrolytic cells and voltaic cells; (d) anode and cathode. *Work Exercises 1, 2, 5, and 6.*

Sections 21-3 through 21-7

2. Describe the general concepts of electrolysis.

Sections 21-2 through 21-5

3. Given the components of the electrodes and observations of what happens at the electrodes when an electrolytic cell is in operation:
   (a) Describe the operation of the cell.
   (b) Write balanced oxidation and reduction half-reactions.
   (c) Write balanced chemical equations for the overall reaction.
   (d) Construct a simplified diagram of the cell (such as Figure 21-4), including designation of anode and cathode, positive and negative electrode, direction of electron flow in the external circuit, and direction of migration of ions within the cell.
   Work Exercises 10 through 15.

Section 21-6

4. Using Faraday's Law of Electrolysis, perform calculations to relate the amount of electricity passing through an electrolytic cell to the amount (mass or gas volume) of a specified reactant consumed or product formed in the cell. Relate this information to the atomic weight of an ion consumed or produced. *Work Exercises 16 through 34.*

Sections 21-7

5. Describe the processes of electrolytic refining of impure metals and electroplating. *Work Exercises 124 and 129.*

Sections 21-8 through 21-10

6. Describe the general concepts of voltaic cells. *Work Exercises 35, 36, 43, and 44.*

Section 21-8

7. Describe standard electrochemical conditions. *Work Exercise 47.*

Sections 21-9 through 21-13

8. Given the components of the electrodes and observations of what happens at the electrodes when a voltaic cell is in operation:
   (a) Describe the operation of the cell.
   (b) Write balanced oxidation and reduction half-reactions.
   (c) Write balanced chemical equations for the overall reaction.
   (d) Construct a simplified diagram of the cell (such as Figure 21-7), including designation of anode and cathode, positive and negative electrode, direction of electron flow in the external circuit, and direction of migration of ions within the cell and in the salt bridge.
   (e) Write and interpret the short notation of voltaic cells.
   Work Exercise 37 through 42, 45, and 46.

Sections 21-11 and 21-14

9. Describe the standard hydrogen electrode. Illustrate how a table of standard electrode potentials (also known as electromotive series or activity series) is constructed and used. *Work Exercise 48 and 49.*

Sections 21-15 and 21-16

10. Assess relative strengths of oxidizing and reducing agents from a table of their standard electrode potentials (the electromotive or activity series). Use these results to assess whether a reaction is spontaneous or nonspontaneous. *Work Exercises 50 through 75.*

Sections 21-17 and 21-18

11. Describe the process of corrosion and methods by which corrosion of a specified metal can be prevented. *Work Exercises 76 and 77.*

Section 21-19

12. Apply the Nernst equation to determine electrode potentials and cell potentials under nonstandard conditions. *Work Exercises 78 through 92.*

Section 21-20

13. Use the Nernst equation to determine concentrations of ions in solution. *Work Exercises 93 through 97.*

Section 21-21

14. Summarize the restrictions on the values of E_{cell}^0, ΔG^0, and the equilibrium constant, K, for: (a) spontaneous reactions, (b) nonspontaneous reactions, and (c) reactions at equilibrium under standard electrochemical conditions. *Work Exercises 98 and 99.*

15. Perform calculations to relate standard cell potentials, standard Gibbs free energy changes, and equilibrium constants for redox reactions. *Work Exercises 100 through 104.*

Sections 21-22 through 21-25

16. Describe and distinguish among primary voltaic cells, secondary voltaic cells, and fuel cells. Give examples, balanced electrode half-reactions, and overall reactions for each. *Work Exercises 105 through 115.*

General

17. Be sure you can recognize and solve various kinds of problems based on the material in this chapter. *Work* **Mixed Exercises** *116 through 125.*

18. Answer conceptual questions based on this chapter. *Work* **Conceptual Exercises** *126 through 131.*

19. Apply concepts and skills from earlier chapters to the ideas of this chapter. *Work* **Building Your Knowledge** *exercises 132 through 138.*

20. Exercises at the end of chapter direct you to sources outside the textbook for information to use in solving them. Work **Beyond the Textbook** *exercises 139 through 142.*

Some Important Terms in This Chapter

Some important new terms from Chapter 21 are listed below. Fill the blanks in the following paragraphs with terms from the list. Each term is only used once. Check the Key Terms list and the chapter reading. Study other new terms, and review terms from preceding chapters (especially Chapter 6) if necessary.

anode	electrolytic (ionic) conduction	primary voltaic cell
cathode	electrolytic cell	reduction
cell potential	faraday	sacrificial anode
corrosion	fuel cell	secondary voltaic cell
electrochemistry	galvanizing	standard cell potential
electrode	metallic conduction	standard electrode
electrode potential	Nernst equation	standard electrode potential
electrolysis	oxidation	voltaic cell

In an oxidation-reduction reaction, one substance undergoes [1]_____, a loss of electrons, while another undergoes [2]_____, a gain of electrons. This transfer of electrons can be used to produce an electric current. The branch of chemistry that deals with the production of electricity by chemical changes and with the chemical changes produced by electric current is called [3]_____. The reacting system is generally contained in a *cell*. Spontaneous chemical reactions produce electricity and supply it to an external circuit from a [4]_____. Electrical energy from an external source causes nonspontaneous chemical reactions to occur in an [5]_____. In these cells, the electric current enters or exits by a conducting material called an [6]_____. The electrode at which reduction occurs, as electrons are gained by some species, is the [7]_____. (Cations often pick up electrons there.) The electrode at which oxidation occurs, as electrons are lost by some species, is the [8]_____. (Anions often lose electrons there.) Generally, electrodes are metal, and the flow of charge through them is called [9]_____. Electric current also flows through melted salts or electrolytic solutions in a cell, a motion known as [10]_____.

In an electrolytic cell, electrical energy from an external source causes nonspontaneous chemical reactions to occur, a process known as [11]_____. The amount of electricity that corresponds to the gain or less of one mole of electrons (6.022×10^{23} electrons) is called a [12]_____, named for Michael Faraday.

A half-cell in which the oxidized and reduced forms of a substance are in their standard states (1 M for aqueous species, 1 atm for gases, pure solids and gases) is referred to as a [13]_____. The electrical potential (E^0) of a standard electrode, as a reduction half-reaction, relative to the standard hydrogen electrode is the [14]_____ of the half-reaction. The corresponding electrical potential (E) of a half-reaction, without the substances involved being a standard conditions is its [15]_____. The electrode potentials of two half-cells are combined to produce the [16]_____ (E^0_{cell}) of the resulting cell. If the reactants and products in the cell are in their standard states, the cell potential is called the [17]_____, (E_{cell}).

A very common consequence of oxidation-reduction reactions is [18]_____, by which metals are oxidized by oxygen (O_2) in the presence of moisture. Two methods used to prevent corrosion are: (a) connecting the metal directly to a [19]_____, a piece of another metal that is more active and, therefore, preferentially oxidized, and (b) [20]_____, coating steel with zinc, a more active metal.

Electrode potentials and cell potentials for concentrations and partial pressures other than standard-state values can be calculated using the [21]_____.

There are three kinds of voltaic cells: (a) the [22]_____, which cannot be "recharged;" (b) the [23]_____, or reversible cell, in which the original reactants can be regenerated; and (c) the [24]_____, in which the reactants are continuously supplied to the cell and the products are continuously removed.

Quotefall

Fit the letters in each vertical column into the squares directly below them to form words across. The letters do not necessarily go into the squares in the same order in which they are listed. Each letter given is used once. A black square indicates a space. Words starting at the end of a line may continue in the next line. Punctuation in the statement is included in the boxes. When all the letters have been placed in their correct squares, you will be able to read the complete quotation across the diagram from left to right, line by line. Hint: It may be helpful at times to fill in small words, like "the," to use up some letters to help in determining other longer words. Some letters have been "seeded" to help you get started. The solution is on page 290.

Puzzle #1:

C	T	O	L	C	R	T	U	O	T	H	A	T	C	E	N	T	S	L	S	C	U	H	H	E	E	G	O	R	O	D	T	X	I	D	A	D	O	R
N	A	R	R	F	I	Y	Y	S	S	O	R	W	N	O	A	S	H	E	A	C	T	T	O	L	R	C	I	E	T	N	T	H	I	S	A	T	I	E
E	C	R	A	D	A	P	S	D	U	R	I	T	I	P	A	A	L	E	T	O	H	R	R	L	Y	S	M	S	S	H	E		E	E	L	L	O	O
F	T	T	O	Y		S	R	B	P	T	A	N	O		O	F	T	E	A	T		T	N	E	U	A	T	O	U	E		O	O	C		E	L	U
N			I				S	E		L	A	O						E		E			D	O		G	H						E			M		
		A									L													L								:						
										N															G										T		
									T																	E											
			P				R												M																		
C														S			R														C			.		

Preliminary Test (Practice many additional Exercises from the text.)

True-False

_____ 1. All electrochemical reactions are oxidation-reduction reactions.

_____ 2. In both electrolytic cells and voltaic cells, reduction occurs at the cathode.

_____ 3. One coulomb of electricity is the amount of charge carried when a current of 1 ampere flows for 1 second.

_____ 4. One faraday of electricity is the amount of charge of Avogadro's number of electrons.

_____ 5. According to Faraday's Law of Electrolysis, the same amount of electricity will deposit 1 mole of any metal from a solution of one of its salts.

_____ 6. The same amount of electricity that would deposit 107.87 g Ag from an Ag^+ solution would also liberate 11.2 L of dry H_2 gas, measured at STP.

_____ 7. One faraday of electricity is required to reduce 1 mole of Ag^+ to Ag.

_____ 8. One faraday of electricity is required to oxidize 1 mole of Ag to Ag^+.

_____ 9. One faraday of electricity is required to reduce 1 mole of Cu^{2+} to Cu.

_____ 10. One faraday of electricity is required to produce 8.00 g of O_2 gas by electrolysis of an aqueous solution.

_____ 11. During the electrolysis of aqueous sodium chloride, the solution gradually becomes more basic.

_____ 12. During the electrolysis of aqueous sodium sulfate, the solution gradually becomes more basic.

_____ 13. A battery is an electrolytic cell.

_____ 14. In a voltaic cell, the electrochemical reaction proceeds spontaneously.

_____ 15. In a voltaic cell, you can just mix the reactants together and then dip electrodes of the appropriate metals into the solution.

Short Answer

1. Electrochemical cells in which nonspontaneous reactions are forced to occur by the input of electrical energy are called _____ cells.

2. Electrochemical cells in which a spontaneous reaction serves as a source of electrical energy are called _____ cells.

3. An electrode that provides a surface at which the oxidation or reduction half-reaction can occur, but which is not itself a reactant or product of the electrochemical reaction, is called a(n) _____.

4. In an electrolytic cell, the electrode at which oxidation takes place is called the _____; in a voltaic cell, the electrode at which oxidation takes place is called the _____.

5. One coulomb is the amount of charge on _____ (a number) electrons.

6. The cell that is used for the commercial electrolytic production of sodium metal is named the _____ cell. In this cell the processes that occur are:

reduction   half-reaction:

_____;

oxidation   half-reaction:

_____;

overall   cell   reaction:

_____.

7. In an electrolysis cell, you observe that 0.5 mole of palladium is deposited by the same current, flowing for the same duration of time, that can also deposit 1 mole of silver. The charge on the palladium ions in this solution is _____.

8. In simple voltaic cells such as those used for measurement, the electrode that gains in weight would be the _____.

9. In discussions of electrochemistry, the initials SHE stand for _____ _____.

10. In the standard hydrogen electrode, the purpose of the platinum metal is to provide ____ _____ __.

11. When the SHE acts as the anode in an electrochemical cell, the half-reaction that takes place at this electrode is _____; when it acts as a cathode the half-reaction is _____.

12. In the cell $Cr|Cr^{3+}(1.0\ M)||H^+(1.0\ M),H_2(1\ atm)|Pt$, the SHE acts as the _____.

13. A voltaic cell consists of a standard hydrogen electrode connected by a salt bridge and a wire to an electrode consisting of a strip of Cd metal dipped into a 1 M solution of $Cd(NO_3)_2$.

    When the cell produces current, the electrons flow through the wire from the _____ electrode to the _____ electrode.

    In this cell, the _____ electrode is acting as the cathode.

14. Liquid bromine _____ (will or will not?) oxidize silver metal to Ag^+ in aqueous solution.

15. The process in which metals are oxidized by O_2 in the presence of moisture is called _____.

16. If, for a given process, E^0_{cell} can be shown to be positive, then ΔG^0 for that process will be _____, K_{eq} will be _____, and the process as written will be _____.

17. In the Leclanché cell, the carbon rod is the _____, at which the half-reaction taking place is _____. The other electrode is provided by the _____, at which the half-reaction taking place is _____.

18. In the lead storage battery, sulfuric acid acts as the _____.

19. During the operation (discharging) of the lead storage battery, the Pb electrode is the _____ and the PbO_2 electrode is the _____.

Multiple Choice

_____ 1. The substances classified as "electrolytes" will conduct electricity when dissolved in water because
   (a) they are polar covalent substances.
   (b) they are nonpolar covalent substances.
   (c) they dissociate in solution to give ions that are attracted to the electrodes having the same signs as those of the ions.
   (d) they dissociate in solution to give ions that are attracted to the electrodes having the opposite signs from those of the ions.
   (e) they cause water to dissociate into H_3O^+ and OH^- ions.

_____ 2. In order to obtain 9,000 coulombs in an electrolytic process, how long would a constant current of 18 amperes be required to flow?

(a) 200 seconds   (b) 500 seconds   (c) 1.6×10^5 seconds

(d) 50 seconds   (e) 0.002 seconds

_____ 3. The electrode in an electrolysis cell that acts as a source of electrons to the solution is called the _____; the chemical change that takes place at this electrode is called _____.

(a) anode, oxidation   (b) anode, reduction

(c) cathode, oxidation   (d) cathode, reduction

(e) Cannot tell unless we know what species is being oxidized or reduced.

The next three questions, 4 through 6, refer to the electrolysis of molten sodium bromide, NaBr. (It may help you to diagram the cell.)

_____ 4. During this electrolysis, sodium ions move

(a) to the anode, which is positively charged.

(b) to the anode, which is negatively charged.

(c) to the cathode, which is positively charged.

(d) to the cathode, which is negatively charged.

(e) through the wire to the battery.

_____ 5. What is the half-reaction that takes place at the cathode?

(a) $2Br^- \rightarrow Br_2 + 2e^-$ (b) $Br_2 + 2e^- \rightarrow 2Br^-$ (c) $Na^+ + e^- \rightarrow Na$

(d) $Na \rightarrow Na^+ + e^-$ (e) $2H_2O + 2e^- \rightarrow 2OH^- + H_2$

_____ 6. What is the half-reaction that takes place at the anode?

(a) $2Br^- \rightarrow Br_2 + 2e^-$

(b) $Br_2 + 2e^- \rightarrow 2Br^-$

(c) $Na^+ + e^- \rightarrow Na$

(d) $Na \rightarrow Na^+ + e^-$

(e) $2H_2O + 2e^- \rightarrow 2OH^- + H_2$

_____ 7. Which of the following is *not* produced in the electrolysis of aqueous sodium chloride?

(a) Na   (b) NaOH   (c) Cl_2 (d) H_2

(e) All of the preceding are produced in this electrolysis.

_____ 8. During the electrolysis of aqueous sodium chloride, the solution gradually

(a) disappears.   (b) gets more basic.   (c) gets more acidic.

(d) dissolves the anode.   (e) dissolves the cathode.

_____ 9. How many coulombs must be supplied in order to plate out 1.07868 g of Ag from an Ag^+ solution?

(a) 96,485   (b) 96,485/1.07868   (c) 96.485

(d) 96,485 × 1.07868   (e) 964.85

_____ 10. How many faradays of electricity are required to perform the reduction of 2.0 g of Sn^{4+} to Sn^{2+}?

(a) 0.017   (b) 0.034   (c) 0.068   (d) 0.008   (e) 2.000

_____ 11. In $Cr_2(SO_4)_3$, chromium is in the +3 oxidation state. What mass of chromium will be deposited by electrolysis of a solution of $Cr_2(SO_4)_3$ for 60 minutes using a steady current of 10 amperes?
(a) 3.25 g   (b) 6.47 g   (c) 17.3 g   (d) 0.187 g   (e) 0.373 g

_____ 12. An aqueous solution of an unknown salt of ruthenium is electrolyzed by a current of 2.50 amperes passing for 50 minutes. This results in the formation of 2.618 g of ruthenium metal at the cathode. What is the oxidation state of ruthenium in this solution?
(a) +1   (b) +2   (c) +3   (d) –2   (e) –3

_____ 13. The electrolysis of an aqueous solution of a cobalt salt proceeds until exactly 8.000 g of O_2 gas has been liberated at the anode. At this time, 29.467 g of Co has been deposited at the cathode. What is the charge on the cobalt ion in the solution?
(a) 1+   (b) 2+   (c) 3+   (d) 1–   (e) 2–

_____ 14. In a voltaic cell, the salt bridge
(a) is not necessary in order for the cell to work.
(b) allows mechanical mixing of the solution.
(c) allows ions to flow from either half-cell into the other half-cell.
(d) is tightly plugged with a firm agar gel in order to keep all ions separate.
(e) drives electrons from one half-cell to the other.

_____ 15. Refer to the table of standard reduction potentials in the text, Table 21-2. The most stable ions of the Group 1B metals are Cu^{2+}, Ag^+, and Au^{3+}. Which one of the following statements regarding the ease of reduction of these ions to the respective metals is true?
(a) Au^{3+} is easier to reduce than Cu^{2+}, which is easier to reduce than Ag^+.
(b) Cu^{2+} is easier to reduce than Au^{3+}, which is easier to reduce than Ag^+.
(c) Au^{3+} is easier to reduce than Ag^+, which is easier to reduce than Cu^{2+}.
(d) Cu^{2+} is easier to reduce than Ag^+, which is easier to reduce than Au^{3+}.
(e) Ag^+ is easier to reduce than Au^{3+}, which is easier to reduce than Cu^{2+}.

_____ 16. Refer to the table of standard reduction potentials in the text, Table 21-2. Which of the following statements about the Group 1B metals is true?
(a) Cu is easier to oxidize than Au.
(b) Au is easier to oxidize than Ag.
(c) Ag is easier to oxidize than Cu.
(d) Au is easier to oxidize than Cu.
(e) Nothing can be decided about ease of oxidation from a table of reduction potentials.

_____ 17. According to the table of standard reduction potentials, which one of the following metals is most easily oxidized?
(a) Cd   (b) Cu   (c) Fe   (d) Ni   (e) Zn

_____ 18. Which one of the following is the strongest oxidizing agent in 1 M aqueous solutions?
(a) Cu^{2+} (b) Ag^+ (c) MnO_4^- (in acid solution)
(d) Ca^{2+} (e) Li^+

_____ 19. Which one of the following is the strongest oxidizing agent? (All solutions are 1 M, gases are 1 atm)

(a) $Cl_2(g)$ (b) $Br_2(\ell)$ (c) $Li(s)$ (d) $Li^+(aq)$ (e) $NO_3^-(aq)$

_____ 20. Would a reaction occur if a piece of pure copper were dipped into a 1 M $FeCl_2$ solution?

(a) No   (b) Yes, forming Fe^{3+} (c) Yes, forming Fe metal

The next five questions, 21 through 25, refer to the following electrochemical cell. Each half-cell contains an aqueous solution. Necessary values of standard electrode reduction potentials may be obtained from Table 21-2 in the text.

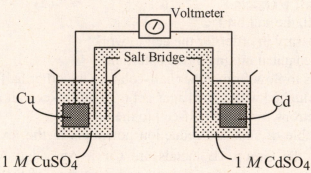

_____ 21. What is the standard cell potential for this cell?

(a) +0.740 V   (b) −0.740 V   (c) +0.066 V   (d) −0.066 V   (e) 0 V

_____ 22. Which of the following statements about the half-cell processes is true?

(a) Cu^{2+} is reduced at the anode.   (b) Cu^{2+} is reduced at the cathode.

(c) Cd^{2+} is reduced at the anode.   (d) Cd^{2+} is reduced at the cathode.

(e) This reaction is not an oxidation-reduction reaction.

_____ 23. When this cell is in operation, electrons flow

(a) through the wire from the Cu electrode to the Cd electrode.

(b) through the wire from the Cd electrode to the Cu electrode.

(c) through the salt bridge from the Cu half-cell to the Cd half-cell.

(d) through the salt bridge from the Cd half-cell to the Cu half-cell.

(e) No other answer is correct, because electrons do not flow in an electrochemical cell.

_____ 24. Which of the following equations describes the net reaction that takes place in this cell?

(a) $Cu + Cd^{2+} \rightarrow Cu^{2+} + Cd$ (b) $Cu + Cd \rightarrow Cu^{2+} + Cd^{2+}$

(c) $Cu^{2+} + Cd^{2+} \rightarrow Cu + Cd$ (d) $Cu^{2+} + Cd \rightarrow Cu + Cd^{2+}$

_____ 25. What is ΔG^0 for the reaction $Cu^{2+} + Cd \rightarrow Cu + Cd^{2+}$?

(a) −71.41 kJ/mol   (b) −142.8 kJ/mol   (c) +142.8 kJ/mol

(d) −34.10 kJ/mol   (e) +71.41 kJ/mol

Answers to Some Important Terms in This Chapter

1. oxidation
2. reduction
3. electrochemistry
4. voltaic cell
5. electrolytic cell
6. electrode
7. cathode
8. anode
9. metallic conduction
10. electrolytic (ionic) conduction
11. electrolysis
12. faraday
13. standard electrode
14. standard electrode potential
15. electrode potential
16. cell potential
17. standard cell potential
18. corrosion
19. sacrificial anode
20. galvanizing
21. Nernst equation
22. primary voltaic cell
23. secondary voltaic cell
24. fuel cell

Answer to Quotefall Puzzle

Puzzle #1:
Faraday's Law of Electrolysis: The amount of substance that undergoes oxidation or reduction at each electrode is directly proportional to the amount of electricity that passes through the cell.

Answers to Preliminary Test

True-False

1. True
2. True
3. True
4. True
5. False. This would depend on the number of electrons necessary to reduce one mole of the metal—that is, on the charge of the metal ion.
6. True. See Table 21-1. This amount of electricity is one faraday. The amount of H_2 liberated is 1.008 g or $\frac{1}{2}$ mole. At STP, one mole of any gas occupies 22.4 L (Chapter 12).
7. True
8. True
9. False. Each Cu^{2+} ion requires 2 electrons in this process. Two faradays would be required.
10. True. The half-reaction is $2H_2O \rightarrow O_2 + 4H^+ + 4e^-$. Thus, 4 faradays are required to produce 1 mole (32.0 g) of O_2. This means that 1 faraday would produce $\frac{1}{4}$ mol O_2, or 8.0 g.
11. True. Remind yourself what the cathode half-reaction is in this cell (Section 21-4).
12. False. At the cathode, OH^- is produced, but H^+ is produced in equal amount at the anode. These neutralize one another (or in the terminology of Chapter 19, the $[H_3O^+]/[OH^-]$ balance is not disturbed).
13. False. A battery is a voltaic cell.
14. True

15. False. The two half-reactions must be forced to take place in two separate places so that the electrons can be transferred externally. If you just mixed the reactants together and dipped the metals into the solution, the same overall redox reaction would occur, but we would not have an electrochemical cell.

Short Answer

1. electrolytic
2. voltaic
3. inert electrode
4. anode; anode
5. 6.241×10^{18}. Avogadro's number of electrons (6.022×10^{23} electrons) corresponds to 96,485 coulombs. So one coulomb would correspond to $\dfrac{6.022 \times 10^{23}}{96,485}$ electrons.
6. Downs

   $Na^+ + e^- \rightarrow Na(\ell)$

   $2Cl^- \rightarrow Cl_2(g) + 2e^-$

   $2Na^+ + 2Cl^- \rightarrow 2Na(\ell) + Cl_2(g)$
7. +2. You should be able to figure this out *without* calculation. This will test whether you really understand the ideas behind those calculations.
8. cathode. This is the electrode at which metal ions are reduced to metal.
9. standard hydrogen electrode
10. a surface at which the electron transfer can take place.
11. $H_2 \rightarrow 2H^+ + 2e^-$. Remember that the anode is the electrode at which oxidation occurs.

    $2H^+ + 2e^- \rightarrow H_2$. Remember that the cathode is the electrode at which reduction occurs.
12. cathode. From the table of standard reduction potentials, you see that H^+ is easier to reduce than is Cr^{3+}, so the SHE is the cathode. At the anode, Cr metal oxidizes to Cr^{3+}.
13. cadmium; hydrogen; hydrogen. From the table of standard reduction potentials, you see that H^+ is more easily reduced than Cd^{2+}. The electrons flow through the wire to the SHE, where they reduce H^+ to H_2. Reduction occurs at the cathode.
14. will. You want to see whether the reaction $Br_2 + 2Ag \rightarrow 2Ag^+ + 2Br^-$ will occur. From Table 21-2 write the half-reactions so that they add to give the reaction we wish to test; remember that when you reverse the half-reaction for $Ag^+|Ag$, you must change the sign of its E^0. Though you multiply the second half-reaction by 2 you *do not* multiply its E^0 value! Then you add the reactions, adding the E^0 values to get E^0_{cell}.

    $$\begin{array}{lll} Br_2 + 2e^- \rightarrow 2Br^- & E^0 & = +1.08 \quad V \\ 2(Ag \rightarrow Ag^+ + e^-) & E^0 & = -0.799 \ V \\ \hline Br_2 + 2Ag \rightarrow Ag^+ + Br^- & E^0_{cell} & = +0.28 \quad V \end{array}$$

    E^0_{cell} for this proposed reaction is positive, so it is spontaneous in the direction written.
15. corrosion
16. negative; greater than 1; spontaneous
17. cathode; $2NH_4^+ + 2e^- \rightarrow 2NH_3 + H_2$; zinc container; $Zn \rightarrow Zn^{2+} + 2e^-$
18. electrolyte

19. anode; cathode

$$Pb(s) + HSO_4^-(aq) \rightarrow PbSO_4(s) + H^+(aq) + 2e^-; \text{ oxidation;}$$
anode

$$PbO_2(s) + 3H^+(aq) + HSO_4^-(aq) + 2e^- \rightarrow PbSO_4(s) + 2H_2O(\ell); \qquad \text{reduction;}$$
cathode

Multiple Choice

1. (d)
2. (b). number of coulombs = (number of amperes) × (number of seconds)

$$\text{time} = \frac{\text{number of coulombs}}{\text{number of amperes}} = \frac{9,000 \text{ coulombs}}{18 \text{ amp}} = 500 \text{ s}$$

3. (d). The gain of electrons is reduction, which occurs at the cathode.
4. (d). Cations migrate to the cathode, which supplies electrons from an external source, and is, therefore, negative.
5. (c). Reduction (the addition of electrons) occurs at the cathode, which eliminates (a) and (d). By far the predominant species present in molten sodium bromide are sodium ions and bromide ions, not elemental bromine or water, which eliminate (b) and (e).
6. (a). Oxidation (the loss of electrons) occurs at the anode, which eliminates (b), (c), and (e). There is little or no elemental sodium in the molten sodium bromide, which eliminates (d).
7. (a). See the discussion in Section 21-4. Remember that the alkali metals are much too active (easily oxidized) to exist in contact with water.
8. (b). The cathode half-reaction in this cell is the reduction of water to H_2. This also gives OH^-. The half-reaction is $2H_2O + 2e^- \rightarrow 2OH^- + H_2$.
9. (e). 1.07868 g of Ag is $\dfrac{1.07868 \text{ g}}{107.868 \text{ g/mol}} = 0.01000 \text{ mol } Ag^+$ to be reduced. Each mole Ag^+ requires 1 faraday or 96,485 coulombs (C). 0.01000 mol × 96,458 C/mol = 964.85 C.
10. (b). The reduction half-reaction is $Sn^{4+} + 2e^- \rightarrow Sn^{2+}$. Thus, 2 faradays are required to reduce 1 mole (118.7 g) of Sn^{4+} to Sn^{2+}. Applying the unit factor $\dfrac{2 \text{ faradays}}{118.7 \text{ g } Sn^{4+} \text{ reduced}}$, we find that 0.034 faraday is required to reduce the desired 2.0 g of Sn^{4+}.
11. (b). The number of coulombs is $60 \text{ min} \times 60 \dfrac{s}{\text{min}} \times 10 \dfrac{C}{s} = 36,000 \text{ C.}$

$$\text{mass} = 36,000 \text{ C} \times \frac{1 \text{ mol } e^-}{96,485 \text{ C}} \times \frac{51.996 \text{ g}}{3 \text{ mol } e^-} = 6.47 \text{ g}$$

12. (c). The number of coulombs is $50 \text{ min} \times 60 \dfrac{s}{\text{min}} \times 2.50 \dfrac{C}{s} = 7,500 \text{ C.}$ If 7,500 coulombs deposits 2.618 grams, then 96,500 coulombs would deposit 33.7 g of ruthenium. This is $\dfrac{33.7 \text{ g}}{101.07 \text{ g/mol}} = 0.333 \text{ mol of Ru deposited by 1 mole of electrons, so each } Ru^{n+} \text{ ion}$ must require three electrons. Thus, the ion is Ru^{3+}.
13. (b). In the anode reaction, $2H_2O \rightarrow O_2 + 4H^+ + 4e^-$, it requires 4 × 96,500 coulombs to liberate 1 mol of 32.0 g O_2. Thus, the reaction involves 1 mole of electrons (8.000 g is ¼ of

32.0 g). Since this liberates $\dfrac{29.467 \text{ g}}{58.933 \text{ g/mol}} = 0.500$ mol Co, each Co^{n+} ion must require two electrons. Thus, the ion is Co^{2+}.

14. (c)

15. (c). The standard electrode potentials for reduction of each of these ions to the free metal are $Au^{3+} = +1.50$ V; $Ag^+ = +0.799$ V; $Cu^{2+} = +0.337$ V. A more positive reduction potential means that reduction is easier to accomplish.

16. (a). Reasoning from the reduction potentials, the hardest reduction to carry out for the Group 1B ions is Cu^{2+}/Cu. Thus, the reverse of this process, the oxidation, would be the easiest of those listed.

17. (e). Each reaction appears as a reduction. The harder the reduction, the easier is the oxidation (the reverse reaction). The half-reaction: $Zn^{2+} + 2e^- \rightarrow Zn$ is hardest to carry out as a reduction. Therefore, its reverse, the oxidation of Zn metal, would be easiest.

18. (c). Referring to the table of reduction potentials, Table 21-2, you see that all species listed are on the left and are being reduced in half-reactions as written. Of the species listed, MnO_4^- in acid solution is most easily reduced. When something is reduced, it is acting as an oxidizing agent. Thus, permanganate ion is the strongest oxidizing agent listed here.

19. (a). Same reasoning as for Question 18. The only wrinkle here is to figure out where Li(s) fits as an oxidizing agent. To act as an oxidizing agent, it would have to be reduced, and we know that alkali metals rarely gain electrons. Thus, Li(s) is a very poor oxidizing agent.

20. (a). Because the Cu metal could only be oxidized ($Cu \rightarrow Cu^{2+} + 2e^-$), Fe^{2+} could not *also* be oxidized (to Fe^{3+}), so answer (b) is eliminated. Might Fe^{2+} be reduced ($Fe^{2+} + 2e^- \rightarrow Fe$)? For this reduction, E^0 is –0.44 V. For the Cu oxidation, E^0 is –0.34 V. Adding these electrode potentials, you see that the only possible reaction, $Fe^{2+} + Cu \rightarrow Fe + Cu^{2+}$, has a standard cell potential of $E^0_{cell} = -0.78$ V. The negative value tells you that the reaction will *not* occur.

21. (a)

$$
\begin{array}{llll}
Cu^{2+} + 2e^- \rightarrow Cu & E^0 & = & +0.337 \text{ V} \\
Cd \rightarrow Cd^{2+} + 2e^- & E^0 & = & +0.403 \text{ V} \\
\hline
Cu^{2+} + Cd \rightarrow Cu + Cd^{2+} & E^0_{cell} & = & +0.740 \text{ V}
\end{array}
$$

22. (b). The half-reactions (above) show that it is the Cu^{2+} ions that are reduced. Reduction occurs at the cathode.

23. (b). The half-reactions (above) show that the Cd loses electrons, which flow through the wire to the copper electrode, where they are gained by the Cu^{2+} ions.

24. (d). See the answer to question 21 above.

25. (b). From the answer to Question 21, E^0_{cell} is +0.740 V for the reaction as written. Convert using $\Delta G^0 = -nFE^0_{cell}$ to find $\Delta G^0 = -142.8$ kJ. This negative value means that this reaction is spontaneous as written. This is also indicated by the positive value of E^0_{cell}.

22 Nuclear Chemistry

Chapter Summary

Up to now, you have studied chemical reactions, in which atoms maintain their elemental identity and interact only through electronic changes. Now turn your attention to the study of nuclear reactions, in which atoms are often converted from one element into another by alterations in their nuclei. This area of chemistry, called **nuclear chemistry,** is the subject of Chapter 22. As you read the chapter introduction, pay close attention to the differences between nuclear reactions and ordinary chemical reactions. A preliminary mention is also made of two important types of nuclear reactions, **nuclear fission** and **nuclear fusion**. You will examine both of these more fully later in the chapter.

Section 22-1 summarizes the main features of atomic nuclei. Nuclear reactions depend on nuclear stability, which in turn is related to the ratio of number of neutrons to number of protons in the nucleus. (These nuclear particles, the neutrons and the protons, are collectively called **nucleons**.) This relationship is explained in Section 22-2. The most stable **nuclides** are those with an even number of protons and an even number of neutrons (146 of them); stable nuclides with an odd number of either nucleon are found less often (103 of them); only five stable nuclides have odd numbers of both protons and neutrons ($^{2}_{1}H$, $^{6}_{3}Li$, $^{10}_{5}B$, $^{14}_{7}N$, and $^{180m}_{73}Ta$). Nuclides having either a number of protons or a number of neutrons (or a sum of the two) equal to one of the so-called **magic numbers**—2, 8, 20, 28, 50, 82, or 126—are exceptionally stable. The plot of the number of neutrons versus number of protons for the stable nuclei is called the **band of stability** (Figure 22-1).

The question of the energy relationships in the nucleus of the atom, the **binding energy** of the nucleus, is addressed in Section 22-3. This topic is related to the release of energy accompanying nuclear reactions such as those that occur in nuclear reactors and in atomic bombs. These will be studied later in this chapter.

The decay or decomposition of unstable nuclei is accompanied by the emission of radiation (**radioactive decay**). The practical reason for carrying out nuclear reactions is often either to make some use of the radiation emitted by nuclear decay or to harness and use the tremendous amounts of energy that are released with many nuclear reactions. You will examine the most common types of radiation emitted in nuclear decay processes in Section 22-4. These various types of radiation may be particles—positive, negative, or neutral—or high-energy electromagnetic radiation, each with a characteristic mass, charge, velocity, and penetrating power as summarized in Table 22-3. Remember that all of these particles or electromagnetic rays are emitted from the **nucleus** of the atom, including the beta particles, which are identical to the extranuclear electrons. The high-energy gamma rays that are often emitted in nuclear decay processes serve to remove excess energy from a nucleus that is in an excited state. This is similar to the emission of light from an atom whose electrons are in an excited configuration.

Each type of decay process causes a change in the atomic number and mass number of the nucleus. These changes are summarized in the following table. Of course, it is not necessary to memorize a table such as this if you just remember the mass number and atomic number for each type of particle that can be emitted or captured (included in some of the symbols in the first column of Table 22-3). Just as for atomic nuclei, the symbol that represents each of these types of radiation can include its atomic number as a left subscript and its mass number as a left superscript.

Process Number	Change in Atomic Number	Change in Mass
beta ($_{-1}^{0}e$) emission	increase by 1	no change
positron ($_{+1}^{0}e$) emission	decrease by 1	no change
alpha ($_{2}^{4}He$) emission	decrease by 2	decrease by 4
proton ($_{1}^{1}H$) emission	decrease by 1	decrease by 1
neutron ($_{0}^{1}n$) emission	no change	decrease by 1
electron capture (K capture)	decrease by 1	no change
gamma ($_{0}^{0}\gamma$) emission	no change	no change

Section 22-5 gives the conventions for writing equations for nuclear reactions. As you study this chapter, you should check that each equation meets the requirements presented in this section:

1. The sum of the mass numbers of the reactants must equal the sum of the mass numbers of the products, and
2. The sum of the atomic numbers of the reactants must equal the sum of the atomic numbers of the products.

Neutron-rich nuclei (those that have a ratio of neutrons to protons too high to be stable) are said to be above the band of stability; these decay so as to decrease the ratio, usually by **beta emission** and less commonly by **neutron emission.** Examples of such nuclei and their characteristic decay processes are discussed in Section 22-6. This section also introduces the writing of equations for nuclear reactions, in a way entirely analogous to the writing of equations for ordinary chemical reactions. In such equations, it is most convenient to write the full **nuclide symbol** for all species involved in the reaction.

Neutron-poor nuclei (with a too-low ratio of neutrons to protons) are described as below the **band of stability**. These nuclei decay by processes that increase this ratio—**positron emission**, **electron capture** (in which an electron from the first, or K, shell is captured by the nucleus), or **alpha emission** (especially for heavier nuclei), as discussed in Section 22-7. All nuclei with *atomic number greater than 83* are radioactive and are below and to the right of the band of stability (Section 22-8) so most such nuclides decay by alpha emission. This includes the heavier representative elements, the most recently produced transition metals, and all of the actinides. Some of these heavy nuclides, especially uranium ($Z = 92$) and elements beyond it, also disintegrate by spontaneous nuclear fission into nuclides of intermediate masses, emitting neutrons as well.

Many studies and applications of nuclear reactions depend on the detection and measurement of the radiation emitted. Several common methods of **detection of radiation** are described in Section 22-9—photographic methods, fluorescence methods, cloud chambers, and gas ionization counters. Each of these methods depends on the properties and energies of the various types of radiation; you should know the general basis of each method of detection. Photographic detection and use of a cloud chamber are convenient methods for detecting the presence of radiation, requiring little in the way of sophisticated equipment, but they are tedious and inaccurate to use for quantitative measurements of radiation intensity and energy. Fluorescence counters such as scintillation counters and gas ionization counters both require electronic circuitry for effective use, but can permit quite sensitive studies of energies and intensities of radiation.

All radioactive decay processes obey temperature-independent **first-order** kinetics (Sections 16-3 and 16-4). Thus, the rates of all such decay processes can be discussed conveniently in terms of the **half-lives** (Section 22-10) of the decaying nuclei. Review the discussion of the order of reactions, the resulting changes in amount of substance present, and the description of reaction rates in terms of the half-life of a reactant (Section 16-4). In the period of one half-life, half of any initial amount of a particular radioactive substance will decay. Remember that the half-life of a substance in a first-order process is constant. Examples of the rate calculations for radioactive decay processes and the interpretations and applications of these calculations are also found in this section.

While some unstable nuclides can attain stability by a single decay process, many must undergo a series of such decays, known as **disintegration series.** Three well-known natural series, beginning with $^{238}_{92}U$, $^{235}_{92}U$, and $^{232}_{90}Th$, respectively, are discussed in Section 22-11. A careful study of these series will give you further practice in understanding the results of emission of various types of radioactive particles.

The steady, predictable decay rates of unstable nuclei (**radionuclides**) and the resulting steady production of radioactivity make them very useful for many applications. Some of these uses are discussed in Section 22-12. The technique of **radioactive dating** depends on the known rates of decay of various long-lived nuclei. Analysis of the amount of carbon-14 remaining in very old materials of plant or animal origin allows dating of such objects when they are less than about 50,000 years old. Similar dating techniques are based on the ratio of potassium-40 to argon-40 (articles up to 1 million years old) or on the ratio of uranium-238 to lead-206 (articles several million years old). The calculations associated with such analyses are, of course, just first-order kinetics calculations. Other uses of radionuclides outlined in this section include applications in medicine, chemical research, agriculture, and industry.

So far in this chapter, you have been investigating processes associated with nuclei that are naturally unstable. It is also possible to bombard nuclei with particles so that they become unstable by collision with the bombarding particle, and then decay. This technique, known as **nuclear bombardment,** can be used to artificially transmute one element to another, as described in Section 22-13. Such processes have resulted in the production of many nuclides that do not occur naturally, including many new elements—technetium, astatine, francium, and the transuranium elements, to name a few.

Most of the nuclear processes studied so far have been ones in which an unstable nucleus emits a particle and is transformed into a nuclide of slightly different mass number or atomic number. Two other kinds of nuclear reactions are discussed in the remainder of this chapter. **Nuclear fission** (Section 22-14) is a process in which an unstable nucleus disintegrates into two nuclei of intermediate masses and one or more neutrons. Nuclear fission is common for isotopes of some elements with atomic numbers above 80. This process is accompanied by the release of a vast amount of energy corresponding to the loss of mass that accompanies the reaction.

The two principal applications of nuclear fission reactions, the **nuclear reactor** and the **atomic bomb**, both depend on establishing a **chain reaction** in which the neutrons from one nuclear disintegration are absorbed by other nuclei. These then disintegrate in turn, each such disintegration releasing energy and more neutrons to propagate the process. In the atomic bomb, this chain reaction is initiated by bringing together more than the minimum amount of material necessary to sustain the chain reaction, called the **critical mass,** and then allowing the reaction to proceed at an uncontrolled rate. (The critical mass is at least 10 kg for ^{239}Pu, 52 kg for ^{235}U, and 60 kg for ^{237}Np.) In the nuclear fission reactor (Section 22-15), the rate of the fission reaction is controlled by other substances that are present to absorb some of the neutrons. The general design features of **light water reactors** are described in this section. Some of the problems associated with the application of nuclear energy to generate power are also discussed.

Nuclear fusion (Section 22-16) is a very energetic process in which two small nuclei are combined to make one nucleus of higher mass. This process is favorable only for the very light nuclei. Nuclear fusion reactions produce even greater amounts of energy than nuclear fission (the hydrogen bomb, for instance). But these reactions are so difficult to contain and control that development of this nearly limitless energy source has been frustrating, for reasons described in this section.

Study Goals

Chapter Introduction and Section 22-1

1. Summarize the characteristics that distinguish between nuclear reactions and ordinary chemical reactions. *Work Exercise 3.*

Section 22-3

2. Compare the calculated mass of an isotope with its observed mass. Relate this mass deficiency to the binding energy of the nucleus. *Work Exercises 4, 5, 7, 8, 14 through 19, 66, 73, 74, 77, and 78.*

Section 22-4

3. Give symbols, identities, and properties of the common types of radiation: (a) beta particles, (b) alpha particles, (c) positrons, (d) neutrons, (e) protons, and (f) gamma rays. *Work Exercises 20, 21, and 41.*

Section 22-5

4. Balance equations for nuclear reactions. *Work Exercises 34, 35, 40 through 43, and 46 through 48.*

Sections 22-2 and 22-4 through 22-8

5. Given a plot of number of neutrons versus number of protons, identify stable nuclei, suggest and give specific examples of the kind of radioactive decay most likely for nuclides that are (a) above, (b) below, and (c) beyond and to the right of the band of stability. *Work Exercises 9 through 13 and 26.*

Section 22-9

6. Describe the following methods for the detection of radiation: (a) photography, (b) fluorescence, (c) cloud chamber, and (d) gas ionization. *Work Exercise 24.*

Section 22-10

7. Relate rate constants for radioactive decays to half-lives. *Work Exercises 50 and 51.*

8. Perform calculations relating the following four variables, calculating the fourth when three are known: (a) initial amount of radionuclide present, (b) specified time, (c) amount of radionuclide remaining after that time, and (d) the rate constant or half-life for the decay. *Work Exercises 52 through 62, 76, and 79.*

Section 22-11

9. In nuclear disintegration processes, be able to relate parent nuclide, daughter nuclide, and type of emission. *Work Exercises 25, 27 through 32, and 34 through 46.*

Section 22-12

10. Give several examples of practical uses of radionuclides. *Work Exercises 22 and 23.*

Section 22-13

11. Describe and illustrate the operation of a cyclotron and a linear accelerator. *Work Exercise 49.*

12. Give specific examples, with balanced equations, of various methods of inducing artificial transmutations of elements. *Work Exercises 34 through 39 and 66.*

Sections 22-14 through 22-16

13. Describe and distinguish between nuclear fission and nuclear fusion. Relate the position of a nuclide on a plot of binding energy per nucleon versus mass number to its tendency to undergo fission or fusion. *Work Exercises 1, 25, and 63 through 65.*

Section 22-14

14. Describe the principles behind the atomic bomb. Tell how ignition of the detonation process is accomplished and how the reaction is sustained. *Work Exercise 67.*

Section 22-15

15. Describe the operation of a light water nuclear reactor. Explain why nuclear fission explosions are impossible in such a setting. *Work Exercises 68, 69, and 71.*

16. Describe the vapor diffusion method currently used for separating fissionable isotopes for use in nuclear reactors. Relate this method to relative rates of diffusion of gas molecules. *Work Exercise 70.*

Section 22-16

17. Give the advantages and disadvantages of nuclear fusion reactors as potential future energy sources. *Work Exercise 72.*

General

18. Answer conceptual questions based on this chapter. *Work* **Conceptual Exercises** *80 through 83.*

19. Apply concepts and skills from earlier chapters to the material of this chapter. *Work* **Building Your Knowledge** *exercises 84 and 85.*

20. Exercises at the end of chapter direct you to sources outside the textbook for information to use in solving them. Work **Beyond the Textbook** *exercises 86 through 91.*

Some Important Terms in This Chapter

Some important terms from Chapter 22 are listed below. Fill the blanks in the following paragraphs with terms from the list. Each term is only used once. Study other new terms and review terms from preceding chapters if necessary.

alpha particle (α)	gamma ray (γ)	nuclides
artificial transmutation	half-life	parent nuclide
band of stability	mass deficiency	positron
beta particle (β)	nuclear binding energy	radioactive dating
chain reaction	nuclear fission	radioactivity
critical mass	nuclear fusion	
daughter nuclide	nuclear reactor	

It was just a little more than one hundred years ago that Henri Becquerel discovered that some atomic nuclei can spontaneously disintegrate, a process known as [1]_____. Two major types of nuclear changes are [2]_____, the splitting of a heavy nucleus into nuclei of intermediate masses with the emission of one or more neutrons, and [3]_____, the combination of light nuclei to produce a heavier nucleus.

Since the term "isotope," properly used, refers only to different forms of the *same* element, different atomic forms of *all* elements are called [4]_____. A graph of the number of neutrons versus the number of protons for stable nuclides reveals a [5]_____ _____. Radioactive nuclides usually lie above (and the left of) or below (and to the right of) this band of stability.

The mass of an atom is not simply the sum of the masses of the protons, neutrons, and electrons that comprise it—it is less. The difference is the [6]_____, Δm. It is the amount of matter that would be converted into energy and released if the nucleus were formed from separate protons and neutrons. Since separating the nucleus into these nucleons would require the addition of this energy, it provides the force that holds the nucleus together, and is known as the [7]_____, BE, and can be calculated from the formula $(\Delta m)c^2$.

There are several types of radiation emitted by radioactive atoms. An [8]_____ _____ consists of two protons and two neutrons, like the nucleus of a helium atom. A [9]_____ is a very fast electron emitted from the nucleus when a neutron decays to a proton and an electron. A [10]_____ is high-energy electromagnetic radiation. A [11]_____ has the mass of an electron but a positive charge, a bit of antimatter. The intensity of the radiation emitted by a radionuclide depends upon its decay rate, which is usually expressed in terms of its [12]_____, $t_{\frac{1}{2}}$, the amount of time required for half of a sample to decay. The radionuclide that decays is called the [13]_____, and the nuclide that is produced is called the [14]_____. The half-life of a radionuclide helps determine its usability in [15]_____, the dating of ancient objects by determining the ratio of amounts of a parent nuclide and one of its decay products present in the objects. Nuclear reactions can also be caused artificially by the bombardment of a nucleus with subatomic particles or small nuclei, a process known as [16]_____.

A radionuclide can achieve sustained nuclear fission, known as a [17]_____, if (a) its nuclei are sufficiently unstable, (b) their fission produces enough neutrons having the right amount of energy, and (c) there is sufficient material (called the [18]_____) contained in a small enough volume. In an atomic bomb this chain reaction is uncontrolled, but controlled nuclear fission reactions, using materials that absorb some of the neutrons, can generate large amounts of heat energy on a [19]_____, and the heat energy can then be converted into electricity.

Preliminary Test

True-False

_____ 1. In a nuclear reaction, one element can be changed into another.

_____ 2. The electrons surrounding the nucleus are usually involved in nuclear reactions.

_____ 3. The actual mass of an atom is equal to the sum of the masses of its constituent particles.

_____ 4. The particles emitted from the nucleus in nuclear reactions are always positively charged.

_____ 5. Only very heavy nuclei are naturally radioactive.

_____ 6. All nuclei except hydrogen are somewhat unstable because of the very close proximity of the positively charged protons in the nucleus.

_____ 7. Nuclei above the band of stability usually decay by nuclear fission.

_____ 8. No naturally occurring nuclides with atomic number less than 8 are radioactive.

_____ 9. All radioactive decay processes follow first-order kinetics.

_____ 10. The eventual product of each of the three common natural radioactive series is a stable isotope of lead.

_____ 11. Because lead is the product of the three common natural radioactive series, we can tell that all isotopes of lead are stable.

_____ 12. Any amount of a fissionable material such as U-235 can explode if heated to a high enough temperature.

_____ 13. Light water nuclear reactors use a large amount of water, both for cooling and as a moderator to slow the neutrons produced in the reaction.

Short Answer

Answer with a word, a phrase, a formula, or a number with units as necessary.

1. Nuclides having either 2, 8, 20, 28, 50, 82, or 126 neutrons are said to have a _____ _____ of neutrons and are found to be exceptionally _____.

2. Above atomic number 20, most stable nuclei have _____ protons than neutrons.

3. A nucleus having a proton to neutron ratio either higher or lower than stable nuclei is said to lie outside the _____.

4. Of the common types of radiation, the ones having positive charge are the _____, the _____, and the _____. (For each one, give the name of the particle and an acceptable symbol.)

5. The types of radioactive emissions that result in the daughter nucleus having a lower atomic number than the parent nucleus are _____, _____, and _____.

6. The type of radioactive emission that results in a lower mass number of the nucleus but no change in atomic number is _____.

7. The type of nuclear emission that produces no change in either mass number or atomic number of the nucleus is _____.

8. The radioactive particles with the lowest penetrating ability are _____.

9. For nuclei *above* the band of stability, the ratio of neutrons to protons is too _____; such nuclei usually decay by _____ emission or by _____ emission.

10. The first several steps of the thorium-232 radioactive decay series involve, in order, the loss of an alpha particle, then a beta particle, then another beta particle; the nuclide symbol for the product after these three steps of decay is _____.

11. All nuclides with atomic number greater than _____ are radioactive.

12. When a nuclear reaction occurs, the particles emitted can have different kinetic energies. The kinetic energy of the emitted particle is equal to the energy equivalent to the _____ difference lost between the reactants and the products, minus the energy associated with any _____ that are later emitted.

For Questions 13 through 19, tell whether the effect of the change would be to increase (I), decrease (D), or leave unchanged (U) the atomic number and the mass number of the nuclide. Tell the magnitude of any increase or decrease.

	Change	Effect on Atomic Number	Effect on Mass Number
13.	alpha emission	_____	_____
14.	beta emission	_____	_____
15.	gamma emission	_____	_____
16.	positron emission	_____	_____
17.	electron capture	_____	_____
18.	proton emission	_____	_____
19.	neutron emission	_____	_____

20. The half-life of lead-210, one of the nuclides in the U-238 series, is 20.4 years. Suppose we start with 100 μg of lead-210 and wait for it to decay. After 20.4 years, the mass of lead-210 present would be _____ μg , after 40.8 years it would be _____ μg , and after 61.2 years it would be _____ μg .

21. The effectiveness of cobalt-60 treatment in arresting certain types of cancer depends on the ability of _____ to destroy tissue.

For each of the next fifteen questions, 22 through 36, fill in the nuclide symbol that correctly completes the equation. Describe each reaction with a phrase, such as beta emission, alpha emission, positron emission, electron capture, neutron emission, gamma emission, nuclear fission, or nuclear fusion. For all reactions *except* those that are nuclear fission or nuclear fusion, tell whether the neutron/proton ratio increases (I), decreases (D), or remains unchanged (U) by the process.

		Change in
Reaction	**Description**	**n/p ratio**
22. $^{1}_{0}n \rightarrow ^{1}_{1}p +$ _____	_____	_____
23. $^{14}_{6}C \rightarrow$ _____ $+ ^{0}_{-1}\beta$	_____	_____
24. $^{17}_{7}N \rightarrow$ _____ $+ ^{1}_{0}n$	_____	_____
25. $^{240}_{94}Pu \rightarrow ^{236}_{92}U +$ _____	_____	_____
26. $^{186}_{73}Ta \rightarrow ^{186}_{74}W +$ _____	_____	_____
27. $^{197}_{80}Hg + ^{0}_{-1}e \rightarrow$ _____	_____	_____
28. _____ $\rightarrow ^{39}_{20}Ca + ^{0}_{-1}\beta$	_____	_____
29. _____ $+ ^{0}_{-1}e \rightarrow ^{22}_{10}Ne$	_____	_____

		Change in
Reaction	**Description**	**n/p ratio**
30. $^{236}_{92}U^{*} \rightarrow ^{236}_{92}U +$ _____	_____	_____
31. $^{223}_{88}Ra \rightarrow$ _____ $+ ^{4}_{2}\alpha$	_____	_____
32. $^{1}_{1}p \rightarrow ^{1}_{0}n +$ _____	_____	_____
33. $^{1}_{1}H + ^{1}_{1}H \rightarrow$ _____ $+ ^{0}_{-1}\beta$	_____	_____
34. $^{3}_{2}He + ^{3}_{2}He \rightarrow ^{4}_{2}He + 2$_____	_____	_____
35. $^{235}_{92}U + ^{1}_{0}n \rightarrow ^{93}_{36}Kr +$ _____ $+ 3^{1}_{0}n$	_____	_____
36. _____ $+ ^{1}_{0}n \rightarrow ^{99}_{40}Zr + ^{139}_{54}Xe + 2^{1}_{0}n$	_____	_____

37. For any radioactive decay, the decaying nuclide is called the _____ nuclide and the product is called the _____ nuclide.

38. The sequence of radioactive decay processes by which some radionuclides attain stability in several steps is called a _____.

39. The method of radioactive dating based on the decay of carbon-14 is applicable only to objects up to about _____ years old.

40. The process in which a nuclide is made unstable by hitting it with a high-energy particle is called _____.

41. All nuclides with mass number greater than _____ undergo nuclear fission spontaneously, whereas those between mass numbers _____ and _____ can be induced to undergo fission by bombarding them with particles of relatively low kinetic energy.

42. The most stable nuclide is _____.

43. A material that is capable of undergoing nuclear fission is said to be _____.

44. A nuclear reaction in which two smaller nuclei combine to form a larger nucleus, with the release of energy, is called a _____ reaction.

45. Most stable, naturally occurring nuclides have an _____ number of protons and an _____ number of neutrons.

Multiple Choice

_____ 1. Which one of the following processes produces two nuclei of intermediate mass from one heavier nucleus?
    (a) alpha emission   (b) nuclear fusion   (c) nuclear fission
    (d) fluorescence   (e) ionization

_____ 2. The observed mass of deuterium, $_1^2H$, is 2.01400 amu. What is the mass deficiency of deuterium?
    (a) 0.0025 amu/atom   (b) 4.0305 amu/atom   (c) 0.126 amu/atom
    (d) 0.127 amu/atom   (e) Deuterium has no mass defect.

_____ 3. What is the nuclear binding energy of one mole of $_{11}^{23}Na$ atoms? Sodium exists in nature as a single isotope, so you may use the observed atomic weight as the observed mass of the Na atom.
    (a) 1.81×10^{10} kJ/mol   (b) 2.70×10^{10} kJ/mol
    (c) 3.92×10^{10} kJ/mol   (d) 9.01×10^9 kJ/mol

_____ 4. Which one of the following is not a commonly used method of radiation detection?
    (a) photographic detection
    (b) detection by fluorescence
    (c) use of a cloud chamber
    (d) use of a gas ionization chamber
    (e) bombardment with positive ions

_____ 5. The radiocarbon method of dating would not be directly applicable to determining the age of which one of the following objects?
   (a) a piece of charcoal from a fire
   (b) a piece of wood
   (c) a piece of animal skin
   (d) a woven straw mat
   (e) a flint arrowhead

_____ 6. Measurement based on which of the following radioactive decay processes would be applicable to age determination for objects of greatest age?
   (a) decay of carbon-14 ($t_{1/2}$ = 5,730 years)
   (b) decay of potassium-40 ($t_{1/2}$ = 1.3 billion years)
   (c) decay of radium-223 ($t_{1/2}$ = 11.4 days)
   (d) decay of uranium-238 ($t_{1/2}$ = 4.5 billion years)
   (e) decay of uranium-235 ($t_{1/2}$ = 710 million years)

_____ 7. The source of high-energy positive ions or high-energy negative ions for nuclear bombardment processes is usually which one of the following?
   (a) a particle accelerator
   (b) a nuclear fission reactor
   (c) a cloud chamber
   (d) an atomic bomb
   (e) a nuclear fusion reactor

_____ 8. Which one of the following experimental methods has been used for the production of elements that do not occur naturally?
   (a) nuclear fusion
   (b) nuclear bombardment
   (c) gas ionization
   (d) natural radioactive decay
   (e) labeling experiments

_____ 9. Which one of the following is *not* a problem associated with the operation of a nuclear fission reactor for the production of electricity?
   (a) possibility of a meltdown
   (b) storage of long-lived radioactive wastes
   (c) a nuclear fission explosion
   (d) thermal pollution
   (e) escape of short-lived radioactive wastes

Answers to Some Important Terms in This Chapter

1. radioactivity
2. fission
3. fusion
4. nuclides
5. band of stability
6. mass deficiency
7. nuclear binding energy

8. alpha particle (α)
9. beta particle (β)
10. gamma ray (γ)
11. positron
12. half-life
13. parent nuclide
14. daughter nuclide

15. radioactive dating
16. artificial transmutation
17. chain reaction
18. critical mass
19. nuclear reactor

Answers to Preliminary Test

Do not just look up or memorize answers—*be sure you could figure them out yourself.*

True-False

1. True. The only nuclear reactions for which this is not true are those in which only a neutron is emitted, and these are not common.
2. False. Most nuclear reactions involve only the nucleus. One kind of nuclear reaction, called electron capture (K capture), involves the capture of an electron from the innermost electronic shell (the K shell) by the nucleus, but this type of nuclear reaction is relatively uncommon.
3. False. Some of the mass, called the mass deficiency, is converted into energy. This is the binding energy that holds the nuclear particles together. See the calculations and discussion in Section 22-3.
4. False. Some are positive, some are negative, and some are neutral. See Section 22-4, and be sure that you are familiar with the general properties of the particles that take part in nuclear decay processes.
5. False. Most very heavy nuclei are radioactive, but so are many light nuclei.
6. False. Many nuclei are nonradioactive, meaning that they are stable and do not decay. It is not entirely understood how the strong coulombic proton-proton repulsive forces in the nucleus are overcome, but it is thought that some of the very short-lived subatomic particles that have more recently been discovered may play a role. See Section 22-1.
7. False. They usually decay by beta emission and less frequently by neutron emission. The only nuclides that spontaneously undergo fission are isotopes of elements with atomic numbers greater than 88. These are found to the right of the band of stability.
8. False. For instance, the chapter mentions radioactive isotopes of carbon and nitrogen.
9. True. They are independent of temperature, and their rates are each proportional only to the concentration of one substance.
10. True. The ^{238}U series ends with ^{206}Pb, the ^{235}U series ends with ^{207}Pb, and the ^{232}Th series ends with ^{208}Pb. [The naturally occurring isotopes of lead are ^{208}Pb (52.4%), ^{206}Pb (24.1%), ^{207}Pb (22.1%), and ^{204}Pb (1.4%). There is a fourth series that is considered "artificial" because the only two isotopes involved that are found naturally are at the end: ^{209}Bi ($t_{1/2} = 1.9 \times 10^{19}$ years) and ^{205}Tl.]
11. False. Several radioactive isotopes of lead occur in these series. See Table 22-4.
12. False. If too little material is present, the reaction cannot become self-sustaining, because too few neutrons are produced. It is necessary to have a critical mass for an explosion. Study the section on the atomic bomb (Section 22-14) and the comments regarding the impossibility of explosion of a nuclear fission reactor (Section 22-15).
13. True. The design and general features of such a light water reactor are described in Section 22-15.

Short Answer

1. magic number; stable. The same principle is found to apply to the numbers of protons in the nucleus and to the sum of protons plus neutrons.

2. fewer. See Figure 22-1. The ratio of protons to neutrons rises from 1:1 to 1:1.2, 1:1.4, and, eventually 1:1.5.

3. band of stability. See Figure 22.1. The dots are the stable nuclei.

4. positron ($_{+1}^{0}\beta$ or $_{+1}^{0}e$); alpha particle (α, $_{2}^{4}\alpha$, or $_{2}^{4}He$); proton ($_{1}^{1}p$ or $_{1}^{1}H$)

5. positron emission; alpha emission; proton emission

6. neutron emission

7. gamma emission

8. alpha particles

9. high; beta; neutron. The first of these, beta emission, is much more common.

10. $_{90}^{228}Th$. You should be able to figure this out without looking at the disintegration series in Table 22-4.

11. 83. Bismuth is the heaviest element that has stable (nonradioactive) isotopes.

12. mass; gamma rays

You should be able to answer questions 13 through 19 without looking up the answer, just from knowing the characteristics of the various kinds of radiation.

13. (D)ecrease by 2;   (D)ecrease by 4
14. (I)ncrease by 1;   (U)nchanged
15. (U)nchanged;   (U)nchanged
16. (D)ecrease by 1;   (U)nchanged
17. (D)ecrease by 1;   (U)nchanged
18. (D)ecrease by 1;   (D)ecrease by 1
19. (U)nchanged;   (D)ecrease by 1

20. 50; 25; 12.5. Radioactive decay is a first-order kinetic process. During any time period equal to one half-life, the amount of material present decreases to one-half of what it was at the beginning of *that* half-life.

21. gamma rays. This is discussed in Example 22-4, which is also an illustration of calculations associated with radioactive decay.

22. $_{-1}^{0}\beta$; beta emission; $\frac{1}{0}$ (undefined) to $\frac{0}{1}$ or 0

23. $_{7}^{14}N$; beta emission; (D)ecreases

24. $_{7}^{16}N$; neutron emission; (D)ecreases

25. $_{2}^{4}\alpha$; alpha emission; (I)ncreases

26. $_{-1}^{0}\beta$; beta emission; (D)ecreases

27. $_{79}^{197}Au$; electron capture; (I)ncreases

28. $_{19}^{39}K$; beta emission; (D)ecreases

29. $_{11}^{22}Na$; electron capture; (I)ncreases

30. $_{0}^{0}\gamma$; gamma emission; (U)nchanged

31. $^{219}_{86}$Rn; alpha emission; (I)ncreases

32. $^{0}_{+1}\beta$; positron emission; (I)ncreases

33. $^{2}_{1}$H; nuclear fusion. (A beta particle is also given off.)

34. $^{1}_{1}$H; nuclear fusion. (Protons are also given off.)

35. $^{140}_{56}$Ba; nuclear fission

36. $^{239}_{94}$Pu; nuclear fission

37. parent; daughter

38. disintegration series or radioactive series

39. 50,000

40. nuclear bombardment

41. 250; 225; 250

42. $^{56}_{26}$Fe

43. fissionable

44. nuclear fusion

45. even; even

Multiple Choice

1. (c). Alpha emission (a) produces a slightly lighter nucleus. Nuclear fusion (b) produces one heavier nucleus from two light nuclei. Fluorescence (d) and ionization (e) are methods for detecting radiation.

2. (a). Solve this problem as in Example 22-1.
   Each atom of $^{2}_{1}$H contains 1 proton, 1 electron, and (2-1) = 1 neutron.
   First, we sum the masses of these particles:
   $$1.0073 + 0.00054858 + 1.0087 = 2.0165 \text{ (rounded to 4 decimal places)}$$
   Then we subtract the actual mass from this "calculated" mass to obtain Δm.
   $$\Delta m = 2.0165 - 2.01400 = 0.0025 \text{ amu/atom}$$
   This mass deficiency could also be expressed as 0.0025 g/mol.

3. (a). First find the mass deficiency, as in Example 22-1.
   Each atom of $^{23}_{11}$Na contains 11 protons, 11 electrons, and (23 – 11) = 12 neutrons.
   First, we sum the masses of these particles:

protons: 11 × 1.0073 amu	=	11.0803 amu
electrons: 11 × 0.00054858 amu	=	0.006034 amu
neutrons: 12 × 1.0087 amu	=	12.1044 amu
	=	23.1907 amu (rounded to 4 decimal places)

   Then we subtract the actual mass from this "calculated" mass to obtain Δm.
   $$\Delta m = 23.1907 - 22.989768 = 0.2009 \text{ amu/atom}$$
   Then convert this to binding energy, as shown in Example 22-2.
   0.2009 amu/atom = 0.2009 g/mol = 2.009×10^{-4} kg/mol
   $BE = (\Delta m)c^2 = (2.009 \times 10^{-4} \text{ kg/mol}) \times (3.00 \times 10^8 \text{ m/s})^2 = 1.81 \times 10^{13} \text{ kg·m}^2/\text{s}^2\text{·mol}$
   $= 1.81 \times 10^{13}$ J/mol $= 1.81 \times 10^{10}$ kJ/mol

4. (e). Bombardment with positive ions is a method for producing artificial transmutations.

5. (e). This method applies only to material that originated as a living organism, either plant or animal.
6. (d). It has the longest half-life.
7. (a). This is discussed in the first part of Section 22-13.
8. (b). See Section 22-13 for a discussion of this topic.
9. (c). Read Section 22-15. Reactors use fuels that have neither the correct composition nor the correct physical arrangement ever to achieve the critical mass necessary for an explosion. The other dangers are real.

23 Organic Chemistry I: Formulas, Names, and Properties

Chapter Summary

Because of carbon's location in Group 4A on the periodic table and its intermediate electronegativity, each carbon atom can form very stable covalent bonds with as many as four other atoms. Thus, carbon has a striking ability to exhibit extensive **catenation** ("chain-making") by bonding to other carbon atoms. As a result, it can form a large number and variety of compounds with carbon chains and rings. The study of such compounds involving C–C or C–H bonds comprises a major branch of chemistry known as **organic chemistry.** Your study of descriptive chemistry continues in Chapters 23 and 24 with some of the major features of this extensive and varied area of chemistry. A more detailed, but still introductory, study of organic chemistry is usually a course of a year or more at the college level, so the present study should be viewed as a short preliminary survey of this field.

Chapter 23 deals with the major classes or "families" of organic compounds, with emphasis on their structures, sources, general properties, and uses. Because of the wide variety of compounds in organic chemistry, it is important to be able to name compounds systematically. Therefore, an introduction to organic nomenclature is an important part of this early study of organic chemistry. Chapter 24 discusses the main classes of reactions that organic compounds undergo, and contains a systematic look at the geometries of organic molecules.

The first two sections of Chapter 23 deal with the **saturated hydrocarbons**, or **alkanes**, which contain only single covalent bonds. This means that each carbon atom is bonded to four other atoms, either carbon or hydrogen, with tetrahedral geometry and sp^3 hybridization (Section 8-7). Section 23-1 introduces the structures and terminology of **homologous series** of compounds, including both the **normal** or unbranched hydrocarbons and the **branched** hydrocarbons. Do not be misled by the term "straight chain" when applied to strings of carbon atoms bonded to each other. Instead, think of such a chain of carbon atoms as "continuous." **Isomers** are different compounds having the same number of atoms of all types, but with differing arrangements. Isomerism, which is very common in organic chemistry, will be discussed in more detail in Chapter 25. The first type to be encountered, called **constitutional** (or **structural**) **isomerism**, is described in Section 23-1.

As noted previously, chemists have developed a systematic way of naming organic compounds. The rules for the nomenclature of saturated hydrocarbons are discussed in Section 23-2. Notice the relation between the name of the parent hydrocarbon and the **alkyl group** obtained by removing one H atom. These alkyl groups can be thought of as building blocks that will appear in many classes of organic compounds. Study carefully the examples for naming organic compounds, and you will find that it is not as complex as it seems at first.

Petroleum consists primarily of hydrocarbons, but it is a mixture of molecules of many different chain lengths. The refining of petroleum involves the separation of this complex mixture into **fractions** that are suitable for various uses. See the Chemistry in Use essay on this topic, page 900.

Hydrocarbons that contain some carbon-carbon bonds other than single bonds are known as **unsaturated hydrocarbons**. There are three classes of such compounds, which are discussed in the next several sections. Section 23-3 describes the structures and naming of the **alkenes**, which contain a carbon-carbon double bond. Because of the lack of free rotation around a double bond, compounds containing such a bond can exhibit **geometric isomerism**. Notice both the differences in properties of such isomeric pairs and the additional terminology needed in the nomenclature—the prefixes *cis-* and *trans-*, which you will use again in Chapter 25. The second class of unsaturated hydrocarbons is **alkynes**, which contain a carbon-carbon triple bond. These are introduced and named in Section 23-4. Sections 23-5 through 23-6 describe the third class of unsaturated hydrocarbons, the **aromatic hydrocarbons**. The origin of this term, which you will see is quite a misnomer, is described in the introductory paragraphs to this segment of the chapter. The bonding in all aromatic compounds is characterized by **delocalization** of π-bonded electrons around a ring (most often a six-membered ring) and consequent lowering of the electronic energy of the molecule. This effect is known as **resonance** (review Sections 7-9 and 9-6). This bonding is described in detail for **benzene**, the simplest aromatic hydrocarbon, in Section 23-5. Other aromatic hydrocarbons, presented in Section 23-6, consist of benzene with one or more hydrogen atoms replaced by alkyl groups (e.g., toluene and the three isomeric xylenes) and the fused-ring aromatics (e.g., naphthalene, anthracene, etc.). The rules for naming derivatives of benzene are presented in this section. Figure 23-13 in Section 23-7 summarizes the classification of hydrocarbons studied in this chapter.

Many organic compounds involve groups of atoms having bonds other than C–C or C–H. Often a given grouping of atoms undergoes similar reactions, no matter in what compound it occurs. Such a group of atoms is often viewed as a potential reaction site, called a **functional group.** The consistent behavior of each functional group enables us to organize large amounts of information. The next major portion of Chapter 23 deals with the most common functional groups and with the resulting characteristic classes of compounds. It is suggested that you organize your study of each of these classes of compounds to emphasize the functional group common to all of the compounds in that class, the system for naming those compounds, and some of their common uses. We sometimes use group symbols to represent unspecified groups of atoms, such as alkyl (R–) or aryl (Ar–, not to be confused with the symbol for argon). This is a commonly used compact notation.

Organic halides, Section 23-8, can be thought of as hydrocarbons in which one or more hydrogen atoms have been replaced by a halogen atom (F–, Cl–, Br–, or I–).

The group of atoms –O–H or –OH (**hydroxyl**), attached to a carbon atom in an aliphatic or aromatic group, is characteristic of two related classes of compounds, **alcohols** and **phenols**. When the hydroxyl group is bonded to an aliphatic carbon atom, the compound is classed as an alcohol (R–OH); when the carbon is aromatic, the compound is a phenol (Ar–OH). Examples of structures and naming of several compounds of these two classes are presented in Section 23-9. Because of electronegativity differences within the molecule, alcohols are polar molecules. Alcohols can be considered as derivatives of water in which one hydrogen atom has been replaced by an alkyl group.

Similarly, **ethers** can be considered to be water derivatives in which both H atoms have been replaced by organic groups. Thus, an ether is a compound in which an oxygen atom is bonded to two organic groups, R–O–R′, where R and R′ ("R prime") may each be either an alkyl or an aryl group. Section 23-10 describes these compounds. They are used for a variety of purposes such as anesthetics, artificial flavors, refrigerants, and solvents.

Two closely related groups of compounds, the **aldehydes** and the **ketones,** both contain the

$$\overset{\displaystyle O}{\overset{\displaystyle \|}{-C-}}$$ group, called the **carbonyl** group. Ketones have two alkyl or aryl groups bonded to the

carbonyl group $$(R-\overset{\displaystyle O}{\overset{\displaystyle \|}{C}}-R',$$ sometimes written as RCOR′), while in aldehydes one of these is

replaced by a hydrogen atom $$(R-\overset{\displaystyle O}{\overset{\displaystyle \|}{C}}-H,$$ sometimes written as RCHO). The structures, nomenclature, and some uses of these two classes of compounds are presented in Section 23-11. The aldehydes are usually called by their common names, which indicate the number of carbon atoms in the compound. Ketones are commonly named according to the two alkyl or aryl groups bonded to the carbonyl group. Many important natural substances are aldehydes and ketones. Their aliphatic or aromatic ends make them mix readily with organic compounds, while the polar carbonyl group makes them miscible with water. As a result, many ketones are useful as solvents, of which the most widely used is acetone, which is often used as fingernail polish remover.

The **amines** are basic compounds that can be considered derivatives of ammonia, NH_3, in which an alkyl or an aryl group has replaced one or more of the hydrogen atoms. These compounds, their structures, and their nomenclature are the subject of Section 23-12.

Carboxylic acids (Section 23-13) contain the **carboxyl group,** $-\overset{\displaystyle O}{\overset{\displaystyle \|}{C}}-OH$, often written as –COOH. Most carboxylic acids are weak acids. You have seen acetic acid, CH_3COOH, many times in weak acid problems or examples. The carboxylic acids can be classed as either **aliphatic** or **aromatic,** depending on the nature of the organic group bonded to the –COOH group. Many carboxylic acids occur in common natural products. The naming and structures of some carboxylic acids are described. Be sure that you understand how we name the common groups that are derived from the carboxylic acids (the **acyl** groups) and how these are related to the parent acid.

Section 23-14 describes the preparation and properties of three important types of derivatives of carboxylic acids, **acyl halides, esters** (organic esters), and **amides.** Each of these can be viewed as resulting from the replacement of the –OH group of the carboxylic acid by another group, such as a halogen atom in the acyl halides or the –OR′ group in the esters. Many commonly recognizable flavors and aromas come from esters and many common fats, oils, and waxes are esters. The Chemistry In Use essay on page 930 relates some dietary aspects of fats and oils to their chemical structures. Amides find use as medicinal agents such as headache and fever remedies, as insect repellents, and as the starting materials for polyamides, of which nylon is an example.

The summary of the classification of compounds according to functional groups in Figure 23-20 will be a great aid in your study of this chapter.

The rest of the chapter is a survey of some important types of reactions of organic compounds. The next three sections deal with three fundamental classes of organic reactions. In a **substitution reaction**, Section 23-16, an atom or a group of atoms attached to a carbon atom is removed and another atom or group of atoms takes its place, so that *no change in the degree of saturation* occurs. This reaction is typical of the saturated hydrocarbons (alkanes and cycloalkanes). Some types of substitution reactions are known by particular names, which indicate the type of group that replaces another group. Examples are halogenation and nitration. In an **addition reaction**, Section 23-17, the number of groups attached to a carbon atom increases, so that the molecule becomes *more nearly saturated*. Again, subcategories are named according to what groups are added—halogenation, hydrogenation, and hydration. The third type of reaction, the **elimination reaction**, Section 23-18, involves a decrease in the number of groups attached to carbon, so that the *degree of <u>unsaturation</u> increases*. Dehydrohalogenation and dehydration reactions are particular types of elimination reactions.

The final section of the chapter discusses **polymerization reactions**, which produce the **polymers** that seem to be everywhere around us. As you learn here, there are two basic types of these reactions. In **addition polymerization**, an addition reaction forms a polymer and no other product. Polyethylene, polypropylene, and natural and synthetic rubbers are formed by this kind of reaction. **Condensation polymerization** is based on condensation reactions, in which molecules combine while eliminating a small molecule, like H_2O or HCl. Polyesters and polyamides (e.g., nylon) are examples of synthetic condensation polymers. The section closes with a discussion of **polypeptides**, which are condensation polymers based on the **peptide bond**. Proteins and other biologically important polypeptides will be discussed further in Chapter 24.

Study Goals

This is quite a long and diverse chapter. Be systematic in your study. Organize your study around the classes of compounds. Look at Figures 23-13 and 23-20 to help you organize the material.

Chapter Introduction
1. Understand the basic ideas that lie behind the study of organic chemistry. *Work Exercises 1 through 4.*

Sections 23-1 and 23-3 through 23-4
2. Know the structures, properties, some sources, and uses that are typical of each type of aliphatic hydrocarbons: (a) alkanes and cycloalkanes, (b) alkenes and cycloalkenes, and (c) alkynes. *Work Exercises 5 through 12, 20, 21, 24, 25, and 31.*

Sections 23-2 through 23-4
3. Know how to name compounds of the classes of hydrocarbons in Study Goal 2. *Work Exercises 13, 14, 16 through 19, 22, and 27 through 30.*

Sections 23-1 through 23-4
4. Understand the types of isomers that are possible for the compounds of the classes of hydrocarbons in Study Goal 2. Be able to name these isomers uniquely. *Work Exercises 13, 14, 16, 23, and 26.*

Sections 23-5 and 23-6
5. Know the structures, properties, nomenclature, some sources, and some uses that are typical of aromatic hydrocarbons. *Work Exercises 32 through 38.*

Section 23-8

6. Know the structures, properties, nomenclature, some sources, and some uses that are typical of alkyl and aryl halides. *Work Exercises 39 through 43.*

Section 23-9

7. Know the structures, properties, nomenclature, some sources, and some uses that are typical of alcohols and phenols. *Work Exercises 44 through 54.*

Section 23-10

8. Know the structures, properties, nomenclature, some sources, and some uses that are typical of ethers. *Work Exercises 55 through 59.*

Section 23-11

9. Know the structures, properties, nomenclature, some sources, and some uses that are typical of aldehydes and ketones. *Work Exercises 64 through 67.*

Section 23-12

10. Know the structures, properties, nomenclature, some sources, and some uses that are typical of amines. *Work Exercises 60 through 63.*

Sections 23-13 and 23-14

11. Know the structures, properties, nomenclature, some sources, and some uses that are typical of carboxylic acids and their derivatives. *Work Exercises 68 through 78.*

Sections 23-16 through 23-18

12. Know the characteristics of some important classes of organic reactions: (a) substitution, (b) addition, and (c) elimination. Know which classes of organic compounds can undergo each type of reaction. *Work Exercises 79 through 94.*

Section 23-19

13. Describe the basic features of polymerization reactions. Recognize various types of polymerization reactions and polymers. *Work Exercises 95 through 110.*

General

14. Apply the ideas of this chapter to unclassified compounds and reactions. Be able to (a) identify the functional groups present, (b) identify the class (or classes) represented, and (c) name the compounds. *Work **Mixed Exercises** 111 through 122.*

15. Answer conceptual questions based on this chapter. *Work **Conceptual Exercises** 123 through 136.*

16. Apply your knowledge about chemical bonding, structure, electronegativity, and stoichiometry to the new material in this chapter. *Work **Building Your Knowledge** exercises 137 through 141.*

17. Exercises at the end of chapter direct you to sources outside the textbook for information to use in solving them. Work **Beyond the Textbook** *exercises 142 through 145.*

Some Important Terms in This Chapter

Some important terms from Chapter 23 are listed below. Fill the blanks in the following paragraphs with terms from the list. Each term is only used once. Study other new terms and review terms from preceding chapters if necessary.

acyl halide	carboxylic acid	hydrogenation reaction
addition polymerization	condensation polymerization	ketone
addition reaction	constitutional isomers	peptide bond
alcohol	copolymer	phenol
aldehyde	dehydration reaction	polyamide
aliphatic hydrocarbons	dehydrohalogenation	polyester
alkanes	elimination reaction	polymer
alkenes	ester	polymerization reaction
alkyl group	ether	saturated hydrocarbons
alkynes	geometric isomers	structural isomers
amide	halogenation reaction	substitution reaction
amine	hydration reaction	unsaturated hydrocarbons
aromatic hydrocarbons	hydrocarbons	

Organic chemistry is the study of the compounds of carbon, except some of the simplest ones. Of the millions of organic compounds that are known, many contain only *carbon* and *hydrogen*. They are called the [1]_____. Some hydrocarbons, benzene being the prime example, contain a delocalized ring of pi bonds, and are known as [2]_____ _____, although some of them don't have pleasant smells. Hydrocarbons that don't have a system of delocalized pi electrons are said to be [3]_____ _____. Some of these aliphatic hydrocarbons have only sigma (single) bonds, and since they cannot add any hydrogen atoms, they are referred to as [4]_____ _____. Hydrocarbons that contain both sigma and pi bonds can add more hydrogen atoms, so they are called [5]_____.

The saturated hydrocarbons are also known as the [6]____._____, compounds in which each carbon atom is bonded to four other atoms. Some of these alkanes have the same molecular formula but they differ in the order in which their atoms are attached, making them different compounds. These are called [7]_____ or [8]_____ (because they have different structures). In the naming of alkanes with complex structures, we imagine taking a simple alkane and removing one hydrogen atom, producing a unit called an [9]_____ (alk*ane* → alk*yl*).

Hydrocarbons that contain at least one carbon-carbon double bond are called [10]_____ (alk*ane* → alk*ene*). The restricted rotation of the double bond can produce compounds whose structures differ only in the arrangements of groups of atoms on the opposite sides of the bond, compounds known as [11]_____. Hydrocarbons that contain at least one carbon-carbon triple bond are called [12]_____ (alk*ane* → alk*ene*).

Several classes of organic compounds have structures closely related to water, which can be written as H−O−H. Some contain the hydroxyl group, −OH. If one or more of the hydrogen atoms of a hydrocarbon are replaced by a hydroxyl, an [13]_____ is produced. Thus, the general formula for an alcohol is R−OH (or HO−R), where R represents an alkyl group. If the hydroxyl group replaces a hydrogen atom on an aromatic ring, the resulting compound is a [14]_____. The general formula is Ar−OH, where Ar represents an aromatic ring. If both

337

of the hydrogen atoms of a water molecule are replaced by alkyl groups, an [15]_____, R—O—R', is produced.

Some other classes of organic compounds contain a carbon atom double-bonded to an oxygen atom (—$\overset{\displaystyle O}{\overset{\|}{C}}$—). In an [16]_____, at least one of the two atoms bonded to the carbon atom is a hydrogen atom (R—$\overset{\displaystyle O}{\overset{\|}{C}}$—H or RCHO). If two alkyl or aryl groups are bonded to the carbon atom, the molecule is a [17]_____ (R—$\overset{\displaystyle O}{\overset{\|}{C}}$—R' or RCOR'). If an alkyl group and a hydroxyl group are bonded to the carbon atom, the compound is a [18]_____ (R—$\overset{\displaystyle O}{\overset{\|}{C}}$—OH or RCOOH). Replacing the hydroxyl group with a halogen, produces an [19]_____ (R—$\overset{\displaystyle O}{\overset{\|}{C}}$—X, where X represents F, Cl, Br, or I). Replacing it with an alkoxy group (oxygen plus an alkyl group), produces an [20]_____ (R—$\overset{\displaystyle O}{\overset{\|}{C}}$—OR' or RCOOR').

An organic compound that consists of an alkyl or aryl group bonded to a nitrogen atom is an [21]_____ (R–NH₂). Alkyl or aryl groups could replace either or both of the two hydrogen atoms bonded to the nitrogen atom. If the carbon atom bonded to the nitrogen atom is also double-bonded to an oxygen atom, the compound is called an [22]_____ (R—$\overset{\displaystyle O}{\overset{\|}{C}}$—NH₂ or RCONH₂).

A chemical reaction in which an atom or group of atoms attached to a carbon atom in an organic compound is removed and replaced by another atom or group of atoms is called a [23]_____. The type of substitution reaction in which a hydrogen atom is replaced by a halogen atom is called a [24]_____. The number of atoms or groups of atoms attached to a carbon atom *increases* in an [25]_____. If a molecule of hydrogen is added, the reaction is called a [26]_____. If a molecule of water is added, the reaction is called a [27]_____. Conversely, the number of atoms or groups of atoms attached to a carbon atom *decreases* in an [28]_____. The elimination reaction in which both a hydrogen atom and a halogen atom are removed from adjacent carbon atoms is called [29]_____. The elimination of both H and OH from adjacent carbon atoms (forming water) is called a [30]_____.

In a [31]_____, many small molecules, called monomers, are combined to form a large molecule, called a [32]_____. There are two kinds of polymerization reactions. In [33]_____, monomers are combined by addition reactions, producing only the polymer and no other product. Often all of the monomer molecules are the same, but when two different monomers are mixed and polymerized, a [34]_____ is formed. In [35]_____, two monomers combine to produce the polymer by eliminating a small molecule every time two of their molecules link (a condensation reaction). When a diol (an alcohol with *two* hydroxyl groups) reacts with a dicarboxylic acid (which has *two* carboxyl groups), each link is an ester, and the polymer produced is a "polymeric ester," or [36]_____. When the links are amides (joining

$$-\overset{\overset{\displaystyle O}{\|}}{C}-$$ with $-NH_2$), the polymer produced is a [37]_____. Proteins are natural polyamides. In polyamides, each amino acid is linked to the next amino acid by a [38]_____, which is produced when the amino group of one amino acid combines with the carboxylic acid group of the next amino acid, with the elimination of a molecule of water.

Preliminary Test

Be sure to practice *many* of the additional textbook exercises.

True-False

_____ 1. Most hydrocarbons are polar.

_____ 2. The carbon atoms in saturated hydrocarbons are best described as sp^3 hybridized.

_____ 3. In straight-chain hydrocarbons the carbon atoms lie in a straight line.

_____ 4. In saturated hydrocarbons, each carbon atom is bonded to either two or three hydrogen atoms.

_____ 5. All alkanes up to C_8H_{18} are gases at room temperature and atmospheric pressure.

_____ 6. Generally, the more branched an alkane is (for the same number of carbon atoms), the lower are its melting and boiling points.

_____ 7. Simple alkanes can exhibit either constitutional isomerism or geometric isomerism.

_____ 8. Simple alkenes can exhibit either constitutional isomerism or geometric isomerism.

_____ 9. Simple alkynes can exhibit either constitutional isomerism or geometric isomerism.

_____ 10. Petroleum consists mainly of hydrocarbons.

_____ 11. Petroleum from all sources is identical, that is, composed of a mixture of the same compounds in the same amounts.

_____ 12. All aromatic hydrocarbons have a pleasant smell.

_____ 13. Many of the common flavoring agents are simple amines.

_____ 14. Alcohols are polar molecules.

_____ 15. Many reactions of carboxylic acids involve displacement of the $-OH$ group by another atom or group of atoms.

_____ 16. All functional groups contain some atom other than carbon or hydrogen.

_____ 17. A single organic compound can have only one functional group.

_____ 18. The most common source of the carbon atoms for the industrial preparation of organic chemicals is carbon dioxide from the air.

_____ 19. The main type of chemical reaction that the alkanes undergo is substitution.

_____ 20. Organic reactions usually form only a single product, so separation and purification is usually not a serious problem.

_____ 21. The reaction in which a nitro group, $-NO_2$, is substituted for a hydrogen atom is called nitration.

_____ 22. Nitration reactions result when a hydrocarbon is mixed with NO_2 gas at high temperature.

_____ 23. Compounds with carbon-carbon double or triple bonds undergo addition reactions much more readily than they undergo substitution reactions.

_____ 24. The alkynes are usually more reactive than the alkenes with respect to addition reactions.

_____ 25. Benzene is an unsaturated hydrocarbon, so it readily undergoes addition reactions like those of the alkenes and the alkynes.

_____ 26. Hydrogenation reactions are substitution reactions in which H atoms are substituted for other groups such as halogen atoms or hydroxyl groups.

Short Answer

1. The existence of the large number of organic compounds is primarily due to the ability of carbon atoms to _____ other carbon atoms.

2. A compound in which there is one continuous chain of carbon atoms, with no branching, is referred to as a straight-chain or _____ compound.

3. Each member of the series of normal saturated hydrocarbons differs from the next by one _____ (name, formula); such a series of compounds is called a _____.

4. The melting points of the normal alkanes generally _____ with increasing carbon content, and their boiling points generally _____.

5. Atomic arrangements that differ only by rotation around single bonds are called different _____ of the same compound.

6. The hydrogen atoms of alkanes are readily replaced by halogen atoms at high temperatures or in the presence of _____.

7. The process of "cracking" of petroleum involves _____ _____ to produce more gasoline.

8. Hydrocarbons containing one carbon-carbon double bond per molecule are called _____.

9. The suffix applied to the names of all alkenes, whether open chain or cyclic, is _____.

10. The suffix applied to the names of all alkanes, whether open chain or cyclic, is _____.

11. The suffix applied to the names of all alkynes, whether open chain or cyclic, is _____.

12. The number of isomers of xylene is _____; these are examples of _____ isomers.

13. The number of isomers of trimethylbenzene is _____.

For each of the following hydrocarbons, either name the hydrocarbon or supply its formula, as required. These formulas should be sufficient to specify the compound uniquely and may be full structural formulas. Write the molecular formula (e.g., C_8H_{18}) for each compound. Classify each compound as an alkane, an alkene, an alkyne, an aromatic hydrocarbon, or derived from one of these by substitution.

Formula	Name	Molecular Formula	Class
14. $H-\underset{\underset{H}{\overset{H}{|}}}{\overset{\overset{H}{|}}{C}}-\underset{\underset{H}{\overset{H}{|}}}{\overset{\overset{H}{|}}{C}}-H$	_____	_____	_____
15. (benzene ring structure)	_____	_____	_____
16. $CH_3-\underset{\underset{CH_3}{\overset{\overset{CH_3}{|}}}}{\overset{\overset{CH_3}{|}}{C}}-CH_2-CH_2-CH_3$	_____	_____	_____
17. _____	2,4-dimethylhexane	_____	_____
18. $H_2C=CHCH_2CH_3$	_____	_____	_____
19. _____	*cis*-2-butene	_____	_____
20. _____	*trans*-2-butene	_____	_____
21. (cyclic structure with H, CH_3, H_2C, CH_2, H_2C-CH_2)	_____	_____	_____
22. (cyclohexane ring structure)	_____	_____	_____

Formula	Name	Molecular Formula	Class
23. _____	o-xylene	_____	_____
24. _____	*p*-xylene	_____	_____
25. _____	acetylene	_____	_____
26. _____	ethyne	_____	_____

27. $CH_3C \equiv CCH_3$ _____ _____ _____

28. _____ anthracene _____ _____

29. $CH_3CH_2CHCH_2CH_2CH_2CH_3$
$\quad\quad\quad | $
$\quad\quad\quad CH_2CH_2CH_3$ _____ _____ _____

30. _____ cyclopentene _____ _____

31. $CH_3CH_2NO_2$ _____ _____ _____

32.
$$\begin{array}{ccc} Br & & Br \\ \diagdown & & \diagup \\ & C{=}C & \\ \diagup & & \diagdown \\ H & & H \end{array}$$
_____ _____ _____

33.
$$\begin{array}{ccc} Br & & H \\ \diagdown & & \diagup \\ & C{=}C & \\ \diagup & & \diagdown \\ H & & Br \end{array}$$
_____ _____ _____

34. _____ *trans*-2,3-dibromo-2-butene _____ _____

35. _____ tetrafluoroethene _____ _____

36. _____ difluorodichloromethane _____ _____

37. _____ Bromobenzene _____ _____

---

38. A compound in which one hydrogen atom of water has been replaced by an organic group is a(n) _____, whereas a compound in which *both* hydrogen atoms of water have been replaced by organic groups is a(n) _____.

39. The simplest aliphatic alcohol is _____; the most common one is _____.

40. Alcohols containing more than one –OH group per molecule are _____ alcohols.

41. Aromatic alcohols as a group are referred to as _____; the simplest member of this group is the compound _____.

42. Alcohols in which the –OH group is bonded to a carbon atom that is in turn bonded to one, two, or three organic groups are called _____ alcohols, _____ alcohols, and _____ alcohols, respectively.

43. The secondary alcohol with the smallest number of carbon atoms is _____ (name, formula).

44. The secondary amine with the smallest number of carbon atoms is _____ (name, formula).

45. The polyhydric alcohol containing two equivalent carbon atoms is _____ _____ (name, formula); this substance is frequently used as _____ .

46. Dilute solutions of phenols are frequently used as _____ .

In each of the following questions, 47 through 64, name the compound or give a formula, as required. For each compound, tell the class of which it is an example.

Formula	Name	Class
47. $CH_3CH_2CH_2OH$	_____	_____
48. _____	phenol	_____
49.	_____	_____
50. _____	pentanoic acid	_____
51. _____	3-chloropentanoic acid	_____
52.	_____	_____
53. _____	methyl butyrate	_____
54. _____	glycerin	_____
55. _____	acetone	_____
56.	_____	_____
57. $CH_3-N-CH_2CH_3$ with H below N	_____	_____
58. _____	aniline	_____
59. _____	N-methylacetamide	_____

Formula	Name	Class
60. $CH_3CH_2-O-CH_2CH_3$	_____	_____
61.	_____	_____
62. _____	oxalic acid	_____

63. _____ _____

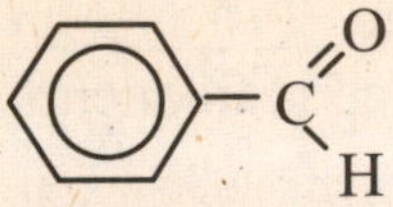

64. _____ _____

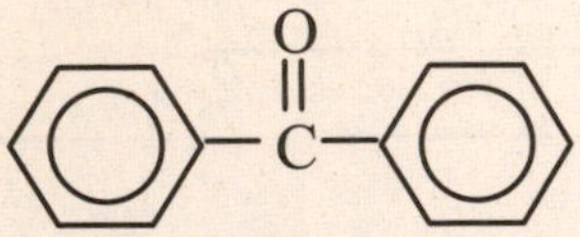

---

65. The three fundamental classes of reactions that are involved in most organic transformations are _____, _____, and _____.

66. The most common kind of reaction of the aromatic ring is _____.

67. The reaction by which a hydrogen atom on the benzene ring is replaced by the group $-NO_2$ is called _____; the resulting compound is named _____.

For each of the following questions, 68 through 75, classify the reaction according to one (or sometimes more) of the following types: aliphatic substitution, aromatic substitution, addition, halogenation, nitration, elimination, dehydration, polymerization.

68. $CH_4 + Cl_2 \rightarrow CH_3Cl + HCl$ _____

69. $CH_3CH=CH_2 + Cl_2 \rightarrow CH_3CH-CH_2$ (with Cl Cl below) _____

70. $CH_3CH=CH_2 + H_2 \rightarrow CH_3CH_2CH_3$ _____

71. $CH_2=CH_2 + HOCl \rightarrow CH_2-CH_2$ (with OH Cl below) _____

72. $CH_2=CH-C=CH_2 \rightarrow \left[-CH_2-CH=C-CH_2- \right]_n$ (with Cl below) _____

73. $CH_3C\equiv CH + 2Br_2 \rightarrow CH_3CBr_2CHBr_2$ _____

74. _____

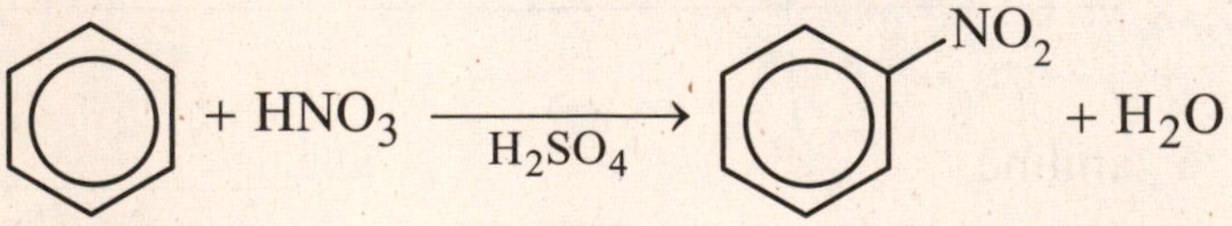

75. _____

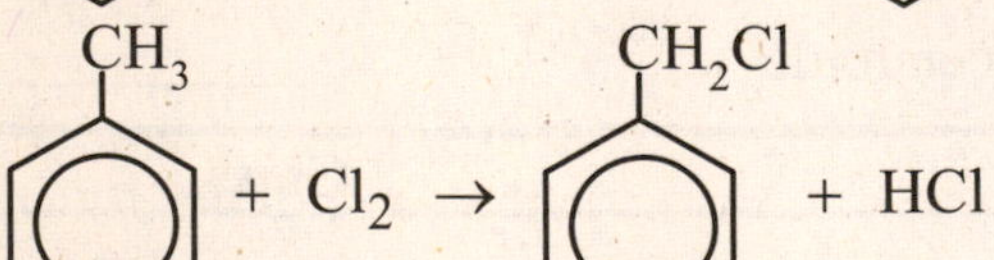

Multiple Choice

____ 1. How many different constitutional isomers are possible with the molecular formula C_3H_7Br? Remember that bromine can form only one covalent bond to another atom.
(a) 1   (b) 2   (c) 3   (d) 4   (e) 5

____ 2. What is the molecular weight of n-pentane?
(a) 44   (b) 70   (c) 42
(d) The same as the molecular weight of 2-methylbutane.
(e) The same as the molecular weight of 2-pentane.

____ 3. In all saturated hydrocarbons, the carbon atoms have which one of the following types of hybrid orbitals?
(a) sp (b) sp^2 (c) sp^3 (d) sp^3d (e) sp^3d^2

____ 4. Substances having the same molecular formula but different structural formulas are called
(a) isotopes.   (b) hydrocarbons.   (c) isomers.
(d) fractions.   (e) free radicals.

____ 5. Which one of the following is an isomer of cyclohexane?
(a) n-hexane   (b) benzene   (c) 1-hexene
(d) 2-methylpentane   (e) 2-methylhexane

____ 6. The possibility of geometric isomerism in some alkenes is due to
(a) the fact that alkenes are saturated hydrocarbons.
(b) the restriction on the orientation of the p orbitals that make up the π bond.
(c) the free rotation about σ bonds.
(d) the acidity of substituted groups.
(e) resonance in these substances.

____ 7. Which one of the following compounds may exist as either of two geometric isomers?
(a) 1,1-dibromoethene   (b) 1,1-dibromoethane   (c) 1,2-dibromoethene
(d) 1,2-dibromoethane   (e) 2-methylpropene

____ 8. Which one of the following compounds is an alkane?
(a) $CH_3CH_2C\!=\!CH_2$ with H   (b) $CH_3CH_2-C-CH_3$ with O (double bond)   (c) CH_3OH
(d) $CH_3CH_2CH_2CH_2Br$ (e) $CH_3CH_2CHCH_2CH_3$ with CH_3

____ 9. The formulas
$$\underset{H}{\overset{H}{\diagdown}}C\!=\!C\underset{CH_3}{\overset{H}{\diagup}}\quad \text{and}\quad \underset{H}{\overset{H_3C}{\diagdown}}C\!=\!C\underset{H}{\overset{H}{\diagup}}\quad\text{represent}$$
(a) geometric isomers.   (b) structural isomers.   (c) the same compound.   (d) ethylene.
(e) None of the preceding answers is correct.

____ 10. What is the number of constitutional isomers of dichlorobenzene?
(a) 1   (b) 2   (c) 3   (d) 4   (e) 5

____ 11. One of the following pairs of compounds is a pair of *constitutional* isomers. Which pair?

(a) $CH_3CH_2CH_2CH_3$ and

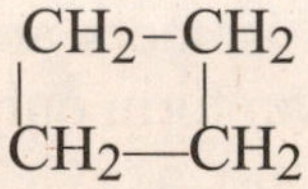

$$\begin{array}{c} CH_2{-}CH_2 \\ | \qquad | \\ CH_2{-}CH_2 \end{array}$$

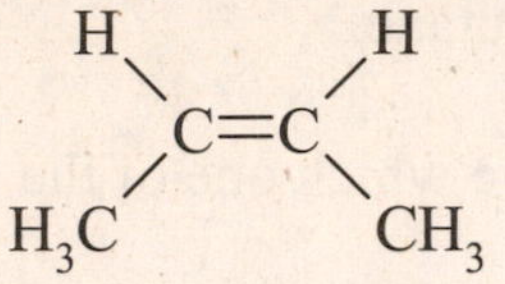

(b) $CH_3CH_2CH_2CH_3$ and

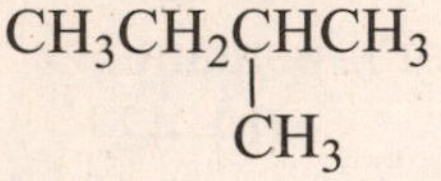

$$\begin{array}{c} CH_3CH_2CHCH_3 \\ | \\ CH_3 \end{array}$$

(c)

$$\overset{H}{\underset{H_3C}{\diagdown}} C{=}C \overset{H}{\underset{CH_3}{\diagup}} \quad \text{and} \quad \overset{H_3C}{\underset{H}{\diagdown}} C{=}C \overset{H}{\underset{CH_3}{\diagup}}$$

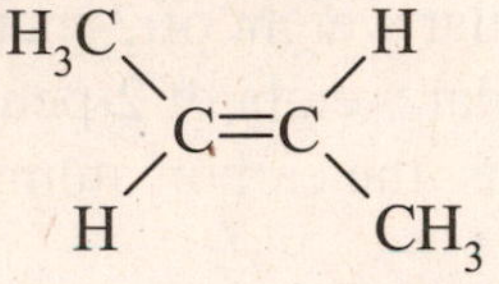

(d) $CH_3CH_2CH_2CH_2CH_3$ and

$$\begin{array}{c} CH_3 \\ | \\ H_3C{-}C{-}CH_3 \\ | \\ CH_3 \end{array}$$

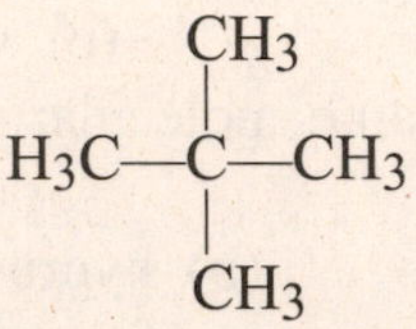

____ 12. Which answer in Question 11 represents a pair of *geometric* isomers?

____ 13. The molecule

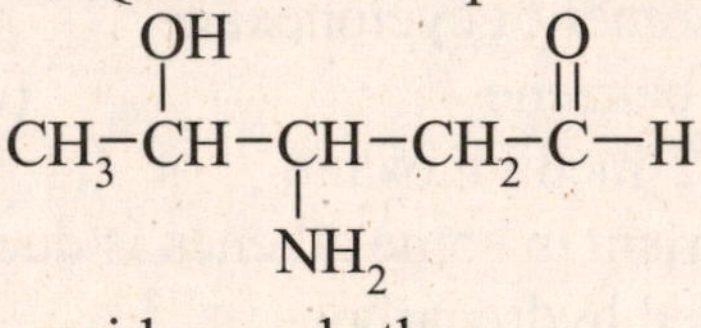

$$CH_3{-}\overset{OH}{\underset{}{CH}}{-}\overset{}{\underset{NH_2}{CH}}{-}CH_2{-}\overset{O}{\overset{\|}{C}}{-}H$$

contains functional groups characteristic of

(a) aldehydes, amides, and ethers. (b) amines, ketones, and esters.
(c) aldehydes, halides, and alcohols. (d) alcohols, aldehydes, and amines.
(e) acids, alcohols, and halides.

____ 14. The group of atoms characteristic of aldehydes is

(a) $-\overset{|}{\underset{|}{C}}{-}\overset{O}{\overset{\|}{C}}{-}\overset{|}{\underset{|}{C}}{-}$ (b) $-\overset{|}{\underset{|}{C}}{-}O{-}\overset{|}{\underset{|}{C}}{-}$ (c) $-\overset{O}{\overset{\|}{C}}{-}O{-}\overset{|}{\underset{|}{C}}{-}$ (d) $-\overset{O}{\overset{\|}{C}}{-}H$ (e) $-C{\equiv}C{-}$

____ 15. The group of atoms that is characteristic of esters is

(a) $-\overset{|}{\underset{|}{C}}{-}NH_2$ (b) $-\overset{|}{\underset{|}{C}}{-}O{-}\overset{|}{\underset{|}{C}}{-}$ (c) $-\overset{O}{\overset{\|}{C}}{-}H$ (d) $-\overset{O}{\overset{\|}{C}}{-}O{-}\overset{|}{\underset{|}{C}}{-}$

(e) $-\overset{|}{\underset{|}{C}}{-}\overset{O}{\overset{\|}{C}}{-}\overset{|}{\underset{|}{C}}{-}$

____ 16. The compound

$$\overset{H_2C{-}CH_2}{\underset{H_2C{-}CH_2}{H_2C{\diagup}\quad{\diagdown}HCOH}}$$

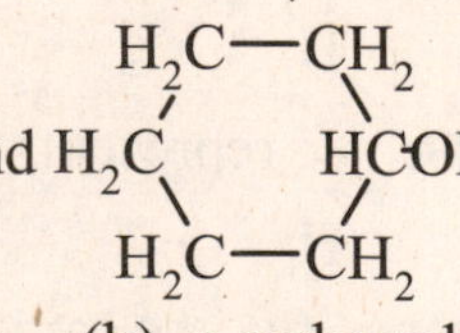

is

(a) an alkane. (b) a carboxylic acid. (c) an alkyne. (d) an alcohol. (e) an amine.

_____ 17. Which of the following is a secondary alcohol?

(a) CH_3OH (b) CH_3CH_2OH (c) 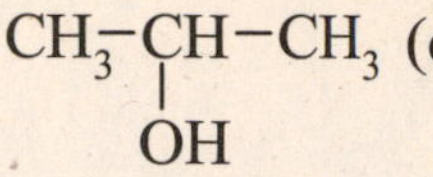 $CH_3-CH-CH_3$ (d) 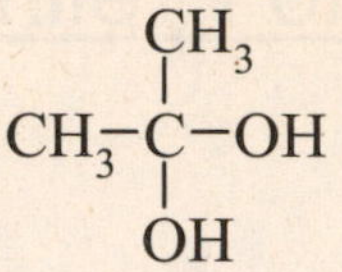 $CH_3-\overset{CH_3}{\underset{OH}{C}}-OH$ (e) 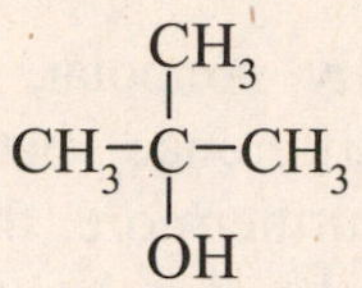
$\underset{OH}{}$

$CH_3-\overset{CH_3}{\underset{OH}{C}}-CH_3$

_____ 18. The structures of malonic acid and fumaric acid are, respectively,

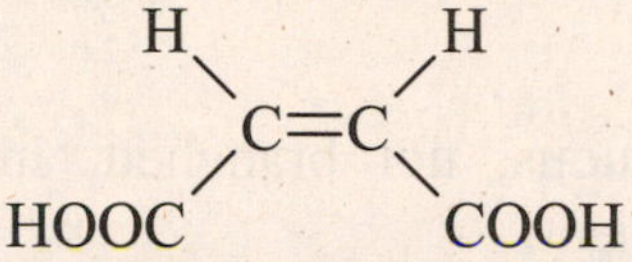

$$\underset{HOOC}{\overset{H}{\diagdown}}C=C\underset{COOH}{\overset{H}{\diagup}} \quad \text{and} \quad \underset{H}{\overset{HOOC}{\diagdown}}C=C\underset{COOH}{\overset{H}{\diagup}}$$

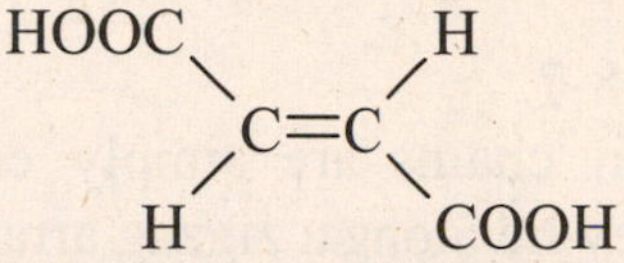

Which of the following terms describes the relationship between these two acids?
(a) a conjugate acid-base pair (b) constitutional isomers   (c) geometric isomers
(d) polymers   (e) dicarboxylic acids

_____ 19. The reaction of methane with chlorine in the presence of ultraviolet radiation can produce
(a) chloromethane.   (b) dichloromethane.
(c) trichloromethane.   (d) carbon tetrachloride.
(e) all of the products listed above.

_____ 20. The nitration of *n*-butane is an example of a(an) _____ reaction.
(a) aromatic   (b) addition   (c) elimination (d) substitution (e) saponification

_____ 21. The reaction by which tetrafluoroethylene is converted into Teflon® is called
(a) fluorination. (b) vulcanization. (c) hydration. (d) addition. (e) polymerization.

Answers to Some Important Terms in This Chapter

1. hydrocarbons
2. aromatic hydrocarbons
3. aliphatic hydrocarbons
4. saturated hydrocarbons
5. unsaturated hydrocarbons
6. alkanes
7. constitutional isomers
8. structural isomers
9. alkyl group
10. alkenes

11. geometric isomers
12. alkynes
13. alcohol
14. phenol
15. ether
16. aldehyde
17. ketone
18. carboxylic acid
19. acyl halide
20. ester
21. amine
22. amide
23. substitution reaction
24. halogenation reaction

25. addition reaction
26. hydrogenation reaction
27. hydration reaction
28. elimination reaction
29. dehydrohalogenation
30. dehydration reaction
31. polymerization reaction
32. polymer
33. addition polymerization
34. copolymer
35. condensation polymerization
36. polyester
37. polyamide
38. peptide bond

Answers to Preliminary Test

True-False

1. False. Carbon-carbon and carbon-hydrogen bonds are essentially nonpolar, since two carbon atoms have the same electronegativity and the difference between the electronegativities of carbon and hydrogen is very small (0.4). Furthermore, these bonds are arranged tetrahedrally around each carbon atom, so that any bond polarities tend to cancel each other.

2. True. See Section 8-7.

3. False. The carbon chains are simply continuous, not branched. In their most stable conformations, they lie along a zigzag arrangement.

4. False. In branched hydrocarbons, some carbon atoms may be bonded to only one hydrogen atom, or even none. Can you write examples of each of these cases of saturated hydrocarbons?

5. False. Only the lightest four normal alkanes are gases at room temperature and atmospheric pressure. The liquefied propane or butane gas that is available in tanks at room temperature is under high pressure.

6. True. This is described in Section 23-1 for the three isomeric pentanes and the five isomeric hexanes.

7. False. They can exhibit only constitutional isomerism, in which the atoms are linked together in a different order. (Cycloalkanes, however, can evidence *cis-trans* isomerism.)

8. True. Restricted rotation around the double bond allows the formation of *cis* and *trans* isomers.

9. False. They cannot exhibit geometric isomerism, because the environment of the triple bond is linear (*sp* hybridization).

10. True. Also present are small amounts of organic compounds containing nitrogen (0.1-2%), oxygen (0.05-1.5%), and sulfur (0.05-6.0%), and trace amounts (<0.1%) of metals such as iron, nickel, copper and vanadium.

11. False. Each oil field produces oil with a characteristic composition.

12. False. Read the introduction to the portion of the chapter on aromatic hydrocarbons. Some have foul odors; others are odorless. Instead, the term is used to describe benzene, its derivatives, and other compounds that exhibit similar chemical properties. Can you describe the bonding feature that all aromatic compounds have in common? See Sections 23-5 and 23-6.

13. False. Most amines have quite unpleasant odors. Many common flavoring agents are organic esters.

14. True. The O−H bond is quite polar.

15. True. Examples of classes of compounds so formed are esters, amides, and acyl halides.

16. False. Any potential reactive site on a molecule is termed a functional group. This can include carbon-carbon double or triple bonds.

17. False. Many organic compounds have several functional groups. You have seen several examples of this in Chapter 23.

18. False. Most organic chemicals are produced from petroleum. This may become a major problem due to decreasing petroleum supplies. Coal is another source of starting materials for organic chemicals.

19. True. The most common reaction is halogenation. See Section 23-16.

20. False. Usually a mixture of products is formed, and often these are isomers of one another. A common method of separation is fractional distillation.

21. True. The reaction involves a mixture of concentrated nitric and sulfuric acids.

22. False. The source of the $-NO_2$ group is usually nitric acid vapor, HNO_3, at high temperature. Aromatic hydrocarbons may be nitrated by a mixture of concentrated HNO_3 and H_2SO_4.

23. True. Addition reactions are the characteristic reactions of the alkenes and the alkynes. See Section 23-17.

24. True. The triple carbon-carbon bond in an alkyne includes two pi bonds, which are sources of electrons.

25. False. The π electrons of the aromatic hydrocarbons (such as benzene) are not available for addition reactions because of delocalization. An addition reaction would disrupt delocalization and the stabilizing effect it has on the molecule. The main reactions of aromatic hydrocarbons such as benzene are substitution reactions.

26. False. Hydrogenation reactions are addition reactions. See Section 23-17.

Short Answer

1. form covalent bonds with. This is known as *catenation*.
2. normal
3. methylene group (CH_2); homologous series
4. increase; increase
5. conformations
6. sunlight (or an ultraviolet light source)
7. breaking of long-chain hydrocarbons into shorter-chain hydrocarbons
8. alkenes
9. -ene
10. -ane
11. -yne
12. 3; structural. Can you draw and name these three isomers?
13. 3. Can you draw and name them?
14. ethane; C_2H_6; alkane
15. benzene; C_6H_6; aromatic hydrocarbon
16. 2,2-dimethylpentane, C_7H_{16}, alkane
17. $CH_3CHCH_2CHCH_2CH_3$; C_8H_{18}; alkane
$\qquad$ $\underset{CH_3}{|}$ $\quad$ $\underset{CH_3}{|}$
18. 1-butene; C_4H_8; alkene
19.
$$\underset{H_3C}{\overset{H}{\diagdown}} C = C \underset{CH_3}{\overset{H}{\diagup}} \text{ ; } C_4H_8 \text{; alkene}$$

20. ; C_4H_8; alkene

21. methylcyclopentane; C_6H_{12}; alkane (cycloalkane)
22. cyclohexane; C_6H_{12}; alkane (cycloalkane)
23. CH_3; C_8H_{10}; aromatic hydrocarbon

24. ; C_8H_{10}; aromatic hydrocarbon

25. $HC\equiv CH$; C_2H_2; alkyne
26. $HC\equiv CH$; C_2H_2; alkyne. This is another name for acetylene. You should be aware of the common, or trivial, names for many of the simple compounds in this and the next chapter.
27. 2-butyne; C_4H_6; alkyne

28. ; $C_{14}H_{10}$; aromatic hydrocarbon

For aromatic hydrocarbons, it is understood that there is one hydrogen atom at each corner. Sometimes they are shown, but it is not necessary to show them explicitly.
29. 4-ethyloctane, $C_{10}H_{22}$; alkane. Remember to name according to the longest chain of carbon atoms, and number so that the substituted group (ethyl, in this case) has the smallest position number possible.

30. ; C_5H_8; alkene (cycloalkene)

31. nitroethane; $C_2H_5NO_2$; substituted alkane
32. *cis*-1,2-dibromoethene; $C_2H_2Br_2$; substituted alkene
33. *trans*-1,2-dibromoethene; $C_2H_2Br_2$; substituted alkene

34. ; $C_4H_6Br_2$; substituted alkene

35. ; C_2F_4; substituted alkene

36. $\text{Cl}-\overset{\overset{\displaystyle F}{|}}{\underset{\underset{\displaystyle F}{|}}{C}}-\text{Cl}$; CF_2Cl_2; substituted alkane. (There is only one isomer.)

37. ; C_6H_5Br; substituted aromatic hydrocarbon

38. alcohol or phenol; ether
39. methyl alcohol (or methanol); ethyl alcohol (or ethanol)
40. polyhydric
41. phenols; phenol
42. primary; secondary; tertiary
43. 2-propanol (or isopropyl alcohol); $CH_3\underset{\underset{\displaystyle OH}{|}}{C}HCH_3$

44. dimethylamine; $CH_3-\underset{\underset{\displaystyle H}{|}}{N}-CH_3$

45. ethylene glycol; $HOCH_2-CH_2OH$; permanent antifreeze
46. disinfectants
47. 1-propanol or normal propyl alcohol; alcohol
48. OH; phenol

   ("Phenol" is the name of the class of compounds and also of the compound itself.)
49. benzoic acid; carboxylic acid
50. $CH_3CH_2CH_2CH_2\overset{\overset{\displaystyle O}{||}}{C}-OH$; carboxylic acid
51. $CH_3CH_2\overset{\overset{\displaystyle Cl}{|}}{C}HCH_2\overset{\overset{\displaystyle O}{||}}{C}-OH$; carboxylic acid. (Note that this is not an acyl chloride.)
52. acetyl chloride; acyl halide (or acid halide)
53. $CH_3CH_2CH_2\overset{\overset{\displaystyle O}{||}}{C}-O-CH_3$; ester
54. $H_2C-\overset{\overset{\displaystyle H}{|}}{\underset{\underset{\displaystyle OH}{|}}{C}}-CH_2$; alcohol (a polyhydric alcohol)
   $\overset{}{\underset{\underset{\displaystyle OH}{|}}{}}\quad\overset{}{\underset{\underset{\displaystyle OH}{|}}{}}$

55. $CH_3\overset{\overset{\displaystyle O}{||}}{C}CH_3$; ketone
56. acetaldehyde (or ethanal); aldehyde
57. ethylmethylamine; amine

58. NH_2; amine

59. $CH_3-\overset{\overset{O}{\|}}{C}-\overset{\overset{H}{|}}{N}-CH_3$; amide

60. diethylether; ether

61. methylphenylether; ether

62. $HO-\overset{\overset{O}{\|}}{C}-\overset{\overset{O}{\|}}{C}-OH$; carboxylic acid (more specifically, a dicarboxylic acid)

63. benzaldehyde; aldehyde

64. diphenylketone; ketone

65. substitution, addition, and elimination. You need to be able to recognize each of these.

66. substitution. Can you describe such a reaction and write an example of it?

67. nitration; nitrobenzene

68. aliphatic substitution, halogenation

69. addition, halogenation

70. addition, hydrogenation

71. addition

72. polymerization

73. addition (actually two steps of addition, as is common for alkynes), halogenation

74. aromatic substitution, nitration

75. aliphatic substitution, halogenation

Multiple Choice

1. (b). Can you convince yourself by drawing them?

2. (d). 2-methylbutane is an isomer of *n*-pentane. (There is no such compound as "2-pentane.")

3. (c). Carbon never forms compounds using the hybrid orbitals indicated in answers (d) or (e) because it does not have any *d* orbitals in its valence shell (the second energy level).

4. (c)

5. (c). You may need to write out structural formulas for each of the answers and figure out the molecular formula (i.e., C_xH_y). Remember that isomers have the same molecular formula.

6. (b)

7. (c). Each of the two carbon atoms double-bonded to each other has two different atoms bonded to it (bromine and hydrogen). If both carbon atoms have the hydrogen atoms on the same side of the double bond, the molecule is *cis*. If the two hydrogen atoms are on opposite sides of the bond, then the molecule is *trans*.

8. (e). An alkane is a hydrocarbon (having hydrogen and carbon atoms only) with only single bonds between the carbon atoms. Answer (a) has a double bond. The other answers have oxygen or bromine atoms.

9. (c). Since one of the double-bonded carbon atoms has two hydrogen atoms, the hydrogen atom on the other carbon atom will always be on the same side of the double bond as one of them and across the double bond from the other.

10. (c). They are the *ortho-*, *meta-*, and *para-* isomers.

11. (d). Pentane and 2,2-dimethylpropane have the same molecular formula (C_5H_{12}). In (a), butane and cyclobutane do not have the same formula. Butane and 2-methylbutane, in (b), also have different formulas. In (c), *cis*- and *trans*-2-butene are geometric isomers.

12. (c). The atoms have the same bonding order, but different arrangements in space.

13. (d). R–OH is characteristic of alcohols, R–NH_2 is characteristic of amines, and R–CHO is characteristic of aldehydes.

14. (d). (a) is characteristic of ketones, (b) of ethers, (c) of esters, and (e) of alkynes.

15. (d). (a) is characteristic of amines, (b) of ethers, (c) of aldehydes, and (e) of ketones.

16. (d). It's cyclohexane with one hydrogen atom replaced by the hydroxyl group.

17. (c). This is 2-propanol, or isopropyl alcohol. It is the substance commonly known as "rubbing alcohol."

18. (c). Each one is a dicarboxylic acid (e), but this does not describe the *relationship* between the two compounds.

19. (e). See the discussion of halogenation reactions in Section 23-16.

20. (d). In order to add a nitro group (–NO_2) to a saturated hydrocarbon like *n*-butane, a hydrogen atom must be removed first.

21. (e). Polymerization is the combination of many small molecules (the monomer, $F_2C=CF_2$) to form large molecules (the polymer).

24 Organic Chemistry II: Shapes, Selected Reactions, and Biopolymers

Chapter Summary

This chapter deals with three important aspects of organic chemistry. In Chapter 23, you learned about some isomeric organic compounds. In the first portion of this chapter, you undertake a more systematic study of the geometric aspects of organic chemistry, a topic known as **stereochemistry**. As you study this subject, pay very close attention to the definitions that are presented in the blue boxes. **Isomers** are substances that have the same number and kind of atoms, but with the atoms arranged differently. Many kinds of isomerism are introduced by first discussing examples of the simplest classes of organic compounds, the hydrocarbons, but all classes of organic compounds can exhibit isomerism.

Section 24-1 discusses **constitutional isomers**, which were introduced (for organic compounds) in Section 23-1. These differ in the *order* in which their atoms are bonded together. **Stereoisomers** (Section 24-2) contain atoms linked together in the same order, but their arrangements in space are different. Geometric isomers differ in the orientation of groups about a plane or direction. The *cis* and *trans* isomers of alkenes (Section 23-3) are examples of geometric isomers. Similarly, groups can be in either a *cis* or *trans* arrangement across a ring. A second kind of stereoisomerism, optical isomerism, depends on the existence of molecules that are nonsuperimposable mirror image structures (**enantiomers**) of each other. Optical isomers are characterized by their ability to rotate the plane of polarized light. Of each pair of optical isomers, one rotates light to the right (**dextrorotatory**) and the other to the left (**levorotatory**). Section 24-3 discusses conformations of a compound, which differ from one another in the extent of rotation about one or more single bonds.

The next few sections discuss some other important organic reactions. You have already seen many examples of Brønsted-Lowry acid-base behavior (Sections 6-9.1 and 10-4). Many organic compounds can act as weak Brønsted-Lowry acids or bases (Section 24-4). The more important organic acids contain carboxyl groups, abbreviated as –COOH, but alcohols and phenols are also weakly acidic compounds. Cations of weak organic bases act as weak acids, in the same way that ammonium salts give acidic solutions (Chapter 18). Substituted amines are usually weaker bases, so their cations are usually stronger acids. Likewise, the amines and the anions derived from weak organic acids are common organic bases.

You have previously encountered many oxidation-reduction (redox) reactions, and you have learned to recognize such reactions by tabulating changes in oxidation numbers. Section 24-5 discusses oxidation-reduction reactions in organic chemistry. In organic chemistry, it is more convenient to talk about oxidation as increase in oxygen content or as decrease in hydrogen content, and reduction as decrease in oxygen content or increase in hydrogen content. Keep these conventions in mind as you study the oxidation and reduction transformations presented in this chapter. The latter part of this section discusses some important aspects of combustion reactions, the most extreme type of oxidation reaction of organic compounds.

In Section 23-14 you learned about several classes of compounds that are structurally related to the carboxylic acids, produced by replacement of the −OH portion of the carboxyl group. In Section 24-6, you learn about some reactions that produce these derivatives—acyl halides, esters, and amides. Section 24-7 describes a reverse reaction, the hydrolysis of an ester to produce the parent alcohol and a salt of the parent acid of the ester. One such reaction, saponification, produces soaps.

The final three sections of Chapter 24 present brief descriptions of three important classes of **biopolymers** (*biological polymers*). **Carbohydrates** (Section 24-8) are composed of **monosaccharides**, or simple sugars. A monosaccharide often contains five or six carbon atoms and either an aldehyde group (an **aldose**) or a ketone group (a **ketose**). Most naturally occurring carbohydrates contain two or more monosaccharide units, linked by **glycosidic bonds**. Examples are amylose (a starch) and cellulose, which differ in the stereochemistry of the glycosidic link.

Polypeptides and **proteins** (Section 24-9) are polymers of amino acids linked by **peptide bonds**. Twenty different amino acids are found widely in nature, most often as the L-isomer (levorotatory). Proteins are large polypeptides with molecular weights usually greater than 5000 g/mol. **Enzymes** are proteins that catalyze specific biochemical reactions.

Nucleic acids (Section 24-10) are more complex polymers that consist of three kinds of monomers—a phosphate group, a simple carbohydrate group, and an organic base. The two types of nucleic acids differ in the carbohydrate groups that they contain; **ribonucleic acid (RNA)** contains ribose, whereas **deoxyribonucleic acid (DNA)** contains 2-deoxyribose. The purine and pyrimidine bases in nucleic acids form pairs with highly specific patterns of hydrogen bonding. The double helix of DNA contains genetic information about an organism. This information is transmitted when living cells of the organism replicate.

Study Goals

Sections 24-1 and 24-2
1. Understand and give examples of various kinds of isomerism in organic chemistry: (a) structural isomers and (b) stereoisomerism (geometric and optical). *Work Exercises 1 through 4 and 6 through 10.*
2. Recognize optical isomers and identify chiral carbons. *Work Exercises 11-16.*

Section 24-3
3. Describe different conformations of organic compounds. *Work Exercises 7 through 9.*

Sections 24-4 through 24-7
4. Describe the characteristics of some important types of organic reactions: (a) acid-base, (b) oxidation-reduction, (c) esterification, and (d) hydrolysis. Know what classes of organic compounds can undergo each type of reaction. *Work Exercises 17 through 41.*

Section 24-4
5. Describe which classes of organic compounds can act as acids and which can act as bases. Understand ranges in strengths of acidic or basic behavior. *Work Exercises 17 through 22, 25, and 28.*

Section 24-5

6. Recognize oxidation and reduction reactions of organic compounds. *Work Exercises 23, 24, 27 and 29 through 34.*

Section 24-6

7. Recognize reactions that form derivatives of carboxylic acids, and predict products of such reactions. *Work Exercises 35 through 39.*

Section 24-7

8. Describe organic hydrolysis reactions and their products. *Work Exercises 40 and 41.*

Sections 24-8 through 24-10

9. Describe the main features of three classes of biopolymers: (a) carbohydrates; (b) polypeptides and proteins; and (c) nucleic acids. Describe the monomeric units and the types of reactions that form these biopolymers. *Work Exercises 42 through 58.*

General

10. Apply the concepts of this chapter to answer questions of unidentified type. *Work **Mixed Exercises** 59 through 64.*

11. Answer conceptual questions based on this chapter. *Work **Conceptual Exercises** 65 through 70.*

12. Apply concepts and skills from earlier chapters to the ideas of this chapter. *Work **Building Your Knowledge** exercises 71 through 76.*

13. Exercises at the end of chapter direct you to sources outside the textbook for information to use in solving them. Work **Beyond the Textbook** *exercises 77 through 79.*

Some Important Terms in This Chapter

Some important terms from Chapter 24 are listed below. Fill the blanks in the following paragraphs with terms from the list. Each term is only used once. There are many additional new terms in this chapter; study others listed in the Key Terms.

achiral	geometric isomers	protein
Brønsted-Lowry acid	glycosidic bond	purine bases
Brønsted-Lowry base	hydrolysis	pyrimidine bases
carbohydrates	monosaccharides	racemic mixture
chiral	nucleic acids	reduction
cis	oligosaccharide	ribonucleic acid
combustion	optical activity	saponification
complementary base-pairing	optical isomers	stereochemistry
conformations	oxidation	stereoisomers
constitutional isomers	peptide	*trans*
deoxyribonucleic acid	polypeptide	
enzyme	polysaccharide	

The study of the three-dimensional aspects of organic structures is known as [1]_____ ("spatial chemistry"). Hydrocarbons exhibit several types of isomerism. Compounds that have the same molecular formula but differ in the *order* in which their atoms are bonded together are [2]_____ or structural isomers. Different compounds composed of molecules in which the atoms are linked together in the same order, but their arrangements in space are different are called [3]_____. There are at least two kinds of stereoisomers, [4]_____ (or *cis-trans* isomers) and [5]_____ (or enantiomers). Geometric isomers are found among cycloalkanes that have substituents on either the same (described at [6]_____) or opposite (described as [7]_____) sides of the ring (top and bottom) and among alkenes that have substituents on the same or opposite sides of the double bond. Optical isomers that are enantiomers are nonsuperimposable mirror images of each other (said to be [8]_____), and they rotate the plane of polarized light in opposite directions, a phenomenon that is called [9]_____. Most chiral molecules contain at least one carbon atom that has four different atoms or groups of atoms bonded to it. A molecule that *is* superimposable with its mirror image is said to be [10]_____. A solution containing equal amounts of the two enantiomers, called a [11]_____, does not rotate a plane of polarized light because the equal and opposite effects of the two isomers exactly cancel. Different shapes of a molecule that result from rotation around one or more single bonds are called [12]_____, and they are *not* isomers.

The acid-base behavior of organic molecules can often be explained by the Brønsted-Lowry acid-base theory. According to this theory, a proton donor is called a [13]_____, and a proton acceptor is called a [14]_____. Oxidation and reduction reactions of organic compounds are also important. [15]_____ of an organic molecule usually involves *increasing* its *oxygen* content or *decreasing* its *hydrogen* content. [16]_____ of an organic molecule usually involves *decreasing* its *oxygen* content or *increasing* its *hydrogen* content. Organic chemists are most often concerned with the *mild* oxidation of an organic compound, which converts it to a different organic compound. However, most organic compounds will also *burn*, reacting with oxygen gas from the air to produce carbon dioxide and water, a reaction known as [17]_____. One other important reaction, particularly in the area of biochemistry, is the splitting of a large molecule into two or more smaller molecules by the addition of the atoms of a water molecule, a process known as [18]_____. Sometimes this process requires the use of a strong base, like NaOH, instead of water (H−OH), and one of the products is then a salt. An example of this is the hydrolysis of esters in the presence of strong bases, producing the salt of a carboxylic acid and an alcohol. If the ester is a fat, the salt produced is a soap, so the process is called [19]_____ (soap-making), even if the ester is not a fat.

Sugars and starches have the general formula $C_n(H_2O)_m$, which makes them appear to consist of hydrated carbons. Therefore, they are often referred to as [20]_____. The carbohydrates are generally composed of molecules of [21]_____, or simple sugars. These molecules of monosaccharides are linked to each other by a C−O bond called a [22]_____. Each molecule of a *di*saccharide contains *two* units of monosaccharide, joined by one glycosidic bond. A molecule of a *tri*saccharide contains *three* units of monosaccharide, joined by two glycosidic bonds. A carbohydrate that contains four to ten monosaccharide units is usually called an [23]_____. Carbohydrates that contain larger numbers of monosaccharide units are called [24]_____.

Similarly, amino acids link together, joined by a *peptide bond*. A *di*peptide contains *two* amino acid units, joined by one peptide bond. A *tri*peptide has *three* amino acid units, joined by two peptide bonds. A small number of amino acids linked together is called a [25]_____. Any number of amino acids (although usually a larger number) linked together is called a [26]_____. One or more polypeptide chains constitute a [27]_____. A protein that catalyzes a biological reaction is an [28]_____.

The third group of important biochemical polymers are the [29]_____, each composed of a phosphate groups, a carbohydrate unit, and one of a small group of organic bases. There are two types of nucleic acids: [30]_____ (RNA), which contains ribose, and [31]_____ (DNA), which contains deoxyribose. Two of the bases used in RNA and DNA are adenine and guanine, which are called the [32]_____ because their structures are very similar to purine. The other two bases are cytosine and thymine (in DNA) or uracil (in RNA), which are called the [33]_____ because their structures are very similar to pyrimidine. Adenine and thymine or uracil each form two hydrogen bonds between them, while guanine and cytosine form three hydrogen bonds between them, an association that is known as [34]_____.

Preliminary Test

Be sure to practice *many* of the additional textbook exercises.

True-False

_____ 1. The different atomic arrangements obtained by rotation around a carbon-carbon single bond represent different compounds.

_____ 2. The different atomic arrangements obtained by rotation around a carbon-carbon double bond represent different compounds.

_____ 3. Saturated hydrocarbons cannot exhibit geometric isomerism.

_____ 4. Reaction of an organic compound with O_2 is usually an exothermic reaction.

_____ 5. Alkoxides of low molecular weight are strong bases.

_____ 6. Ethers are very reactive compounds.

_____ 7. Some of the reactions of alcohols are similar to those of water.

_____ 8. Many reactions of carboxylic acids involve displacement of the –OH group by another atom or group of atoms.

_____ 9. Most carboxylic acids are weak acids.

_____ 10. The conversion of an alcohol to a ketone is an oxidation.

_____ 11. The conversion of an alcohol to an aldehyde is an oxidation.

Short Answer

1. Compounds that differ only in the order in which their atoms are bonded together are called _____.

2. Compounds in which the atoms are linked together in the same order but with different arrangements in space are called _____.

3. Compounds that differ only in the spatial orientation of atoms or groups about a plane or direction are called _____.

4. A compound that is *not* superimposable with its mirror image is said to be _____.

5. An equimolar solution of D-glucose and L-glucose is called a _____ mixture.

6. The eclipsed and staggered forms of ethane are different _____.

7. When a hydrocarbon undergoes complete combustion the two products are _____ and _____.

8. As the chain length increases for the normal alkanes, the heat of combustion _____ on a molar basis and _____ on a mass basis.

9. In a process analogous to the reaction of active metals with water to form hydroxides, active metals react with alcohols to form _____. In both of these reactions, the gas _____ is liberated.

10. The reactions of alcohols with organic acids result in the formation of _____.

11. The hydrolysis of esters in the presence of strong soluble bases, a process which (with some esters) produces soaps, is called _____.

12. Aldehydes can be prepared by the oxidation of _____ alcohols, whereas ketones can be prepared by the oxidation of _____ alcohols.

For each of the following questions, 13 through 20, classify the reaction according to one (or sometimes more) of the following types: neutralization, displacement, esterification, saponification, combustion, oxidation, reduction, hydrolysis.

13.
$$\begin{matrix} CH_2OH \\ | \\ CHOH \\ | \\ CH_2OH \end{matrix} + 3HNO_3 \rightarrow \begin{matrix} CH_2-ONO_2 \\ | \\ CH-ONO_2 \\ | \\ CH_2-ONO_2 \end{matrix} + 3H_2O$$ _____

14. $CH_3-\overset{\overset{O}{\|}}{C}-OCH_2CH_3 + NaOH \rightarrow$

$CH_3-\overset{\overset{O}{\|}}{C}-O^-\ Na^+ + CH_3CH_2OH$ _____

15. $3C_2H_4 + 2KMnO_4 + 4H_2O \rightarrow$

$\begin{matrix} 3CH_2-CH_2 + 2MnO_2 + 2KOH \\ |\quad\ | \\ OH\ OH \end{matrix}$ _____

16. $2CH_3CH_2OH + O_2 \xrightarrow{Cu} 2CH_3\overset{\overset{O}{\|}}{C}H + 2H_2O$ _____

17. $CH_3-\overset{\overset{\displaystyle OH}{|}}{CH}-CH_3 \xrightarrow[300\,°C]{Cu} CH_3\overset{\overset{\displaystyle O}{\|}}{C}CH_3$ _____

18. $2CH_3CH_2CH_2\overset{\overset{\displaystyle O}{\|}}{C}H + O_2 \rightarrow 2CH_3CH_2CH_2\overset{\overset{\displaystyle O}{\|}}{C}OH$ _____

19. [benzene]—$CH_2CH_3 \xrightarrow[\text{heat, } H^+]{K_2Cr_2O_7}$ [benzene]—$C\overset{\nearrow O}{\underset{\searrow OH}{}}$ _____

20. $2C_8H_{18} + 25O_2 \rightarrow 16CO_2 + 18H_2O$ _____

Multiple Choice

____ 1. Which one of the following *cannot* exist in isomeric forms?
(a) C_3H_8 (b) C_4H_{10} (c) C_5H_{12} (d) C_6H_{14} (e) C_7H_{16}

____ 2. Vegetable oils can be converted to fats by _____ in the presence of a catalyst under high pressures and at high temperatures.
(a) substitution   (b) addition   (c) elimination   (d) vulcanization   (e) hydrogenation

____ 3. Members of which of these classes of compounds generally act as the weakest bases?
(a) alkoxides   (b) aliphatic amines   (c) phenoxides   (d)   aromatic amines
(e) None of the other answers is correct, because organic compounds cannot be bases.

____ 4. The reaction of propanoic acid with _____ produces propionyl chloride, CH_3CH_2COCl.
(a) sodium chloride   (b) hydrochloric acid   (c) hypochlorous acid
(d) chlorine   (e) phosphorus pentachloride

____ 5. The reaction of 1-propanol with $K_2Cr_2O_7$ in acidic solution produces _____.
(a) propyne   (b) propene   (c) propanal   (d) propanone   (e) cyclopropanone

____ 6. The reaction of methylamine with hydrochloric acid produces _____.
(a) ammonia   (b) ammonium methylate   (c)   methylammonium chloride
(d) chloromethane   (e) chloromethylamine

> You will need much more practice in recognizing types of reactions and predicting their products. Answer many of the Exercises from the end of the text chapter.

Answers to Some Important Terms in This Chapter

1. stereochemistry
2. constitutional isomers
3. stereoisomers
4. geometric isomers
5. optical isomers
6. *cis*
7. *trans*
8. chiral
9. optical activity
10. achiral
11. racemic mixture
12. conformations
13. Brønsted-Lowry acid
14. Brønsted-Lowry base
15. Oxidation
16. Reduction
17. combustion
18. hydrolysis
19. saponification
20. carbohydrates
21. monosaccharides
22. glycosidic bond
23. oligosaccharide
24. polysaccharide
25. peptide
26. polypeptide
27. protein
28. enzyme
29. nucleic acids
30. ribonucleic acid
31. deoxyribonucleic acid
32. purine bases
33. pyrimidine bases
34. complementary base-pairing

Answers to Preliminary Test

True-False

1. False. These are just different conformations of the same compound. See Section 24-3.
2. True. These are geometric isomers. See Sections 23-3 and 24-2.
3. False. Substituted cycloalkanes can exist as *cis* and *trans* isomers. Figures 24-1 and 24-2 show examples of this.
4. True. Remember that the term exothermic means that the process liberates heat. Another way of making the statement that appears in this question is "Most organic compounds burn." The production of heat energy is sometimes the reason we carry out these reactions.
5. True. They hydrolyze (react with water) to form the parent alcohol and a strong soluble base. This reaction is described in Section 24-4.
6. False. They burn very readily but otherwise are not particularly reactive, much like alkanes.
7. True. For example, they react with reactive metals to form strong bases and hydrogen gas.
8. True. Examples of classes of compounds so formed are esters, amides, and acyl halides. See Section 24-6.
9. True. See, for instance, the K_a values for common carboxylic acids, listed in Table 24-1.
10. True. See Section 24-5.
12. True. See Section 24-5.

Short Answer

1. structural isomers. See the definition at the beginning of Section 24-1.
2. stereoisomers. See the definition at the beginning of Section 24-2.
3. geometric isomers. See the description in the first part of Section 24-2.
4. chiral

5. racemic. Such a solution does not rotate a plane of polarized light because the equal and opposite effects of the two isomers exactly cancel.
6. conformations. These are *not* isomers. Be sure you know the distinction. Study Section 24-3.
7. CO_2; H_2O (the oxides of carbon and hydrogen)
8. increases; decreases. See the comparisons in Table 24-3.
9. alkoxides; hydrogen (H_2).
10. esters
11. saponification
12. primary; secondary
13. esterification (inorganic)
14. saponification, hydrolysis of an ester
15. oxidation, addition
16. oxidation, dehydrogenation
17. oxidation, dehydrogenation
18. oxidation
19. oxidation
20. oxidation (combustion, complete combustion)

Multiple Choice

1. (a). $CH_3CH_2CH_3$ can only be arranged one way.
2. (e). The addition of hydrogen (hydrogenation) converts unsaturated oil molecules to saturated fat molecules. See Section 23-17.
3. (d). See the ranking of base strengths at the end of Section 24-4.
4. (e). This method of forming acyl chlorides is described at the beginning of Section 24-6.
5. (c). Oxidation of a primary alcohol under the conditions stated produces an aldehyde. See Section 24-5, under the heading "Oxidation of Alcohol."
6. (c). This is a Brønsted-Lowry acid-base reaction involving the organic base methylamine and the strong acid HCl. See Section 24-4.

25 Coordination Compounds

Chapter Summary

Many compounds are produced by **coordinate covalent bond** formation. This is a process in which one or more atoms or groups of atoms provide pairs of electrons to be shared with another atom, forming a covalent bond *without* having the same number of electrons provided for sharing by each bonded atom. As you saw in Section 10-10, this process can be described in acid-base terms, using the Lewis acid-base theory. Transition metal ions have vacant d orbitals, as well as vacant s and p orbitals, so they often act as Lewis acids, reacting by coordinate covalent bond formation with molecules or ions that have unshared electron pairs. The compounds or polyatomic ions thus formed are referred to as **coordination compounds**, **coordination complexes**, or **complex ions**. Some representative metal ions can also bond in this way (e.g., Al^{3+}, Sn^{2+}, Sn^{4+}) but even these often use d orbitals in their formation of coordination compounds. In Chapter 25, you shall study some aspects of the formation, nomenclature, structures, and properties of coordination compounds.

Section 25-1 introduces this topic, with many examples to illustrate the importance and some of the history of coordination chemistry. Recall that the stability of these complexes with respect to dissociation is described by the value of K_d, the **equilibrium constant** for the dissociation reaction.

Several transition metals whose hydroxides are amphoteric can dissolve in an excess of strong soluble base. Ammonia, however, is such a weak base that it cannot dissolve most amphoteric hydroxides by the usual method of forming hydroxo complexes. Yet the hydroxides of some metals do dissolve in an excess of aqueous ammonia, to form soluble **ammine complexes** (Section 25-2).

Several important terms in coordination chemistry are defined and illustrated in Section 25-3. A molecule or ion that acts as the Lewis base in complex formation is called a **ligand**. Within a ligand, the atom that actually provides the electron pair is the **donor atom**. A ligand is termed **monodentate** or **polydentate** (or, alternately, **bidentate**, **tridentate**, and so on) to describe the number of different donor atoms that a ligand can use simultaneously. The metal atom or ion and its ligands (but not any uncoordinated ions) constitute the **coordination sphere**. The **coordination number** is the number of donor atoms (not ligands) to which the metal is bonded. Study the examples in Table 25-5, which illustrate the use of these terms.

Likewise, a knowledge of and careful adherence to the rules for naming coordination compounds (Section 25-4) along with much practice with examples and exercises, will enable you to write formulas from names and vice versa. Notice that the naming of the compound sometimes must specify the oxidation state of the metal. You should review the rules for assignment of oxidation state in polyatomic molecules and ions (Section 5-7). It will help you to remember that the sum of the oxidation states of atoms in a neutral molecule (and also a neutral ligand) is equal to *zero* and that the sum of the oxidation states of the atoms in a polyatomic ion (or in an ion acting as a ligand) is equal to the *charge* on that ion.

Once you know the coordination number of the metal in a coordination compound, you can use the **VSEPR theory** (do you recall what these letters mean?—see Section 8-2) to predict the arrangement of donor atoms in the coordination sphere—the structure of the complex. Review the VSEPR theory (Sections 8-5 through 8-12) before you study Section 25-5, which deals with the structures of coordination compounds. The heart of this discussion is Table 25-7, which is similar to Table 8-4 in the earlier applications of VSEPR theory. Please notice, though, that in coordination chemistry there may be different structures possible for the same coordination number, especially 4 (tetrahedral or square planar) and 5 (trigonal bipyramidal or square pyramidal). Each of the different geometries around the metal ion corresponds, of course, to a different set of hybrid orbitals.

One point of confusion for some students involves the way structural formulas of coordination compounds are represented. Look, for instance, at the representation of the hexachlorostannate(IV) ion, at the beginning of Section 25-1. The lines that are drawn from one Cl^- ligand to another (the blue parallelogram, a square drawn in perspective) are not bonds, but are drawn merely to help visualize the shape of the coordination sphere and the arrangement of the ligands. The lines drawn from each Cl^- to the tin(IV) ion are bonds (coordinate covalent bonds). Similar comments apply to the common representations of square planar, trigonal bipyramidal, and square pyramidal structures.

The next portion of the chapter, Sections 25-6 and 25-7, deals with the existence of **isomers**, which are substances having the same number and kinds of atoms, but arranged differently. There are several ways in which the atoms can differ in their arrangements, so several different classes of isomers exist. (Compare these with the types of isomers found in organic molecules, Sections 24-1 to 24-3.) These are broadly classified into two major categories. (1) **Structural (constitutional) isomers** involve differences in the bonding arrangement of different donor atoms on the same ligand or differences in bonding beyond a single coordination sphere. Section 25-6 breaks the topic of structural isomers into four subclasses—**ionization (ion-ion exchange) isomers**, **hydrate isomers**, **coordination isomers**, and **linkage isomers**—with several examples of each. (2) The differences among **stereoisomers** involve only one coordination sphere and the same ligands and donor atoms. The two subclasses of stereoisomers, known as **geometric (*cis-trans*) isomers** and **optical isomers** (or **enantiomers**) are described in Section 25-7. The following distinctions between structural isomers and stereoisomers may help you to recognize the broad categories. In structural isomers, the pattern or order of bonds is different, so you should be able to find at least one atom that is bonded to different atoms in the isomers. The rules already given for naming compounds serve to distinguish among various structural isomers. In stereoisomers, on the other hand, the order of bond linkages is the same, but the spatial arrangements of the atoms are different. If this is due to the atoms being linked in different relative positions on the central metal atom, the isomers are geometric isomers; if it is due only to the "handedness" of the entire arrangement, they are optical isomers. As you find out in your study of stereoisomerism, additional terms are used to describe such isomers. Be sure you understand and can use properly such terms, as *cis-*, *trans-*, **dextro-**, **levo-**, and **racemic**.

The rest of the chapter is concerned with the bonding in coordination compounds. Section 25-8 presents the very successful **crystal field theory**, illustrated with applications to complexes with coordination number 6. The crystal field theory takes the opposite view that the bonding is entirely (or at least predominantly) ionic and that the *d* orbitals of the metal, which in an isolated ion would be at the same energy, have their energies disturbed (**degeneracy** is removed) by the presence of the ligands. The amount by which these energies change, called the **crystal field splitting**, depends on the nature and arrangement of the ligands. These ideas are illustrated for the octahedral $[CoF_6]^{3-}$ and $[Co(CN)_6]^{3-}$ ions. The magnetic nature of a complex, that is, whether it is diamagnetic or paramagnetic (and the degree of paramagnetism), is often used to propose or verify the bonding description for that complex. As you study these sections, you should become familiar with the meanings and uses of such terms as e_g, t_{2g}, **crystal field splitting**, **low spin complex**, **high spin complex**, and Δ_{oct}.

The great variety of colors exhibited by transition metal coordination compounds can be interpreted in terms of the crystal field theory, as in Section 25-9. According to this interpretation, the different crystal field strengths of various ligands cause different e_g vs. t_{2g} splitting. Thus, to promote an electron from one d orbital to another, the various complexes must absorb light of different energies, that is, different colors. The common ligands are ranked in order of increasing crystal field strength, based on a study of the colors of light absorbed. This ordered list of ligands is called the **spectrochemical series**.

Study Goals

It may help you to write out a brief summary of the ideas of each Study Goal, summarizing the applicable portions of your class notes and text reading. Be sure to answer many of the suggested Exercises related to the Study Goals.

Section 25-1

1. Account for the tendency of the d-transition metals to form coordination compounds. *Work Exercises 1 and 3.*
2. Describe the experiments and reasoning of Werner that allowed him to determine which species of a complex were in the metal's coordination sphere, and therefore, what were the true formulas for coordination compounds. *Work Exercises 2, 9, and 10.*

Section 25-2

3. Write equations that account for the solubility of water-insoluble hydroxides in aqueous ammonia by formation of ammine complexes. *Work Exercises 16 through 19.*

Section 25-3

4. Distinguish among the terms ligand, donor atom, polydentate, monodentate, chelate complex, coordination number, and coordination sphere. *Work Exercises 4 through 8 and 11 through 15.*

Section 25-4

5. Know the names and formulas of common ligands as well as their donor atoms.
6. Know the general IUPAC rules for naming coordination compounds. Name coordination compounds if given their formulas, and supply formulas if given names. *Work Exercises 20 through 29.*

Sections 25-5

7. Sketch structures of coordination compounds of the types ML_2, ML_4 (square planar and tetrahedral), ML_5 (trigonal bipyramidal and square pyramidal), and ML_6. *Work Exercises 30 and 31.*

Sections 25-6 and 25-7

8. Distinguish between structural isomers and stereoisomers. *Work Exercise 38.*

9. Describe and give examples of the following kinds of isomers: (a) ionization, (b) hydrate, (c) coordination, (d) linkage, (e) geometric, and (f) optical. Name the isomers, using appropriate prefixes where necessary. *Work Exercises 32 through 37 and 40 through 43.*

Section 25-7

10. Describe and relate the following: (a) *dextro* and *levo* isomers, (b) racemic mixture, (c) plane polarized light, (d) polarimeter, and (e) optical activity. *Work Exercise 39.*

Section 25-8

11. Draw and distinguish between the t_{2g} and e_g sets of orbitals. *Work Exercise 47.*

12. Describe with words and diagrams the crystal field theory as it applies to octahedral coordination compounds. Explain the reversal in relative energies of t_{2g} and e_g orbitals for the two cases. *Work Exercise 44.*

13. Distinguish between and give examples of low spin and high spin complexes. Explain the relationship between electron pairing energy and Δ_{oct} with respect to these kinds of complexes. *Work Exercises 45 through 51.*

Sections 25-8 and 25-9

14. Describe what is meant by the terms "strong field ligand" and "weak field ligand." Describe the spectrochemical series and its relationship to the colors of coordination compounds. *Work Exercises 47 and 48.*

General

15. Answer conceptual questions based on this chapter. *Work* **Conceptual Exercises** *52 through 57.*

16. Apply concepts and skills from earlier chapters to the ideas of this chapter. (Review appropriate sections of Chapters 2, 3, 4, 5, 8, 10, and 20.) *Work* **Building Your Knowledge** *exercises 58 through 65.*

17. Exercises at the end of chapter direct you to sources outside the textbook for information to use in solving them. Work **Beyond the Textbook** *exercises 66 through 71.*

Some Important Terms in This Chapter

Some important terms from Chapter 25 are listed below. Fill the blanks in the following paragraphs with terms from the list. Each term is only used once. Study other new terms and review terms from preceding chapters if necessary.

coordinate covalent bond	e_g orbitals	optical isomers
coordination compound	geometric isomers	spectrochemical series
or complex	high spin complex	stereoisomers
coordination isomers	hydrate isomers	strong field ligand
coordination number	ionization isomers	structural (constitutional)
coordination sphere	ligand	isomers
crystal field theory	linkage isomers	t_{2g} orbitals
donor atom	low spin complex	weak field ligand

If a Lewis base donates a share in an electron pair and a Lewis acid accepts it, a [1]_____ is formed. When the Lewis acid involved in the coordinate covalent bond is a *d*-transition metal ion that has empty *d* orbitals to accept shares of electron pairs, a [2]_____ may be formed. The Lewis base in this complex is called a [3]_____. An atom of the ligand that donates a share in an electron pair to a metal is called a [4]_____. The number of these donor atoms that coordinate with the metal atom or ion in the complex is referred to as the [5]_____. The metal atom or ion plus the ligands surrounding it make up the [6]_____ (not to be confused with the *coordination number*).

Isomers are different compounds that have the same molecular formula. They can be divided into two major categories: (a) [7]_____, which have different bonds, involving either more than one coordination sphere or different donor atoms on the same ligand (and, therefore, different structures); and (b) [8]_____, which have the same chemical bonds, but different spatial arrangements. There are four types of structural isomers: (a) [9]_____, also known as ion-ion exchange isomers, which result from the exchange of ions inside and outside the coordination sphere; (b) [10]_____, which differ in the location of water molecules inside or outside the coordination sphere; (c) [11]_____, which involve exchange of ligands between a complex cation and a complex anion of the same coordination compound; and (d) [12]_____, in which a ligand bonds to a metal ion through different donor atoms. The two kinds of stereoisomers are: (a) [13]_____ _____, also known as *cis-trans* isomers, in which ligands are arranged in different ways on two sides of a rigid structure; and (b) [14]_____, or enantiomers, which are nonsuperimposable mirror images of each other.

The theory of bonding in transition metal complexes that treats metals ions and ligands as point charges is the [15]_____, a purely ionic model. According to crystal field theory, the $d_{x^2-y^2}$, and d_{z^2} orbitals, which are directed along the *x*, *y*, and *z* axes, are referred to as [16]_____. The d_{xy}, d_{yz}, and d_{xz} orbitals, which lie between the axes, are called the [17]_____. As the ligand donor atoms approach the metal ion along the axes to form octahedral complexes, the electrons on the ligands repel electrons in e_g orbitals more strongly than they repel those in t_{2g} orbitals, separating the two sets of orbitals into two levels of energy. The energy separation between the two sets is *crystal field splitting energy*,

Δ_{oct}, which is proportional to the *crystal field strength* of the ligands, how strongly the ligand electrons repel the electrons on the metal ion. Small values for the crystal field splitting energy are caused by [18]_____, such as F^-, and large splitting energies result from [19]_____, such as CN^-. A certain amount of energy is required to pair electrons by bringing two negatively charged particles into the same region of space, the *electron pairing energy* (*P*). Hund's Rule (Section 4-18), a helpful guide in determining the electron configurations of atoms, is a consequence of the pairing energy (*P*). If the energies of the *d* orbitals are not split by the ligands, each of the orbitals would be occupied singly before any pairing occurred. Similarly, if the splitting energy (Δ_{oct}) is *smaller* than the pairing energy (*P*), the electrons occupy all five orbitals singly before any pairing. Then, once all the *d* orbitals are half-filled, additional electrons pair with electrons in the t_{2g} orbitals. Consequently, the complex has the same number of *unpaired* electrons on the metal atom or ion as when the metal is uncomplexed, which is *higher* than it would be otherwise, and the complex is called a [20]_____. Conversely, if the splitting energy is *greater* than the pairing energy, the electrons require less energy to pair in the t_{2g} orbitals than to occupy the higher energy e_g orbitals. More electrons are paired than on the uncomplexed metal atom or ion, so a *low* number are unpaired, and the complex is called a [21]_____
_____. Thus, a weak field ligand, which produces a small splitting energy, tends to yield a high spin complex, and a strong field ligand, which produces a large splitting energy, tends to yield a low spin complex. One of the physical evidences of crystal field splitting is the color of most transition metals compounds, which is produced by electronic transitions from t_{2g} orbitals to e_g orbitals. The wavelength of the color observed is directly related to the value of the splitting energy (Δ_{oct}) which is determined, in turn, by the field strength of the ligands. Weak field ligands, which produce small crystal field splittings and high spin complexes, absorb lower-energy (longer-wavelength) light. Strong field ligands, which produce large crystal field splittings and low spin complexes, where possible, usually absorb higher-energy (shorter-wavelength) light. By examining the visible spectra of many complexes, chemists have been able to arrange common ligands in order of increasing crystal field strengths:

$$I^- < Br^- < Cl^- < F^- < OH^- < H_2O < (COO)_2{}^{2-} < NH_3 < en < NO_2{}^- < CN^-$$

This list is called the [22]_____.

Preliminary Test

True-False

_____ 1. The higher the value of K_d for a complex, the more stable the complex is.

2. The properties of transition metal ions that enable them to form coordination complexes are their magnetism and their colors.

_____ 3. Whenever a metal is bonded to four donor atoms in a coordination complex, the coordination geometry must be tetrahedral.

_____ 4. Even though the atoms are arranged differently in geometric isomers, the properties of these compounds are alike.

_____ 5. When we say that a substance is exhibiting optical activity, we mean that it is colored.

_____ 6. Ligands must be neutral molecules.

_____ 7. Coordination complexes in which the metal is bonded to six donor atoms have octahedral coordination geometry.

_____ 8. The crystal field theory can be applied to complexes only when they occur in crystals.

_____ 9. In an isolated metal atom or ion, the five d orbitals with the same principal quantum number all have the same energy.

_____ 10. When a metal atom or ion is bonded to several ligands, the five metal d orbitals with the same principal quantum number all have the same energy.

_____ 11. The ion Fe^{3+} is referred to as a d^5 ion.

_____ 12. In the crystal field theory description of coordination compounds, ligands that are more electronegative lead to smaller crystal field splitting.

_____ 13. All octahedral complexes of metal ions with d^5 configuration are low spin complexes.

Short Answer

1. A molecule or ion that provides an electron pair for coordinate covalent bond formation is called a Lewis _____; a molecule or atom that accepts the electron pair is called a Lewis _____.

2. The name of the chemist who was responsible for much of the early development of the ideas of coordination chemistry was _____.

3. The molecules or ions that act as Lewis bases in the formation of coordination compounds are called _____.

4. Ligands that can bond through only one donor atom at a time are called _____.

5. Ligands that can bond through two or more donor atoms simultaneously are called _____.

6. A molecule or ion that can act as a ligand in a coordination complex usually has at least one _____.

7. The portion of a compound consisting of the metal ion and its coordinated ligands, but no uncoordinated ions, is called the _____.

_____

For each of the next fifteen questions, 8 through 22, provide either the formula or the name of the ligand, as required, and classify each ligand as monodentate, bidentate, and so on. When a ligand name is required, be sure that you name it as a ligand, and not as a free molecule or ion. Dot formulas are not required.

Formula	Name	Classification
8. NH_3	_____	_____
9. _____	cyano	_____
10. _____	chloro	_____
11. $H_2NCH_2CH_2NH_2$	_____	_____

369

12. _____   diethylenetriamine (dien)   _____

13. PH_3 _____   _____

14. _____   carbonyl   _____

15. Br^- _____   _____

16. SO_4^{2-} _____   _____

17. _____   oxo   _____

18. _____   hydroxo   _____

19. $^{2-}O_2C-CO_2^{2-}$ _____   _____

20. OH_2 _____   _____

21. NO_2^- (bonded
    through the N)   _____   _____

22. _____   nitrato   _____

23. A ligand that can bond through three donor atoms simultaneously is called _____.

24. When a metal has a coordination number of 2 in a complex, the coordination geometry is _____.

25. When a metal has a coordination number of 6 in a complex, the coordination geometry is _____.

26. The two mirror image forms of a molecule are called _____ of one another; one is identified by the term _____, while the other is called _____.

27. The number of geometric isomers possible for the square planar metal complex with formula MX_2Y_2 is _____. (Note: M is the metal and X and Y are each monodentate ligands.)

28. The number of geometric isomers possible for the tetrahedral metal complex with formula MX_2Y_2 is _____. (Note: The symbols have the same meaning as in Question 27.)

For each of the next twelve questions, 29 through 40, provide either the formula or the name of the complex. For each complex, give the oxidation state (Ox. State) of the metal and the coordination number (Coord. No.) of the metal. In naming, do not worry about stereoisomerism, unless sufficient information is explicitly shown.

	Formula	Name	Ox. State	Coord. No.
29.	$[Cu(en)Br_2]$	_____	____	____
30.	_____	hexamminecobalt(III) ion	____	____
31.	_____	hexacyanocobaltate(III) ion	____	____
32.	$[Fe(CO)_5]$	_____	____	____

33. $[CuCl_5]^{3-}$ _____

34. _____ cis-diamminedichloroplatinum(II) ____ ____

35. _____ trans-diamminedichloroplatinum(II) ____ ____

36. $[Cr(OH_2)_2Br_2(NH_3)_2]^+$ _____ ____ ____

37. _____ cis-tetraamminediaquacobalt(III) ion ____ ____

38. _____ trans-tetraamminediaquacobalt(III) ion ____ ____

39. _____ tetraamminedichloroplatinum(IV) chloride ____ ____

40. _____ triamminetrichloroplatinum(IV) chloride ____ ____

---

41. The name of the complex ion $[AlCl_6]^{3-}$ is _____; in this complex, the oxidation state of aluminum is _____.

42. The name of the theory of bonding in coordination compounds that is discussed in this chapter is the _____ theory.

43. An experiment that is used to distinguish between inner and outer orbital complexes is to determine how _____ the complex is.

44. According to the crystal field theory, the five originally equivalent d orbitals on a metal become nonequivalent under the influence of the array of ligands. If there are six ligands in an octahedral arrangement, these d orbitals then occur in two groups; the two groups are called, collectively, the _____ orbitals (a set of three orbitals) and the _____ orbitals (a set of two orbitals). For this octahedral arrangement of ligands, the _____ set is at lower energy, with the two sets being separated by an amount of energy designated _____.

45. According to the crystal field theory, the crystal field splitting is larger for ligands that are more able to _____ into vacant metal orbitals; for octahedral complexes, this is expressed quantitatively as a _____ value of Δ_{oct}.

46. Low spin octahedral complexes exist only for metal ions with configurations ____, ____, ____, and ____.

47. The elements whose hydroxides dissolve in excess aqueous ammonia by the formation of soluble ammine complexes are the metals of the _____, _____, _____, and _____ families.

48. According to the crystal field description of bonding, the color that an octahedral complex of a particular metal exhibits depends on the value of _____, which in turn depends on the _____ of the ligands.

49. The ranking of a set of ligands according to increasing crystal field strength, as determined by the study of the colors of several of their complexes, is called the _____.

Multiple Choice

_____ 1. What is the coordination number of cobalt in $[Co(NH_3)_6]^{3+}$?
  (a) 2   (b) 3   (c) 4   (d) 6   (e) 9

_____ 2. What is the coordination number of cobalt in $[Co(NH_3)_6]Cl_3$?
  (a) 2   (b) 3   (c) 4   (d) 6   (e) 9

_____ 3. What is the oxidation state of cobalt in $[Co(NH_3)_6]^{3+}$?
  (a) 0   (b) +2   (c) +3   (d) +4   (e) +6

_____ 4. What is the coordination number of cobalt in $[Co(en)_2Cl_2]^{+}$?
  (a) 2   (b) 3   (c) 4   (d) 6   (e) 9

_____ 5. Which one of the following types of isomers is not an example of structural isomerism?
  (a) ionization isomers   (b) optical isomers   (c) hydrate isomers
  (d) hydrate isomers   (e) coordination isomers

_____ 6. What is the number of geometric isomers possible for the complex $[Pt(NH_3)_2Cl_4]$?
  (a) 1   (b) 2   (c) 3   (d) 4   (e) 6

_____ 7. What is the number of geometric isomers possible for the complex $[Pt(NH_3)_3Cl_3]^{+}$?
  (a) 1   (b) 2   (c) 3   (d) 4   (e) 6

_____ 8. Which of the following is correctly referred to as a d^8 ion?
  (a) Ti^{3+} (b) Ru^{3+} (c) Ni^{2+} (d) Ag^{+} (e) Cr^{3+}

Answers to Some Important Terms in This Chapter

1. coordinate covalent bond
2. coordination compound
  or complex
3. ligand
4. donor atom
5. coordination number
6. coordination sphere
7. structural (constitutional) isomers

8. stereoisomers
9. ionization isomers
10. hydrate isomers
11. coordination isomers
12. linkage isomers
13. geometric isomers
14. optical isomers
15. crystal field theory

16. e_g orbitals
17. t_{2g} orbitals
18. weak field ligands
19. strong field ligands
20. high spin complex
21. low spin complex
22. spectrochemical series

Answers to Preliminary Test

Do not just look up answers. *Think about the reasons for the answers.*

True-False

1. False. Remember that K_d is the equilibrium constant for the *dissociation* of the complex. A higher value for K_d means that the numerator (product concentration) is higher, which

would correspond to *less* stable complexes. Section 20-6, part 3, discusses these dissociation constants in quantitative terms.

2. False. They form complexes so readily because of the availability of vacant *d* orbitals. Properties such as magnetism and color are a *result* of the bonding in the complexes, not a *cause* of that bonding.

3. False. Many tetracoordinate metal atoms form square planar complexes.

4. False. A different arrangement of atoms gives different properties. Many examples of this are given in the text chapter, especially regarding the color of the complex.

5. False. Read the second part of Section 25-7.

6. False. Many common ligands are anions, and a few are cations.

7. True

8. False. The term "crystal field" has to do with the symmetry of the arrangements of the ligands and the resulting effects on symmetry of the metal orbitals. Similar symmetry arrangements frequently occur in crystals.

9. True

10. False. In the terminology of the crystal field theory, the degeneracy is removed by the arrangement of the ligands around the metal. In the terminology of the valence bond theory, some of the *d* orbitals become involved, along with *s* and *p* orbitals, in the formation of hybrid orbitals.

11. True. Fe^0 has the outer electronic configuration $4s^2 3d^6$; on formation of the Fe^{3+} ion, iron has the configuration $4s^0 3d^5$.

12. True. See the discussion of $[Co(CN)_6]^{3-}$ versus $[CoF_6]^{3-}$ in Section 25-8.

13. False. They can be either high spin or low spin, depending on the nature of the ligands. See Section 25-8.

Short Answer

1. base; acid. Review Sections 10-10 and 25-1.

2. Alfred Werner. In fact, for many years coordination complexes were commonly called Werner complexes. The amminechloroplatinum complexes described in Section 25-1 are now often referred to as Werner complexes.

3. ligands

4. monodentate

5. polydentate

6. unshared pair of electrons

7. coordination sphere

8. ammine; monodentate

9. CN^- (with the carbon acting as the donor atom); monodentate

10. Cl^-; monodentate

11. ethylenediamine (or en); bidentate

12. $NH_2-CH_2-CH_2-\underset{\underset{H}{|}}{N}-CH_2-CH_2-NH_2$; tridentate

13. phosphine; monodentate

14. CO (with carbon as the donor atom); monodentate

15. bromo; monodentate

16. sulfato; monodentate
17. O^{2-}; monodentate (M=O) or bidentate (M–O–M)
18. OH^-; monodentate
19. oxalato; bidentate
20. aqua; monodentate
21. nitro; monodentate
22. NO_3^- (bonded through an oxygen); monodentate
23. tridentate
24. linear
25. octahedral
26. optical isomers (or enantiomers); *dextro*; *levo*
27. 2
28. 1. No matter how you arrange the ligands, the two X ligands will be adjacent, as will be the two Y ligands.
29. dibromoethylenediaminecopper(II); +2; 4
30. $[Co(NH_3)_6]^{3+}$; +3; 6
31. $[Co(CN)_6]^{3-}$; +3; 6
32. pentacarbonyliron(0); 0; 5
33. pentachlorocuprate(II) ion; +2; 5. Note the use of the "-ate" ending when the total complex ion has a negative charge.
34. 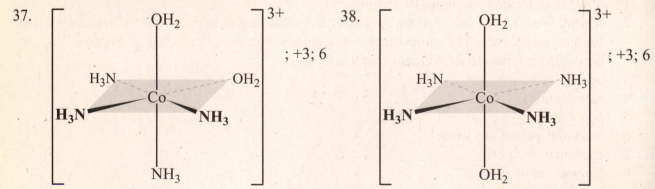 ; +2; 4

35. ‖ EMBED
   ChemDraw.Document.6.0 ; +2; 4

36. Diamminediaquadibromochromium(III) ion; +3; 6. Notice the application of the rule about alphabetizing ligand names. There is no particular convention about the order in which the ligands are shown in the formula.

37. ; +3; 6

38. ; +3; 6

39. $[Pt(NH_3)_4Cl_2]Cl_2$; +4; 6
40. $[Pt(NH_3)_3Cl_3]Cl$; +4; 6
41. hexachloroaluminate ion; +3. The "-ate" ending is used because the complex ion is negatively charged. You do not specify the oxidation state of aluminum by including "(III)" in the name, because this is the only oxidation state that aluminum has in compounds.
42. crystal field

43. paramagnetic
44. t_{2g} ; e_g ; t_{2g} ; Δ_{oct}
45. donate electrons; larger
46. d^4; d^5; d^6; d^7
47. cobalt; nickel; copper; zinc
48. Δ_{oct}; crystal field strength. This is discussed in some detail in Section 25-9.
49. spectrochemical series

Multiple Choice

1. (d)
2. (d). The coordination sphere is enclosed in the square brackets so you know that the chloride ions are not bonded to cobalt. Only the nitrogen atoms of the six ammonia molecules (ammine ligands) are bonded to cobalt. This contains the complex ion in Question 1.
3. (c). Each ammine ligand has a total oxidation state equal to zero (it's neutral). The sum of oxidation states in a polyatomic ion must be equal to the charge on the ion, so cobalt must be in an oxidation state of +3.
4. (d). Each (en) ligand is bidentate, containing two donor atoms.
5. (b). Optical isomers are a type of stereoisomer.
6. (b). Complexes of the type MA_4B_2 can exist in two isomeric forms, *cis* and *trans*.
7. (b). Complexes of the type MA_3B_3 can exist in two isomeric forms, *mer* (for *meridian*) and *fac* (for *facial*).
8. (c). Remember that when transition metal atoms form ions the two *s* electrons are lost first, after which some *d* electrons may be lost also.

26 Metals I: Metallurgy

Chapter Summary

Earlier in the textbook, the emphasis was on developing a fundamental understanding of chemistry through broad concepts and basic principles. In recent chapters, however, you have studied in more detail the specific reactions and properties of various elements and groups of elements—the general area of chemistry called **descriptive chemistry.** As you study the rest of the chapters, keep in mind the basic principles that you have studied previously. Go back whenever necessary to review appropriate earlier chapters or sections. In this way, the principles you have learned will become more meaningful and useful as you see how they relate to dozens of specific properties, reactions, and processes. For instance, in this chapter you use the ideas of oxidation states, redox reactions, anions and cations, acidity and basicity, amphoterism, basic salts, densities, melting points, solubility rules, crystallization, melting point depression, electrolysis, activity series, relative thermodynamic stability, K_{sp} values, and complex ion formation. In your study of each of these descriptive chemistry chapters, it will probably help you to organize the material in your own words, emphasizing the general reactions or processes of the chapter, and especially the similarities and differences of the various elements or compound types considered.

In Chapter 26, you will study metals from several aspects—the importance of metals and other trace elements in life processes, the distinctions between properties of metals and nonmetals, the occurrence of metals, the methods of obtaining and using free metals, and conservation of metals. You should recall (see Study Goal 1) the general distinguishing properties, both physical and chemical, of **metals** and **nonmetals.** Review (Chapter 5) the locations in the periodic table of general classes of elements and be familiar with the terminology of representative elements, transition elements, and so on. As pointed out in Section 26-1, some metals have sufficiently high positive reduction potentials (see Section 21-16 for a review of what this means) that they can occur in nature in the uncombined **native state.** Most metals are so easily oxidized that they are found in nature only in compounds. When this is true, they are always in positive oxidation states and methods for obtaining the free metals must involve **reduction.** The terms **ore, mineral**, and **gangue**, which will be used many times later in the chapter, are introduced in this section. We can classify ores according to the anion associated with the metal. Methods of extraction of the metals depend on which compounds of the metal are present (Table 26-1). This also depends on the solubilities of the compounds. Review the solubility guidelines, Section 6-1, part 5 and Table 6-4.

The next portion of the chapter, Sections 26-2 through 26-4, is concerned with the general principles and techniques of **metallurgy**—the commercial purification and use of free metals. The central chemical process in metallurgy is often one of reduction of the metal from a positive oxidation state in a compound. However, metallurgy also involves separating the components of very complex heterogeneous mixtures. As described in Section 26-2, **pretreatment** of ores to remove the gangue (pronounced "gang") may take advantage of differences in densities and hydrophilic properties (as in the **flotation** method), or converting the metal compound to a more easily reduced form, often the oxide (as in the **roasting** of sulfides), or both. A second stage of the metallurgical process is the reduction of the metal ores to the free metals (Section 26-3). This can be done in a variety of ways, either by oxidation of the anion away from the metal as in the roasting of cinnabar to produce mercury, by chemical reduction of the ore (which is often an oxide at this stage) with coke or carbon monoxide, by chemical reduction with hydrogen or another reducing agent, or by electrolytic reduction. The general reduction processes are summarized in Table 26-2, with specific applications cited in Table 26-3. The reduction process provides impure metals, which are then **refined** or purified by the methods described in Section 26-4. Metals are often used in the form of **alloys**, in which they are mixed with other elements to give desired changes in their physical and chemical characteristics.

Sections 26-5 through 26-9 discuss in detail the metallurgies of five important metals—magnesium, aluminum, iron, copper, and gold. As you study these sections, remember that the metallurgical process described for each metal must involve both (1) processes by which compounds eventually produce the free metals and (2) removal of other unwanted substances, including unwanted components of the ores and the other products of the reduction reactions and other chemical processes. One thing you should recognize is that tremendous amounts of energy are used for many of these processes, both for melting huge amounts of material and sometimes for electrolytic reduction.

The Chemistry in Use essay at the end of the chapter discusses some important practical aspects of the utilization, conservation, and recycling of metals.

Study Goals

Review Sections 4-10, 5-8, 5-9, and 10-11; Introduction to Chapter 26
1. Summarize the differences in physical and chemical properties of metals and nonmetals. *Work Exercise 1.*

Sections 26-1 through 26-4
2. Relate the electronic structures of metals to their tendencies to be found combined or uncombined in nature and to the methods that must be used to obtain them from their ores. *Work Exercises 2 through 6.*

Section 26-1
3. Know the anions that are commonly found combined with metals in ores and give examples of ores containing each. *Work Exercise 8.*

Sections 26-2 through 26-4
4. Describe the general sequence of steps used in obtaining metals from their ores. Discuss and illustrate these processes in detail for specific metals. Write balanced equations for important reactions. *Work Exercises 7, 9 through 17, and 19 through 21.*

Sections 26-5 through 26-9
5. Describe in detail the separation processes and chemical reactions involved in the metallurgies of Mg, Al, Fe, Cu, and Au. *Work Exercises 22 through 41.*

General (Review Chapters 2, 3, 15, and 21)

6. Answer conceptual questions based on this chapter. *Work* **Conceptual Exercises** *42 through 45.*

7. Carry out calculations relating metallurgy to the concepts of earlier chapters. *Work* **Building Your Knowledge** *exercises 46 through 58.*

8. Exercises at the end of chapter direct you to sources outside the textbook for information to use in solving them. Work **Beyond the Textbook** *exercises 59 through 63.*

Some Important Terms in This Chapter

Some important new terms from Chapter 26 are listed below. Fill the blanks in the following paragraphs with terms from the list. Each term is only used once. Check the Key Terms list and the chapter reading. Study other new terms, and review terms from preceding chapters if necessary.

alloying	metallurgy	roasting
cast iron	native ore	slag
flotation	ore	smelting
gangue	pig iron	steel
Hall-Héroult process	refining	zone refining

Some of the first elements to be discovered by ancient people were metals of low chemical activity (positive reduction potentials), which are often found in the uncombined free state as [1]_____. Other metals are found in the earth as compounds (minerals) contained within a natural deposit (an [2]_____), mixed with large amounts of sand, soil, clay, rock, and other material, which are collectively called [3]_____.

The science of commercially extracting metals from their ores and preparing them for use is known as [4]_____. Metallurgists use different methods to treat ores, depending on the ore. For example, hydrophilic and hydrophobic particles, which differ in their attraction to water, can be separated by the [5]_____ process. Some sulfides can be converted to oxides by [6]_____, which is heating below the melting point of the sulfide in the presence of oxygen from air. Similarly, there are various methods for obtaining the pure metal from the ore. Reduction of metal ores to the free metals, called [7]_____, depends upon the strength of the bonds in the mineral. During smelting, much of the unwanted gangue is eliminated as [8]_____, often in the form of silicates. Once the elemental, but impure, metal is obtained, it must undergo [9]_____, or purifying. When extremely pure metals are needed, they may be purified by [10]_____, in which a bar of the metal is passed through an induction heater, causing impurities to move along in the melted portion. After refining, a metal may be mixed with other substances (usually other metals) to modify its properties, a process known as [11]_____.

Aluminum is such an active metal that to obtain the elemental metal, its oxide must be reduced by electrolysis. The melting point is reduced by adding cryolite (a mixture of NaF and AlF_3), and the molten mixture is electrolyzed at 1000°C with carbon electrodes, a procedure known as the [12]_____.

Iron is an especially important metal. The iron obtained directly from smelting in a blast furnace, known as [13]_____, contains impurities, such as carbon. If it is remelted, run into molds, and cooled, it becomes [14]_____, which is brittle because it contains much iron carbide, Fe_3C. Pure iron is silvery in appearance, quite soft, and of little practical use. If some carbon is left and other metals (Mn, Cr, Ni, W, Mo, and V) are added, the mixture becomes stronger and is known as [15]_____.

Preliminary Test

As in other chapters, this test will check your understanding of basic concepts and types of calculations. Be sure to answer *many* of the additional textbook exercises.

True-False

Mark each statement as true (T) or false (F).

_____ 1. There are more metallic elements than nonmetallic elements.
_____ 2. Metallic character decreases with increasing numbers of electrons in the valence shell.
_____ 3. For elements with the same number of electrons in the valence shell, metallic character decreases with increasing numbers of electron shells.
_____ 4. Metals usually have higher electrical conductivity than nonmetals.
_____ 5. Most metals are solids at room temperature.
_____ 6. Metals are usually fairly electronegative.
_____ 7. Free metals are often good oxidizing agents.
_____ 8. Most metals are found free in nature.
_____ 9. Roasting of sulfide ores is one of the most environmentally clean methods presently used in metallurgy.
_____ 10. The metallurgy of all metals involves reduction of a compound to the free metal.
_____ 11. Gold often occurs in the native state, so all that is ever done to recover it is to carry out a mechanical separation from the other components of the ore.

Short Answer

Answer with a word, a phrase, a formula, or a number with units as necessary.

1. When the oxides and hydroxides of metals dissolve in water, they give solutions that are _____ or _____.
2. The most common ores of metals, except those in which the element occurs in the native state, have the metal combined with one or more of the following anions: _____, _____, _____, _____, _____, _____.
3. The concentrating of an ore before the reduction stage is referred to as _____ of the ore.
4. In the pretreatment of ores in the cyclone separator, the property that is used to effect the separation is differences in _____.

5. The pretreatment technique in which hydrophobic materials are separated from other components of a mixture by suspending them in water is called the _____ method; this method often involves coating the particles with _____ and blowing a stream of _____ through the suspension.

6. In the pretreatment process known as roasting, metal sulfides are converted to _____ by _____.

7. An example of a metal that can be obtained from the sulfide ore by the simple process of roasting is _____; the sulfide ore from which it is obtained is called _____ and has the formula _____.

8. The reduction of metal compounds to the free metal, especially by chemical means, is called _____.

9. The formula of a carbonate that is converted to the oxide by heating to drive off carbon dioxide is _____.

10. The very active metals are usually reduced by _____ methods.

11. The coke that is used in reduction processes in metallurgy is _____.

12. An example of a reaction in which hydrogen is used as the reducing agent in the production of a metal from an oxide is _____.

13. The name of the common ore of aluminum is _____. The commercial process most widely used for the production of aluminum from this ore is called the _____ process.

14. The relatively large amount of sand, soil, clay, rock, and other material that accompanies the minerals in ores is called _____.

15. In the blast furnace process for the metallurgy of iron, limestone is added with the crushed ore. The purpose of this limestone (the flux) is to react with the silica in the gangue to form calcium silicate, which melts at these high temperatures. This molten material is referred to as _____.

16. Two metals that are commercially produced by the electrolytic reduction of the molten chloride salts are _____ and _____.

17. In the commercial electrolysis cell for refining of metals, the purified copper is deposited at the _____.

18. In the commercial electrolytic process for the refining of copper, the net cell reaction is _____.

Each of the following equations, Questions 19 through 32, represents a reaction that was discussed in Chapter 26. Each reaction is used in the production of some metal. For each equation, tell what metal or metals M could represent. The possible metals are Al, Au, Ca, Cu, Fe, Hg, Mg, Na, Ni, Pb, and Zn.

19. $MCO_3(s) \xrightarrow{\text{heat}} MO(s) + CO_2(g)$

20. $M(OH)_2(s) \xrightarrow{\text{heat}} MO(s) + H_2O(g)$

21. $2MS(s) + 3O_2(g) \xrightarrow{\text{heat}} 2MO(s) + 2SO_2(g)$

22. $M_2S(s) + O_2(g) \xrightarrow{\text{heat}} 2M(s) + SO_2(g)$

23. $MS(s) + O_2(g) \xrightarrow{heat} M(\ell) + SO_2(g)$

24. $MO_2(s) + 2C(s) \xrightarrow{heat} M(\ell) + 2CO(g)$

25. $MO(s) + C(s) \xrightarrow{heat} M(s) + CO(g)$

26. $M_2O_3(s) + 3CO(g) \xrightarrow{heat} 2M(\ell) + 3CO_2(g)$

27. $M_2O_3(s) + 3C(s) \xrightarrow{heat} 2M(\ell) + 3CO(g)$

28. $2M_2O_3(\ell) \xrightarrow[electrolysis]{heat} 4M(\ell) + 3O_2(g)$

29. $2MCl(\ell) \xrightarrow[electrolysis]{heat} 2M(\ell) + Cl_2(g)$

30. $2MCl_3(\ell) \xrightarrow[electrolysis]{heat} 2M(\ell) + 3Cl_2(g)$

31. $4M(s) + 8CN^-(aq) + O_2(g) + 2H_2O(\ell) \rightarrow 4[M(CN)_2]^-(aq) + 4OH^-(aq)$

32. $Ca(OH)_2(s) + M^{2+}(aq) \rightarrow Ca^{2+}(aq) + M(OH)_2(s)$

Multiple Choice

_____ 1. Which one of the following metals is most likely to occur in the native state?
   (a) platinum   (b) sodium   (c) magnesium (d) iron   (e) mercury

_____ 2. Which one of the following metals is produced in large quantities by treatment of its salts found dissolved in sea water?
   (a) sodium   (b) magnesium (c) mercury   (d) gold   (e) iron

_____ 3. Which one of the following is *not* a method for reduction of ores?
   (a) electrolysis   (b) reaction with coke   (c) reaction with hydrogen
   (d) roasting   (e) zone refining

_____ 4. Which one of the following metals is produced primarily by electrolytic reduction of its molten oxide?
   (a) gold   (b) iron   (c) copper   (d) platinum   (e) magnesium

_____ 5. The purification process known as zone refining depends on the fact that
   (a) many impurities will not fit into the lattice of a pure crystalline substance.
   (b) different substances have different densities.
   (c) some metals are more easily reduced than others.
   (d) some salts are soluble in water.
   (e) some particles are wet by oil but not by water.

_____ 6. Which one of the following metals finds use in lightweight structural alloys because of its low density?
   (a) gold   (b) iron   (c) copper   (d) platinum   (e) magnesium

_____ 7. Purification of which one of the following metals involves precipitation of a slightly soluble hydroxide?

(a) gold　　　　(b) iron　　　　(c) copper　　　　(d) platinum　　　　(e) magnesium

_____ 8. In the metallurgy of iron, the metal obtained from the separation of the slag and the molten metal is called

(a) gold　　　　(b) iron　　　　(c) copper　　　　(d) platinum　　　　(e) magnesium

_____ 9. The purpose of adding sand to the roasted copper ore in the reverberatory furnace is to

(a) oxidize the copper oxide to copper metal.

(b) provide structural strength to the copper metal.

(c) aid in the melting for the subsequent electrolytic refining of copper.

(d) react with the limestone impurity to form a molten silicate glass.

(e) cause the metal to conduct so it can be electrolyzed.

Answers to Some Important Terms in This Chapter

1. native ore
2. ore
3. gangue
4. metallurgy
5. flotation

6. roasting
7. smelting
8. slag
9. refining
10. zone refining

11. alloying
12. Hall-Héroult process
13. pig iron
14. cast iron
15. steel

Answers to Preliminary Test

Do not just look up answers. *Think about the reasons for the answers.*

True-False

1. True. See Chapter 4 and Section 26-1.
2. True. Remember that metallic character decreases going left to right across a period in the periodic table.
3. False. Remember that metallic character increases going down a group in the periodic table. If you need help to remember this trend, look at Group 4A, which goes from carbon (a nonmetal) through silicon and germanium (metalloids) to tin and lead (metals). Review Section 4-10.
4. True
5. True. Mercury is a liquid at room temperature, and cesium and gallium melt slightly above room temperature. All other metals have higher melting points.
6. False. To be quite electronegative means to have a tendency to gain electrons; metals tend to lose electrons, having low ionization energies.
7. False. Because metals tend to lose electrons, they are easily oxidized; hence, they act as reducing agents.

8. False. Only a few are sufficiently inactive that they are found in the native state. Can you name some of these metals?
9. False. Read Section 26-2 regarding the SO_2 formed in the process.
10. False. Some metals occur in the free or native state, and require no reduction.
11. False. As described in Section 26-9, additional gold can be recovered by amalgamation with mercury followed by distillation of the mercury, or by oxidation in the presence of cyanide ion to form the soluble complex ion $[Au(CN)_2]^-$, followed by electrolytic reduction.

Short Answer

1. basic; amphoteric
2. oxide; sulfide; chloride; carbonate; sulfate; silicate (sometimes phosphate). (These answers could appear in any order.)
3. pretreatment
4. density
5. flotation; oil; air
6. oxides; heating in the presence of oxygen
7. mercury; cinnabar; HgS
8. smelting. In such processes, the metal is often obtained in molten (liquid) form.
9. $CaCO_3$. This conversion is used both in the production of calcium (Section 26-3) and to produce CaO for use in the production of magnesium (Section 26-5).
10. electrochemical (or electrolytic). You may wish to reread Section 21-3 regarding the production of sodium metal by electrolysis of molten sodium chloride in the Downs cell. This method is used to produce the metals of Groups 1A and 2A, as well as some others.
11. impure carbon
12. $WO_3 + 3H_2 \rightarrow W + 3H_2O$. See Table 26-3. This method is also used for molybdenum.
13. bauxite; Hall-Héroult
14. gangue. Can you pronounce this word?
15. slag
16. magnesium; sodium. These two have been discussed—magnesium in Section 26-9 and sodium in Section 21-3. As pointed out in Table 26-3, this method is also used for calcium and the other active Group 1A and 2A metals.
17. cathode
18. none. Copper is oxidized from the impure electrode (anode) and redeposited at the cathode.

As you think about the reactions in Questions 19 through 32, remember to take into account the stoichiometry in identifying the metals. For example, even though the Group 2A metals are produced by electrolysis of their molten chloride, neither the equation in Question 29 nor that in Question 30 could represent this process, because these metal chlorides would have the formula MCl_2.

19. Ca. See Section 26-2.
20. Mg. See Section 26-2.
21. Zn, Pb, Ni. See Table 26-3 and Section 26-3 (Ni).

22. Cu. See Table 26-3.
23. Hg. See Table 26-3.
24. Sn. See Section 26-3.
25. Zn. See Table 26-3.
26. Fe. See Table 26-3.
27. Fe (Both this reaction and that in Question 26 take place in the blast furnace.)
28. Al. See Table 26-3.
29. Na (as well as the other Group 1A metals). See Table 26-3.
30. Al. See Section 26-6.
31. Au. See Section 26-9.
32. Mg. See Section 26-5.

Multiple Choice

1. (a). This is discussed in Section 26-1. The way to tell is by comparing reduction potentials for the cations of these metals. Metals with higher reduction potentials are less active and are more likely to occur in the native state. The list of metals inactive enough to occur in the native state is sufficiently short that you should remember which metals are included.
2. (b) (Section 26-5)
3. (e). This is a method for purifying metals once they have been reduced.
4. (e). The metallurgy of magnesium is discussed in detail in Section 26-5.
5. (a)
6. (e)
7. (e). The addition of $Ca(OH)_2$ precipitates the less soluble $Mg(OH)_2$.
8. (b). The slag is less dense than the molten iron, so it floats on the surface of the liquid iron.
9. (d). The calcium silicate that is formed also dissolves the iron(II) oxide impurity and floats on the surface of the molten copper(I) sulfide.

27 Metals II: Properties and Reactions

Chapter Summary

In this descriptive chemistry chapter, you study some properties and reactions of the **representative metals** and of selected **d-transition metals**. The elements that have predominantly metallic properties include: all members of Group 1A (the alkali metals) and Group 2A (the alkaline earth metals), as well as some of the lower (heavier) members of Groups 3A, 4A, and 5A. Each group of metals is discussed in this chapter from three aspects—(1) **properties and occurrence**, (2) **reactions**, and (3) **uses of the elements and their compounds**.

Descriptive chemistry contains much detail, but the textbook presents it so that you can learn it by a systematic approach. The following suggestions may help you to organize and remember the substance of this chapter. First, do not first read the entire chapter in detail, front to back—go over it lightly at first, just to see the main points of the presentation. Second, outline the chapter in detail for yourself, perhaps one group of elements at a time, noting particular examples and more details of the main ideas that were pointed out. Third, try to remember something about the occurrence of the elements in each group—at least which are the most common and which are quite rare—and the major sources of the elements of the group. Fourth, as you study each group, try to see the trends in the general chemical and physical properties, such as those summarized in Table 27-1 for the alkali metals. Such trends and similarities are more important to remember and to understand than the particular numerical values (especially now, with nearly "ubiquitous access to the information superhighway"). For example, it is not as important to remember that sodium metal melts at 97.5°C as it is to recall that all alkali metals melt at relatively low temperatures and that the melting points decrease with increasing atomic number—and to relate these observations to atomic properties. Also, be sure that you see and understand exceptions to these trends. Often, even significant exceptions are systematic enough to help you remember them. An example is found in the somewhat anomalous properties of the first members of many groups and the resulting **diagonal similarities** discussed in the latter part of Section 27-2. Try to organize your study of the specific chemical reactions of the elements of each group along these same general lines. In the discussion of each group of elements, a table summarizes the reactions of the metals of that group (e.g., Table 27-2 for the alkali metals), but you also need to remember when one or two members of a group display a different reactivity than other elements in the group. The interesting Chemistry in Use essay "Trace Elements and Life" early in this chapter discusses the biological importance of a few elements, most of them metals.

The **alkali metals, Group 1A**, are discussed in Sections 27-1 through 27-3. You should be able to name these metals without reference to the text. All of these melt at relatively low temperatures and are relatively low-density metals whose properties are determined by their ns^1 valence shell configuration. All are very electropositive, with low first ionization energies and very negative reduction potentials. These metallic properties all increase going down the group.

Thus, all alkali metals are very reactive elements, often reacting so vigorously and with evolution of so much heat that they can be dangerous, as in their reactions with water. They are never found free in nature, but occur in many stable salts, all with the metals in the +1 oxidation state. In their reactions (Section 27-2), the alkali metals are all strong reducing agents. The oxides and hydroxides of all of these metals are strongly basic. In your study of Group 1A, be sure to notice that the physical and chemical properties of lithium are anomalous in nearly all respects, presumably as a consequence of its very small size and resulting high charge density. Some of the many uses of alkali metal compounds, especially those of sodium, are listed in Section 27-3.

The metals of **Group 2A**, the **alkaline earth metals** (can you name them?) are discussed in Sections 27-4 through 27-6. All of these metals have the valence shell configuration ns^2. They are not as reactive as the alkali metals, but are still reactive enough that they are never found free in nature. The lighter elements, especially Be, have more tendency to form covalent substances than any of the alkali metals. Calcium and magnesium are the especially abundant members of the group. As you see in Section 27-5, the types of reactions of these elements are similar to those of the alkali metals, except for stoichiometry. Notice especially the trends in basic and amphoteric nature of the oxides of Groups 1A to 3A, summarized in this section. Several important uses of compounds of the Group 2A metals, especially of calcium and magnesium, are mentioned in Section 27-6.

Each of the other groups of elements in this chapter is a post-transition metal group—i.e., it is to the right of the transition metals in the periodic table. Each of these groups ranges from predominantly nonmetallic elements at the top of the group to the metallic ones at the bottom. Only elements with significantly metallic properties are discussed in this chapter. The corresponding discussion of the nonmetallic elements of these groups appears in Chapter 28.

Groups 3A and 4A each consist of elements having properties that range from nonmetallic to metallic. Periodic trends in these two groups are presented in Section 27-7. You will see that gradations in properties within the post-transition metal groups are more extreme than they are in Group 1A or 2A. All Group 3A metals can exhibit +3 oxidation state, but with predominantly covalent bonding. The elements of this group that are to the right of a transition metal series (Ga, In, Tl) also exhibit increasingly stable +1 oxidation states going down the group, often with marked ionic character. This is due to the tendency, increasing down the group, for the elements to lose or share only their outer p electrons; the s electrons remain neither ionized nor shared. This **inert s-pair effect** is common to all post-transition metals. The properties and chemistry of **aluminum**, the most important of the post-transition metals, are also presented in this section. As you will see, many of its important properties depend on its intermediate, almost metalloid, character. Hydrolysis reactions (Chapter 18) play an important role in the aqueous chemistry of aluminum.

In the latter portion of Chapter 27, you survey the major aspects of the properties and chemistry of the **d-transition metals.** There are many of these elements and each has a quite varied chemistry, so you will not study all of the elements in detail. That would (and does!) require several quite large volumes. Rather, your concern here is with summarizing the major trends in properties and behavior of these elements. You will look more closely at a few compounds of only one of them, chromium.

The d-transition elements (usually called just "the transition metals") are the ten columns of elements located between Groups 2A and 3A in the periodic table. Their general properties are summarized in Section 27-8.

When a group of elements has such a widely varied chemistry as do the transition elements, you must organize the reactions or compounds of the elements in some way. As you have seen before, a very convenient organization is in terms of the **oxidation state** of the element. You should pay attention to the discussion in Section 27-9 with regard to the existence of several oxidation states for each transition metal and the relative reducing and oxidizing tendencies of the element in its various oxidation states. It will help to remember that in its simple ions, each of the transition metals loses its outer s electrons (or sometimes electron, as in Cu^+), and may also lose one or more d electrons and to observe that, for most odd-numbered groups, at least some of the most important oxidation states are odd, whereas for even groups, several of the principal oxidation states are even (though this is by no means an infallible rule but only a guide to your memory). As is very clear from Table 27-7, the tendency to form several different oxidation states is much higher near the middle of each series of transition metals than near the ends. Moving down within a group, higher oxidation states become more stable and common, which is an opposite trend from that observed among the representative elements. Other trends regarding correlations of acidity, basicity, and covalent character with oxidation state and with position in the group are parallel to those already seen for the representative elements.

As mentioned earlier, the chemistry of the transition metals is far too vast and too varied to cover in detail in any one chapter. However, it is illustrative to look a little more closely at one of these metals. The next section does this, going into some detail regarding the reactions and properties of the oxides and hydroxides of chromium (Section 27-10). The hydroxides of transition metals in high oxidation states are acidic, with acidity increasing with oxidation state; you saw this earlier for the representative nonmetals. Compounds with chromium in its highest oxidation state (+6) are very strong oxidizing agents.

Study Goals

Review Sections 4-10, 4-18, and 5-9; Introduction to Chapter 27
1. Review the general physical and chemical characteristics of metals. Know where the representative metals are located in the periodic table. Be able to write electronic configurations for metals and their ions. *Work Exercises 1 through 5 and 13 through 15.*

Sections 27-1 through 27-3
2. Account for the trends in periodic properties of the alkali metals, Group 1A. Give several examples of typical reactions of the Group 1A metals. Know some of the uses of the alkali metals and their compounds. *Work Exercises 7 through 10, 12, 16 through 19, 21, 22, 29, 31, 32, and 40.*

Section 27-2
3. Describe, with examples, diagonal similarities. *Work Exercises 25, 26, and 31.*

Sections 27-4 through 27-6
4. Account for the trends in periodic properties of the alkaline earth metals, Group 2A. Give several examples of typical reactions of the Group 2A metals. Know some of the uses of the alkaline earth metals and their compounds. *Work Exercises 8, 9, 11 through 16, 18 through 20, 23, 24, 28, 30, 32, 45, and 46.*

Section 27-7

5. Summarize and illustrate, with properties and reactions, the trends in metallic and nonmetallic characteristics of the post-transition metals. Summarize and illustrate the trends in basic and acidic characteristics of their oxides, hydroxides, and salts. Know some of the properties and uses of aluminum.

Section 27-8

6. Summarize the general properties of the *d*-transition metals. Explain why the properties of successive *d*-transition metals in one period vary less dramatically than those of representative metals. *Work Exercises 33 through 35.*

Section 27-9 (Review Sections 4-18 through 4-20)

7. Write out the electron configurations of the first row (Period 4) *d*-transition metals and their common ions. Explain why the Period 5 and Period 6 *d*-transition metals contain more exceptions to the expected, or Aufbau, electron configurations than do the Period 4 metals. *Work Exercises 36 and 37.*

Section 27-9

8. Know trends in oxidation states, occurrence, and stabilities for the *d*-transition metals. *Work Exercises 39 and 41 through 43 and 47.*

Section 27-10

9. Show how the properties and reactions of the oxides and hydroxides of chromium demonstrate the ideas of this chapter. Be familiar with the structures, properties, and reactions of compounds of chromium. *Work Exercise 44.*

General

10. Answer conceptual questions based on this chapter. *Work* **Conceptual Exercises** *49 through 55.*

11. Apply principles learned earlier to perform calculations concerning the representative metals, the *d*-transition metals, and their compounds. Review earlier chapters as needed—especially Chapters 2, 3, 12, 15, 19, 20, and 21. *Work Exercise 33 and* **Building Your Knowledge** *exercises 56 through 62.*

12. Exercises at the end of chapter direct you to sources outside the textbook for information to use in solving them. Work **Beyond the Textbook** *exercise 63 through 65.*

Some Important Terms in This Chapter

Some important terms from Chapter 27 are listed below. Fill the blanks in the following paragraphs with terms from the list. Each term is only used once. Check the Key Terms list and the chapter reading. Study other new terms, and review terms from preceding chapters if necessary.

acidic anhydride	basic anhydride	oxidation state
alkali metals	diagonal similarities	post-transition metals
alkaline earth metals	inert *s*-pair effect	transition metals
amphoteric		

This chapter describes metallic elements in four regions of the periodic table:

a. the [1]_____, in Group 1A,
b. the [2]_____, in Group 2A,
c. the [3]_____, in Groups 3A, 4A, and 5A, below the stair-step division that separates the metals from the nonmetals (and below the metalloids), and
d. the [4]_____, also known as *d*-transition elements, in Groups 3B, 4B, 5B, 6B, 7B, 8B, 1B, and 2B (if you don't get too technical).

It's been observed that the elements of the second period (Li, Be, and B, particularly) have several properties that are more like those of the third period elements in the next group to the right (Mg, Al, and Si), a characteristic known as [5]_____.

The location of an element on the periodic table largely determines the real (or apparent) charge an atom of the element has in a compound, known as the [6]_____ or oxidation number, which, in turn, has a large influence on some of the properties of the compound. Metallic representative elements generally lose all of their valence shell electrons in forming a positive ion, but the heavier post-transition metals often exhibit two oxidation states. One results from losing all of their valence electrons, but the other from losing only the *p* valence electrons. This is called the [7]_____ because the two *s* valence electrons are not ionized or shared for the lower charge. The transition metals often exhibit several oxidation states by losing one or more *d* electrons in addition to the two *s* valence electrons. For those that form more than one oxide, like chromium,

a. the oxide in which the metal is in the lowest oxidation state tends to be a [8]_____, reacting with water to form a base,
b. the oxide in which the metal is in the highest oxidation state tends to be an [9]_____, reacting with water to form an acid, and
c. the oxides in which the metal is in an intermediate oxidation state tend to be [10]_____, reacting with both acids and bases.

Preliminary Test

True-False

Mark each statement as true (T) or false (F).

_____ 1. All transition elements are metals and all representative elements are nonmetals.
_____ 2. Of the alkali metals (Group 1A), only the heavier ones are sufficiently inactive to be found free in nature.
_____ 3. Alkali metals (Group 1A) can easily form either 1+ or 2+ ions because they have such low ionization energies.
_____ 4. Francium is such a rare element that its properties have not been well characterized.
_____ 5. Because they are easily oxidized by atmospheric oxygen, the alkali metals (Group 1A) are usually stored under water.
_____ 6. In many ways, the compounds of lithium resemble those of magnesium more than they do those of the other alkali metals (Group 1A).
_____ 7. Because they have ns^1 configurations in their valence shell, alkali metals (Group 1A) generally show the +1 oxidation state and never the +2.

_____ 8. Because they have ns^2 configurations in their valence shell, alkaline earth metals (Group 2A) frequently show either the +1 or the +2 oxidation state in their compounds.

_____ 9. All alkali metals (Group 1A) react readily with water at room temperature to produce hydroxides and hydrogen.

_____ 10. All alkaline earth metals (Group 2A) react readily with water at room temperature to produce oxides or hydroxides and hydrogen.

_____ 11. Because of its ease of oxidation, aluminum is rarely used in the pure state.

_____ 12. The Group 3A metals are sufficiently less reactive than those of Groups 1A and 2A that they (Group 3A) frequently occur free in nature.

_____ 13. All of the representative metals are solids at room temperature.

_____ 14. Oxides and hydroxides of an element in a higher oxidation state are always more acidic and covalent than oxides and hydroxides of the same element in a lower oxidation state.

_____ 15. Nonmetallic character _decreases_ as the oxidation state of a particular metal _increases_.

_____ 16. The _d_-transition metals are generally more reactive than the Group 1A and 2A representative elements.

_____ 17. The _d_-transition metals discussed are in B groups of the periodic table.

Short Answer

Answer with a word, a phrase, a formula, or a number with units as necessary.

1. The names and symbols of the metals of Group 1A are _____, _____, _____, _____, _____, and _____.

2. As a common group name, the elements of Group 1A are referred to as the _____ _____.

3. The most common human trace element deficiency is a dietary shortage of _____.

4. The metal that is present in insulin and many enzymes is _____.

5. The names and symbols of the metals of Group 2A are _____, _____, _____, _____, _____, and _____.

6. As a common group name, the elements of Group 2A are referred to as the _____ _____.

7. The names and symbols of the metals of Group 3A are _____, _____, _____, and _____.

8. The names and symbols of the metals of Group 4A are _____ and _____.

9. The name and symbol of the metal of Group 5A are _____.

10. The two most commonly occurring metals of Group 1A (name, symbol) are _____ _____ and _____.

11. The melting points of the alkali metals _____ (increase or decrease?) going down the group.

12. All of the metals of Group 1A have quite _____ (positive or negative?) reduction potentials, which tells us that they are very _____ (easy or difficult?) to reduce to the metallic state.

13. Denoting the metal by M, the formula of the normal oxides of the Group 1A metals is _____, the formula for the peroxides of these metals is _____, and the formula for their superoxides is _____.

14. The only Group 1A metal (name, symbol) that combines with nitrogen to form a nitride is _____.

15. When burned in limited O_2, the heavier alkali metals form compounds called _____ _____, having the formula _____ (using M for the metal); in the presence of excess O_2, they form compounds called _____, which have the formula _____.

16. The formulas and uses of two compounds of sodium are _____ _____ (formula,   use)   and _____ _____(formula, use).

17. Aqueous solutions of the oxides and hydroxides of all Group 1A elements are strongly _____.

18. In general, oxides of the representative metals become decreasingly acidic going _____ across the periodic table and going _____ within a group of the periodic table.

19. One use of beryllium is _____.

20. The substance that is used to coat the gastrointestinal tract for obtaining x-ray photographs is _____.

21. The metal used in electrical wiring is primarily _____, but homes built between 1965 and 1973 may contain _____ wiring.

Each of the equations in Questions 22 through 29 represents a reaction of one or more of the representative metals (M), from Groups 1A and 2A, as discussed in Chapter 27. For each equation, tell for which metal or metals of Groups 1A and 2A the reaction is typical. Some reactions may be limited to only one or a few of the metals of the groups; if so, you should indicate that. Some other reactions may be applicable to metals from more than one group. Be careful of the stoichiometry.

22. $4M + O_2 \rightarrow 2M_2O$ (limited O_2) _____

23. $6M + N_2 \rightarrow 2M_3N$ _____

24. $M + O_2 \rightarrow MO_2$ _____

25. $2M + 2H_2O \rightarrow 2MOH + H_2$ _____

26. $M + 2H_2O \rightarrow M(OH)_2 + H_2$ _____

27. $M + X_2 \rightarrow MX_2$ (X = halogen) _____

28. $2M + X_2 \rightarrow 2MX$ (X = halogen) _____

29. $M_2O + H_2O \rightarrow 2[M^+ + OH^-]$ _____

Questions 22 through 29 emphasize only a few of the general reaction types of the Group 1A and 2A elements. Tables 27-2 and 27-4 in the text include more of them. In addition, the reactions that may be specific to one or a few metals should also be studied, e.g., the many reactions demonstrating acidic or basic behavior of oxides, hydroxides, and salts.

30. The electron configuration of Ti is _____.
31. The electron configuration of V is _____.
32. The electron configuration of Cr is _____.
33. Because many transition metals and metal ions have one or more unpaired electrons, they are _____.
34. The three common oxidation states of chromium are ____, ____, and ____.
35. An example of a compound in which chromium has an oxidation state of +2 is _____.
36. An example of a compound in which chromium has an oxidation state of +3 is _____.
37. Two common ions in which chromium has an oxidation state of +6 are _____ and _____ (name and formula for each).
38. Chromate and dichromate ions in solution exist in a pH-dependent equilibrium, for which the equation is _____.
39. A common (but potentially hazardous) laboratory glassware cleaning solution is made by adding concentrated _____ to a concentrated aqueous solution of _____.

Multiple Choice

____ 1. Which one of the following oxides is most *basic*?
   (a) B_2O_3 (b) Ga_2O_3 (c) BeO   (d) CaO   (e) K_2O
____ 2. Which one of the following oxides is most *acidic*?
   (a) B_2O_3 (b) Ga_2O_3 (c) BeO   (d) CaO   (e) K_2O
____ 3. Which one of the following is not a substance that includes calcium?
   (a) quicklime   (b) gypsum   (c) slaked lime   (d) mortar   (e)   baking soda
____ 4. Which one of the following is the most abundant metal in the earth's crust?
   (a) aluminum   (b) sodium   (c) silicon   (d) oxygen   (e) iron
____ 5. What metal has the largest liquid-state temperature range of any element?
   (a) gallium   (b) mercury   (c) aluminum   (d) potassium   (e) bromine
____ 6. The tendency for the post-transition metals to retain their *s* electrons nonionized or unshared is referred to as
   (a) the diagonal similarity.   (b) the heat of hydration.   (c) the inert *s*-pair effect.
   (d) amphoterism.   (e) stoichiometry.

_____ 7. Which one of the following is not a general property of the transition elements?
   (a) They form many complex ions.
   (b) They often exhibit multiple oxidation states.
   (c) They are metals.
   (d) They form their most stable ions by attaining a noble gas configuration.
   (e) Their ions and compounds are often colored.

_____ 8. The colors exhibited by many transition metal ions and compounds are caused by
   (a) light absorption of wavelengths in the visible region of the spectrum.
   (b) light emission of wavelengths in the visible region of the spectrum.
   (c) inability to form positive ions.
   (d) the different color of electrons in a *d*-orbital from those in a *p*-orbital.
   (e) the large number of oxidation states available to these elements.

_____ 9. Which one of these 3*d*-transition elements exhibits the largest oxidation state?
   (a) Mn   (b) Cr   (c) Fe   (d) Zn   (e) Ni

_____ 10. Which one of the following oxides of chromium is the strongest oxidizing agent?
   (a) CrO (b) Cr_2O_3 (c) CrO_3

_____ 11. Which one of the following oxides of chromium is most acidic?
   (a) CrO (b) Cr_2O_3 (c) CrO_3

Answers to Some Important Terms in This Chapter

1. alkali metals
2. alkaline earth metals
3. post-transition metals
4. transition metals
5. diagonal similarities
6. oxidation state
7. inert *s*-pair effect
8. basic anhydride
9. acidic anhydride
10. amphoteric

Answers to Preliminary Test

Do not just look up answers. *Think about the reasons for the answers.*

True-False

1. False. It is true that all transition elements are metals, but not all representative elements are nonmetals. All of the elements of Groups 1A and 2A, as well as several elements of Groups 3A, 4A, and 5A have pronounced metallic character.
2. False. All alkali metals are far too reactive to occur in the native state. The heavier ones are even more active than the lighter ones.
3. False. These metals have low *first* ionization energies but very high *second* ionization energies. Thus, they do not form 2+ ions by ordinary chemical means. Think about what electron configuration would need to be disturbed for one of these metals to go from a 1+ to a 2+ ion.
4. True

5. False. They react with water to form hydrogen in a manner ranging from "readily" for Li, through "vigorously" for Na, to "explosively" for K, Rb, and Cs. They are stored under anhydrous nonpolar liquids such as mineral oil.

6. True. Do you know why this is so? (Do not answer "diagonal similarity"—this is just another way of saying that it is so, but not *why* it is.)

7. True

8. False. Alkaline earth metals do not exhibit the +1 oxidation state, because their second ionization energies are also quite low.

9. True

10. False. The Group 2A metals are sufficiently less reactive than those of Group 1A that only the heavier alkaline earth metals react with water at room temperature. Magnesium reacts with water only as steam. Even when red hot, beryllium will not react with pure water. See Section 27-5.

11. False. In fact, the easily formed, hard, unreactive, transparent oxide coating protects the underlying metal from further oxidation.

12. False. These are still so reactive that they never occur in the native state. Aluminum occurs mainly in aluminosilicate minerals (from which it cannot easily be extracted) and as the oxide. Gallium, indium, and thallium occur mainly as the sulfide, but are much rarer than aluminum.

13. True. Mercury is the only metal that is liquid at room temperature. It is in Group 2B, so it is not a representative metal.

14. True. To help you remember and understand this, read Section 18-11 again. There you learned that the hydrolysis of small, highly charged metal ions (that is, the action of their hydrates as acids) is due to the strong pull of the highly charged metal for the electrons in the water molecule bonded to the metal.

15. False. In terms of both the acidity and the oxidizing capability of its oxides and hydroxides, the nonmetallic character of a metal *increases* as its oxidation state increases.

16. False. The Group 1A metals, as a group, are the most reactive metals, followed by the Group 2A metals.

17. True. The *d*-transition metals are found in Groups 3B, 4B, 5B, 6B, 7B, 8B, 1B, and (ostensibly) 2B, according to the CAS group numbering system.

Short Answer

1. lithium (Li); sodium (Na); potassium (K); rubidium (Rb); cesium (Cs); francium (Fr)

2. alkali metals

3. iron. See "Chemistry in Use: Trace Elements and Life."

4. zinc. See "Chemistry in Use: Trace Elements and Life."

5. beryllium (Be); magnesium (Mg); calcium (Ca); strontium (Sr); barium (Ba); radium (Ra)

6. alkaline earth metals

7. aluminum (Al); gallium (Ga); indium (In); thallium (Tl). Boron (B) is not considered to have significant metallic properties.

8. tin (Sn); lead (Pb). Carbon (C) and silicon (Si) are not metals. Germanium (Ge) is a metalloid.

9. bismuth (Bi). Nitrogen (N) and phosphorus (P) are nonmetals. Arsenic (As) and antimony (Sb) are metalloids.

10. sodium (Na); potassium (K)
11. decrease
12. negative; difficult. Review the meaning of reduction potentials, Section 21-14.
13. M_2O; M_2O_2; MO_2. Do you remember which metals form each of these compounds and under what conditions? See Table 27-2.
14. lithium (Li)
15. normal oxides, M_2O; superoxides, MO_2. Remember that only K, Rb, and Cs form superoxides.
16. Many possible answers to this question appear in the second part of Section 27-3.
17. basic
18. right to left; down
19. as windows for X-ray tubes. See the photo on page 868.
20. $BaSO_4$
21. copper; aluminum. These homes (about 1.5 million of them) are 55% more likely to have one or more electrical connections reach "fire hazard" condition than homes wired with copper. Aluminum wiring in itself, when properly installed, is not dangerous. But if it has not been properly installed, the connections—where the wires join the outlets and switches—can present a fire hazard. For further information, see the U.S. Consumer Product Safety Commission Booklet CPSC#516 (http://www.cpsc.gov/CPSCPUB/PUBS/516.pdf).
22. Group 1A (all metals). See Table 27-2.
23. Group 1A (Li only). Magnesium also reacts with nitrogen to form nitrides, but the stoichiometry is different. Can you write that equation and the formulas of the Group 2A nitrides (using M)?
24. Group 1A (K, Rb, Cs only, in the presence of excess O_2—these are the Group 1A superoxides); Group 2A (Ba only—barium peroxide, BaO_2). This general equation is also typical of Group 4A metals forming their normal oxides. Notice that the formulas for the various oxides (normal, per-, and super-) differ for different groups.
25. Group 1A. (The heavier alkali metals undergo this reaction in a dangerously explosive manner)
26. Group 2A (Ca, Sr, Ba at room temperature, Mg only at high temperature and can dehydrate to the oxide MgO, Be not at all)
27. Group 2A. See Table 27-4.
28. Group 1A. See Table 27-2.
29. Group 1A. See Section 27-2.
30. $[Ar]3d^24s^2$
31. $[Ar]3d^34s^2$
32. $[Ar]3d^54s^1$. This irregularity in electron configuration is related to the special stability of the d^5 configuration (half-filled). Copper is an analogous exception to the usual order, because of the stable d^{10} configuration.
33. paramagnetic. The relationship between unpaired electrons and paramagnetism was first discussed in Chapter 4. This property is a useful aid in verifying electron configurations.
34. +2; +3; +6
35. Some examples are seen in Table 27-8, including CrO, $Cr(OH)_2$, and various salts such as $CrCl_2$.

395

36. See Table 27-8. Examples are Cr_2O_3, $Cr(OH)_3$, and various salts such as $CrCl_3$.

37. chromate (CrO_4^{2-}); dichromate ($Cr_2O_7^{2-}$)

38. $2CrO_4^{2-} + 2H^+ \rightleftharpoons Cr_2O_7^{2-} + H_2O$

39. H_2SO_4; $K_2Cr_2O_7$

Multiple Choice

1. (e). Trends in acidic and basic character of oxides were discussed in Section 5-9.
2. (a). Trends in acidic and basic character of oxides were discussed in Section 5-9.
3. (e). Can you write the formula for each of these substances?
4. (a). Both silicon and oxygen are more abundant than aluminum in the earth's crust, but neither is classified as a metal.
5. (a). See Section 27-7.
6. (c). See Section 27-7.
7. (d). Very few transition metal ions have noble gas configurations. Can you name some that do?
8. (a). Transition metal ions absorb visible light due to the transition of electrons between two closely spaced d orbitals. The colors of gemstones are often due to such transitions. Zinc ions are colorless because their $3d$ orbitals are completely filled, which is why the Group 2B elements are occasionally not included among the d-transition metals.
9. (a). For the d-transition metals, the *maximum* oxidation state is given by a metal's group number (except for the Group 8B metals in the fourth period). See Table 27-7. Other lower oxidation states are also found, giving manganese one of the largest numbers of different oxidation states. In addition to those listed in the table, +5 and +6 have also been reported.
10. (c). Remember that strengths of oxidizing agents increase with increasing oxidation state for the same element.
11. (c). See Table 27-8.

28 Some Nonmetals and Metalloids

Chapter Summary

Chapter 28 briefly presents the descriptive chemistry of some representative **nonmetallic** elements and some **metalloids.** First, you will learn about the group of quite unreactive elements at the extreme right of the periodic table, the **noble gases** (Sections 28-1 and 28-2). Group 7A, the **halogen group**, which consists of very reactive nonmetals, is the subject of Sections 28-3 through 28-7. Sulfur and other heavier elements from Group 6A are discussed in Sections 28-8 through 28-12. The two elements at the top of Group 5A, nitrogen and phosphorus, are covered in Sections 28-13 through 28-17. The chapter closes (Section 28-18) with a brief section on silicon, from Group 4A.

Because they are rare on earth (formerly called the rare gases) and they are not present in any natural compounds (very low reactivity), the noble gases were discovered rather recently. The discovery, isolation, and occurrence of these gases are described in Section 28-1. It is interesting to observe that argon, a noble gas, is actually the third most abundant element in the earth's atmosphere (about one percent), many times more abundant there than elemental hydrogen. (Of course, hydrogen is very abundant on earth, but most of it appears in compounds, mostly water.) Section 28-1 also gives the physical properties of the noble gases. Because of their very low intermolecular attractions, all have low melting and boiling points, which increase with increasing atomic number (i.e., going down the group). Several uses of the noble gases are given in Table 28-2. As you might expect, most of these uses depend on the relative chemical inertness of these elements. Some applications involve the emission of light under the influence of an electric field. Some of the uses of helium are based on its unusually low density. Hydrogen, the only gas that is less dense than helium at the same conditions, has many disadvantages because of its high reactivity, especially with oxygen.

In the 1960s it was discovered that some of these elements, previously thought to be totally inert to chemical reaction, could form compounds. It should be noted, however, that the only known reactions of the noble gases are with extremely powerful oxidizing agents. Their compounds all involve the noble gas elements in positive oxidation states. The first compounds of noble gases to be made, and those best characterized to date, are the **xenon-fluorine compounds** (Section 28-2). In the known compounds of this class, XeF_2, XeF_4, and XeF_6, the hybrid orbitals involve Xe d orbitals, which accounts for the deviation from the octet rule.

Compounds of radon and krypton have also been made, but until recently no compounds of argon, neon, or helium, the lighter members of the group, were known. In 2000, however, chemists at the University of Helsinki, Finland, produced molecules of argon fluorohydride (HArF). The molecules were trapped in a matrix of solid argon at 7.5 K. Heating above 17 K causes the molecules to decompose into argon and hydrogen fluoride.

The next few sections concern the properties, reactions, and compounds of the elements of Group 7A, the **halogens.** As is described in Sections 28-3 and 28-4, the halogens, except astatine, are distinctly nonmetallic, displaying a greater similarity in properties than any other group except the noble gases and possibly the Group 1A metals. Table 28-4 emphasizes the quite regular gradation in properties of the halogens. In most of their compounds, these elements exhibit the -1 oxidation state, as well as the $+1$, $+3$, $+5$, and $+7$ states, except for fluorine.

The occurrence, production, and uses of the individual halogens are detailed in Section 28-4. As you would expect from their quite high reactivities, these elements do not occur free in nature. They appear most commonly as halide salts, containing X^- (X represents a halogen). **Fluorine** is produced mostly by the electrolysis of molten potassium hydrogen fluoride (KHF_2). The free element is used as a powerful oxidizing agent. Many fluorocarbons (compounds involving C–F bonds) are extremely stable, leading to a variety of uses as refrigerants, lubricants, plastics, aerosol propellants, coating agents, and in preventing tooth decay. **Chlorine,** produced primarily by electrolysis, is used extensively in extractive metallurgy, as a disinfectant, in making chlorinated hydrocarbons, and in bleaching agents. **Bromine,** one of the two liquid elements at room temperature, is a very volatile, corrosive substance. It is produced as the free element primarily by extraction from sea water or brine wells and by displacement of Br^- by Cl^- (already discussed in Section 6-8). Compounds of bromine find use as light-sensitive substances (such as silver bromide) in photographic emulsions, as a sedative, and as a soil fumigant. **Iodine,** a volatile black crystalline solid, is obtained from dried seaweed or from $NaIO_3$, an impurity in nitrate deposits. Its biological importance in the function of the thyroid gland was discussed in the "Trace Elements and Life" essay in Chapter 27. Iodine and its compounds are used as antiseptics and germicides.

Section 28-5 describes some reactions of the free halogens. These are sufficiently strong oxidizing agents to react with most other elements. Oxidizing strengths range from F_2, a very vigorous oxidizing agent, to I_2, a quite mild one. The reactions with iron and copper detailed early in this section exemplify this, but you can see the same trend by studying the remarks in this section. Notice that the displacement reaction with other halogens, $X'_2 + 2X^- \rightarrow 2X'^- + X_2$, is limited to X being lighter than X', consistent with the order of oxidizing power of the diatomic free elements.

The most important and generally useful compounds of the halogens are the **hydrogen halides, HX**. The preparation and properties of these compounds are summarized in Section 28-6. These compounds can be prepared by direct reaction of the elements H_2 and X_2, but at greatly varying rates, $F_2 > Cl_2 > Br_2 > I_2$. Hydrogen halides, which are polar covalent compounds, react with water to give the **hydrohalic acids**, of which HF(aq) is a weak acid and HCl(aq), HBr(aq), and HI(aq) are strong. The strengths of the last three, which cannot be discriminated in water, are seen to increase going down the group. Some uses of the hydrohalic acids are also described in this section.

Section 28-7 is concerned with the other important class of inorganic halogen compounds, the **oxoacids (ternary acids)** and their salts. You may wish to review the naming of ternary acids and their salts, Section 6-4. These compounds are conveniently organized and studied according to the oxidation state of the halogen. Note that fluorine rarely, if ever, exists in an oxidation state other than -1 in a compound; this is because the most electronegative element in a compound is always assigned the negative oxidation state (Section 5-7) and fluorine is the most electronegative element. Stability of higher oxidation states increases with increasing halogen size and atomic number.

The next several sections present the three most common heavier Group 6A elements: **sulfur, selenium,** and **tellurium**. (Polonium, a very rare radioactive element, is not well characterized by chemists.) There is considerable variation in the properties of this group of elements, much more so than for the halogens. This is particularly true for groups that are further from the extreme ends of the periodic table. By looking at the group properties presented in Table 28-6, you can see that, although most properties undergo a trend within the group, there is a big jump in properties is between oxygen (which has no valence shell d orbitals) and sulfur (which does). For instance:

1. Oxygen is the only gaseous element in Group 6A—the other members of the group are solids.
2. The boiling point of oxygen is more than 600 degrees lower than that of sulfur, while the differences between the heavier members of Group 6A are much less.
3. The electronegativity of oxygen is one full unit higher than sulfur, while the electronegativities of the other elements of Group 6A differ by only 0.6 units altogether.
4. Oxygen exhibits almost exclusively the −2 oxidation state, while for the other members of the group, +2, +4, and +6 are also common.

You can see other such irregularities in this table and as you study the chapter in more detail. Recall that the chemistry of the most nonmetallic Group 6A element, oxygen, was discussed in Section 5-9.

The occurrence, properties, and uses of these elements are discussed in Section 28-8. Because of its ability to bond to itself, sulfur can exist in several forms in each of the three physical states. As you learn here, the properties of the two solid and the various liquid forms of sulfur depend on the extent of S−S bonding present. Selenium, a rare element, can also exhibit several different solid forms and exists in various molecular forms in the vapor phase. Several uses of elemental selenium are mentioned, including some that depend on its sensitivity to light. Not surprisingly, selenium, like the even rarer tellurium, occurs mainly in sulfide ores.

The general reactions of the Group 6A nonmetals are summarized in the table in Section 28-9. As you study this table, notice that not all elements of the group undergo all of these reactions. All of the Group 6A elements form covalent hydrides H_2E (E represents a Group 6A element), with E in the −2 oxidation state. Section 28-10 is concerned with the preparation, properties, and reactions of the heavier Group 6A hydrides, both as compounds and in aqueous solution. The acidity of the Group 6A hydrides increases going down the group, just as for the hydrogen halides (Section 28-6).

All of the heavier Group 6A elements form oxides, Section 28-11. The most important of these are the dioxides, EO_2, and the trioxides, EO_3. The tendency toward increasing metallic character going down a group is shown dramatically by the stable dioxides of the 6A elements. As has been mentioned before, SO_2 is a serious atmospheric pollutant that is formed by the combustion of sulfur-containing fossil fuels and by the roasting of sulfide ores. Some of the chemistry of this noxious, dangerous gas and some methods for removing it from flue gases are discussed in this section.

The oxides of the Group 6A elements are acid anhydrides: (1) the dioxides, EO_2, which dissolve in water to give the "-ous" acids, and (2) the trioxides, EO_3, which are the anhydrides of the "-ic" acids. Section 28-12 is concerned with the chemistry of the oxoacids of sulfur, sulfurous and sulfuric acids and their salts, the sulfites and sulfates, respectively.

In Group 5A, sometimes referred to as the **nitrogen family**, the properties of the elements range from the quite nonmetallic **nitrogen** and **phosphorus**, through predominantly nonmetallic arsenic and more metallic **antimony** (both metalloids), to metallic **bismuth**. The latter was discussed with the other representative metals in Chapter 23. In this chapter we learn more about the two lightest nonmetallic members of this group: nitrogen and phosphorus.

As you can see in Table 28-7, the elements of this group show the expected trends in properties. All of the elements of Group 5A, including nitrogen, show an unusually large range of oxidation states in their compounds. Thus, their chemistry is quite varied and you will find it convenient to organize your study of these elements by oxidation state of the Group 5A elements. These elements also show chemical trends with which you are probably familiar by now. For instance, observe the regular variation from acidic to basic properties of the oxides formed from Group 5A elements in the same oxidation state. This trend is illustrated by the oxides N_2O_3, P_4O_6, As_4O_6, Sb_4O_6, and Bi_2O_3.

Although nitrogen is the most abundant element in the atmosphere, it occurs in the form of its compounds as only a minor fraction of the earth's crust (Section 28-13). All living matter contains nitrogen, especially in proteins and in nucleic acids. In the nitrogen cycle, atmospheric nitrogen is converted by bacteria into a form that is chemically available to higher living systems. Nitrogen in the form of N_2 is one of the most stable elements, except for the noble gases and the noble metals. It does, however, have a very rich and interesting chemistry, forming a wide variety of compounds. The versatility of the chemistry of nitrogen is seen in the observation that it can exhibit all oxidation states from –3 to +5.

Ammonia, NH_3, is commercially and chemically the most important of the binary compounds of nitrogen with hydrogen (Section 28-14). The industrial production of ammonia by the Haber process was discussed in detail in Section 17-7 in connection with the concepts of chemical equilibrium and kinetics. This would be a good time for you to review that discussion. As you have seen (Chapters 10 and 18), aqueous solutions of ammonia are weakly basic and form **ammonium salts** when neutralized by acids. Ammonia can also act as a Lewis base, both with metal ions (Chapter 25) and with other electron-pair acceptors. **Amines**, structurally related to ammonia but with one or more hydrogen atoms replaced by organic groups, are mentioned in this section. These have already been encountered in Section 18-4 as weak bases and more generally in the chapters on organic chemistry, Sections 23-12 and 24-4.

The variety of the chemistry of nitrogen is further shown by the number of oxides it forms, with nitrogen having all positive oxidation states from 1 to 5 in these binary compounds. The preparations, properties, reactivities, and uses of some of these, as well as their interconversions, are described in Section 28-15. These compounds and their rather involved chemistry are of great environmental significance in atmospheric processes, especially those related to the production of photochemical smog (see the Chemistry In Use essay, page 1075).

In its **oxoacids** and their salts (Section 28-14), nitrogen displays oxidation states +1 (**hyponitrous acid** and **hyponitrites**), +3 (**nitrous acid** and **nitrites**), and +5 (**nitric acid** and **nitrates**). Of these, the latter two are the most important. (There is also a pernitric acid, HNO_4, but it is very unstable.) The commercial production of nitric acid by the **Ostwald process** is described in this section. Much of the useful chemistry of nitric acid depends on its oxidizing power. You should observe that the reduction products of nitric acid in such reactions depend on the concentration of this acid, with increasing acid concentration generally leading to products with nitrogen in higher oxidation states. Oxidation of metals produces solutions of the metal ions, whereas oxidation of nonmetals yields solutions of oxoacids of the nonmetal. The use of nitrates and nitrites as food additives, as well as the possible danger from this use, are also discussed in Section 28-16.

The occurrence, production, and a few uses of phosphorus are the topic of Section 28-17. Phosphorus is an essential nutrient to all living organisms. The natural phosphorus cycle that produces soluble phosphorus compounds is extremely slow, so the largest commercial use of phosphorus is in **fertilizers.** The production of superphosphate of lime also represents the biggest single use of sulfuric acid.

To close this brief look at the nonmetals, you study silicon and a few of its compounds (Section 28-18). Silicon is the second most abundant element in the earth's crust, occurring in vast quantities in **silica** and the **silicate minerals**. It does not occur free in nature. Pure silicon has a structure like that of diamond, but it is less dense and less hard than diamond. Uses of elemental silicon include semiconductors, transistors, solar cells, additives to steel and aluminum alloys, and as the source of silicon in silicone polymers (polymerized siloxanes or polysiloxanes, $[R_2SiO]_n$, where R is an organic group). Even though silicon is in the same group as carbon, it has quite different chemical properties. (1) Silicon-silicon double bonds (Si=Si) tend to be unstable. (2) Silanes (containing silicon-silicon single bonds, Si-Si, and similar to alkanes) are very reactive compounds, unlike alkanes. (3) Silicon has $3d$ orbitals available for bonding, allowing it to share more than four electron pairs.

The only important oxide of silicon is silicon dioxide (silica). However, unlike carbon dioxide, which is a gaseous molecular substance, silicon dioxide is a polymeric substance that is a solid at room temperature. The two familiar natural forms of silicon dioxide, quartz and flint, are described in this section. Section 28-18 discusses the widely occurring **silicates**. These include a large variety of compounds, all having SiO_4 tetrahedra linked into chains, sheets, or three-dimensional networks, but with metal ions occupying spaces between the tetrahedra. These silicate minerals can adopt a wide variety of structural arrangements. This section also includes a brief discussion of the formation of **glass** and the dependence of its properties on other elements that may be present with the silicates.

Study Goals

Section 28-1
1. Give a few practical uses of the noble gases and the physical and chemical properties on which they depend. *Work Exercises 1 through 5.*

Section 28-2
2. Describe some compounds of xenon, their formation and the bonding of the noble gas element in each. *Work Exercises 6 through 8.*

Section 28-3
3. Describe the bonding in the free halogens. Describe the trends in such properties of the halogens as electronegativities, ease of oxidation, atomic and ionic radii, polarizabilities, and melting and boiling points. Give reasons for the trends. *Work Exercises 9 through 12.*

Section 28-4
4. Describe the occurrence, production, and uses of the halogens. *Work Exercises 13 and 14.*

Section 28-5

5. Give several typical reactions of the free halogens, and know how the tendency of the halogens to undergo these reactions varies with position in the group. *Work Exercises 15 through 19.*

Section 28-6

6. Give several methods of preparation of the hydrogen halides and some of their properties.

7. Compare and explain the trends in acid strengths of the hydrogen halides, both in anhydrous form and in aqueous solution. Describe some uses of these solutions.

Section 28-7 (Review Section 6-4)

8. Know the names and formulas of the halogen oxoacids and their salts and draw representative structures of perhalate, halate, halite, and hypohalite ions. *Work Exercises 20, 23, 24, and 25.*

Section 28-7

9. Give representative preparations and reactions of the perhalic, halic, halous, and hypohalous acids. Explain the trend in acid strengths of the chlorine oxoacids. *Work Exercises 26 and 28.*

Sections 28-8 and 28-9

10. Describe and account for the trends in properties of the Group 6A elements given in Table 28-6. Summarize the occurrence, properties, uses, and general reactions of sulfur, selenium, and tellurium. *Work Exercises 29 through 33.*

Sections 28-10 and 28-11

11. Draw Lewis formulas for compounds of S, Se, and Te. Describe the structures and bonding in such compounds. *Work Exercises 34 and 35.*

Sections 28-9 through 28-10

12. Write equations for the preparations of the Group 6A hydrides. Compare the properties of the 6A hydrides, including the acidic nature of their aqueous solutions. *Work Exercises 36 and 38.*

Section 28-11

13. Compare the structures, properties, preparations, and reactions of the Group 6A dioxides and trioxides. *Work Exercises 34, 36, 39, 41, and 42.*

Section 28-12

14. Draw representative structures of sulfur oxoacids and their salts, and write equations for their preparations: (a) sulfurous acid and sulfite salts; (b) sulfuric acid and sulfate salts. Be familiar with the properties and reactions of these acids and salts. *Work Exercises 34, 35, and 42.*

Section 28-13

15. Compare the Group 5A elements, accounting for the unusual stability of elemental nitrogen and the differences in its chemical properties from those of other 5A elements. *Work Exercises 45 through 47.*

16. Summarize the occurrence, properties, preparation, and importance of nitrogen. *Work Exercises 45 and 49.*

17. Describe the nitrogen cycle in nature. *Work Exercise 48.*

Sections 28-14 through 28-16

18. Give Lewis formulas and structures and assign oxidation states for some compounds of nitrogen. Know the preparations and some reactions of these compounds. *Work Exercises 52 through 54, 56, 60, and 62 through 64.*

Section 28-14 (Review Section 17-7)

19. Describe the structure and bonding of ammonia and the Haber process for its preparation. Describe (with equations) ammonia as a Brønsted-Lowry base and as a Lewis base. Know the important reactions of ammonia and its salts. *Work Exercises 50, 53, 56, and 58.*

Section 28-14

20. Compare the properties of liquid ammonia and liquid water. *Work Exercises 58 and 59.*
21. Draw the general structural formulas for typical amines. Know their most important chemical property.

Section 28-15; Chemistry In Use essay, page 1075

22. Describe the role of nitrogen oxides in the problem of photochemical smog. *Work Exercise 65.*

Section 28-16

23. Describe, with equations, the preparations of the following oxoacids of nitrogen and their salts: (a) nitrous acid and nitrite salts, (b) nitric acid and nitrate salts. Give Lewis formulas, structures, and uses of these compounds. *Work Exercises 61 through 66.*

Sections 28-17

24. Summarize the occurrence, production, and uses of phosphorus and some of its compounds. *Work Exercises 46, 47, 49, 55, 57, 66, and 67.*

Sections 28-18

26. Describe the physical properties and production (with equations) of silicon. Know some reactions of the free element.
27. Compare the structure of silicon dioxide (silica) with carbon dioxide. Understand reasons for differences in structures and properties. Describe the occurrence and some uses of silicon and the silicates.

General

28. Answer conceptual questions based on this chapter. *Work* **Conceptual Exercises** *69 through 81.*
29. Apply concepts and skills from earlier chapters to the ideas of this chapter. *Work* **Building Your Knowledge** *exercises 82 through 89.*
30. Exercises at the end of chapter direct you to sources outside the textbook for information to use in solving them. Work **Beyond the Textbook** *exercises 90 through 94.*

Some Important Terms in This Chapter

Some important terms from Chapter 28 are listed below. Several of them were first used in earlier chapters. Fill the blanks in the following paragraphs with terms from the list. Each term is only used once. Check the Key Terms list, the chapter reading, or in preceding chapters.

acid anhydride	ionization energy	PANs
bond energy	nitrogen cycle	photochemical oxidants
contact process	noble gases	photochemical smog
displacement reaction	nonmetals	silica
Frasch process	Ostwald process	silicates
Haber process	oxoacids	ternary acids
halogens		

This chapter is about the metalloids and most of the elements above and to the right of them on the periodic table. These elements exhibit poor electrical conductivity, are good heat insulators, have no metallic luster, and are brittle and nonductile in the solid state. They are called [1]_____, because they are not metals. The first group of these elements, Group 8A, tend not to react to form compounds at all. Therefore, they were once, and still occasionally are, called the *inert gases*, but since some highly reactive compounds have been made from some of them, they are now called [2]_____. (This means they'll react, but they really don't want to—somewhat like the noble metals.)

The next group studied is Group 7A, elements that form many of our simple salts, which is the basis of the group name, the [3]_____. The chemical behavior of the halogens is determined by their energy characteristics. In particular, a large amount of energy is required to remove an electron from a neutral atom of one of these elements, a property known as the [4]_____. Each of these elements, as the free element, consists of diatomic molecules. Energy is required to break the bond that holds the two atoms together, which is known as the [5]_____. Differences in their chemical reactivity allows us to obtain the less reactive members of this group as the free element; a more active member of this group will displace a less active element from one of its compounds by a [6]_____. The elements of this group produce many of the most important acids used in chemistry. In addition to the binary acids, HX, they produce a number of acids that consist of three elements, called [7]_____, also known as [8]_____ because the third element is oxygen.

Sulfur is the most important element of Group 6A. It is mined along the U.S. Gulf Coast by the [9]_____, in which the sulfur underground is melted by hot water, forced out by compressed air, and blown into giant rectangular forms, where it cools and solidified. Sulfur dioxide reacts with water to produce sulfuric acid, so it is an [10]_____. It is oxidized in the presence of a catalyst to sulfur trioxide, a procedure known as the [11]_____.

In Group 5A, the most common element is nitrogen. It makes up 75% of the mass of the air and is vital to life. The complex series of reactions by which nitrogen is slowly but continually cycled from the atmosphere to water and earth to plant and animal life to water and earth and back to the atmosphere is known as the [12]_____. Industrially rather unreactive elemental nitrogen is converted to ammonia by a chemical procedure called the [13]_____. In urban areas, colorless NO is often oxidized by ultraviolet radiation to brownish NO_2, producing a haze called [14]_____.
Further chemical reactions produce ozone, aldehydes, ketones, and peroxyacyl nitrates

(abbreviated [15]_____), which damage rubber, plastics, and plant and animal life and are very irritating to the eyes and throat. Collectively they are known as [16]_____. Nitric acid, a very important industrial chemical, is produced by a procedure known as the [17]_____.

The most common compound of silicon is silicon dioxide, known as [18]_____. Most of the earth's crust is made of silica and compounds made from it, called [19]_____.

Preliminary Test

Be sure to answer additional textbook Exercises. Many of these are indicated in the Study Goals.

True-False

Mark each statement as true (T) or false (F).

_____ 1. All of the noble gases are chemically inert.

_____ 2. Most of the nonmetals are found at or near the right-hand end of the periodic table.

_____ 3. All of the noble gases are gaseous at all conditions of temperature and pressure.

_____ 4. In their compounds, the noble gases always obey the octet rule.

_____ 5. The noble gases react only with very strong reducing agents.

_____ 6. Reactivity of the noble gases increases going down the group.

_____ 7. All of the halogens exist as stable diatomic gases at room temperature and atmospheric pressure.

_____ 8. Even though they are in the same group, the halogens display widely varying properties.

_____ 9. Generally, small ions are harder to polarize than large ones.

_____ 10. The halogens are generally nonmetallic in their properties.

_____ 11. Because of their strengths as oxidizing agents, halogens always show negative oxidation states in their compounds.

_____ 12. Because of its strength as an oxidizing agent, fluorine cannot be produced by electrolysis.

_____ 13. The most abundant natural sources of the halogens are halide salts.

_____ 14. Halogens are so reactive that their binary compounds with other elements are all ionic.

_____ 15. The elements of Group 6A are much more alike than are the elements of Group 7A.

_____ 16. All of the elements of Group 6A form covalent compounds in which they exhibit the oxidation state –2.

_____ 17. All of the elements of Group 6A form covalent compounds in which they exhibit the oxidation state +6.

_____ 18. The hydrides of all of the Group 6A elements are often used as perfumes in soaps and cosmetics.

_____ 19. Aqueous solutions of hydrogen sulfide, hydrogen selenide, and hydrogen telluride are acidic.

_____ 20. The metallic character of the Group 6A elements studied in this chapter increases in the order: sulfur < selenium < tellurium.

_____ 21. Sulfur is too reactive to occur as the free element, so it is found exclusively in compounds.

_____ 22. The major source of sulfur in naturally occurring compounds is as metal sulfides.

_____ 23. Pure sulfur is always bright yellow, whether solid, liquid, or gas.

_____ 24. Selenium is so rare that it has no significant practical use.

_____ 25. All Group 6A hydrides are liquid at room temperature and atmospheric pressure.

_____ 26. Sulfur dioxide is a gas at room temperature and atmospheric pressure.

_____ 27. Sulfur trioxide is a gas at room temperature and atmospheric pressure.

_____ 28. Sulfurous acid is a diprotic acid.

_____ 29. Sulfuric acid has never been isolated in pure form.

_____ 30. Nitrogen, N_2, is a very reactive molecule because it contains a highly reactive unsaturated triple bond.

_____ 31. Like O_2, N_2 is paramagnetic.

_____ 32. No other element exhibits more oxidation states than does nitrogen.

_____ 33. Amines can be considered as derivatives of ammonia.

_____ 34. Gaseous nitrogen oxide, NO, is paramagnetic.

_____ 35. The oxidizing power of the nitrogen oxides _de_creases as the oxidation state of the nitrogen _in_creases.

_____ 36. The atmospheric reaction in which nitrogen dioxide reacts with water vapor to produce nitric acid and nitrogen oxide is a disproportionation reaction.

_____ 37. Nitrogen can form so many oxides because of the availability of _d_ orbitals on nitrogen for bonding.

_____ 38. Phosphorus can exist in more than one solid form.

_____ 39. Phosphorus is so reactive that it must be stored under oil to keep it from reacting with water vapor in the air.

_____ 40. Because of its relatively low reactivity, most of the silicon in the earth's crust occurs in the native, or uncombined, form.

_____ 41. Like carbon dioxide, SiO_2 is a gas.

_____ 42. Because of the availability of _d_ orbitals, silicon can form some stable species in which its coordination number exceeds four.

Short Answer

Answer with a word, a phrase, a formula, or a number with units as necessary.

1. The noble gas whose uses often depend on its low density is _____.
2. There are more compounds of _____ than of any other noble gas.
3. The order of decreasing electronegativity of the halogens is _____.
4. The halogens have _____ standard reduction potentials.
5. The order of increasing strength of the halogens as oxidizing agents is:
   _____.
6. All halogens have the outer-shell electron configuration _____.
7. The order of increasing atomic size of the halogen atoms is _____.
8. The least metallic halogen is _____; the most metallic halogen is _____.

9. As a result of the _____ (greater or lesser?) ease of polarizability of a larger electron cloud, compounds containing iodide ions display _____ (greater or lesser?) covalent character than those containing fluoride ions.

10. The binary compounds of hydrogen and a halogen are called _____.

11. The oxidation states exhibited by the halogens in the halogen oxoacids and their salts are: ____, ____, ____, and ____.

12. The hydrogen halides, arranged in order of increasing acidity in aqueous solution are: _____.

13. The general formula of the alkali metal hypohalites is _____; one of their main uses is as _____.

14. The oxoacids of chlorine, arranged according to increasing acidity, are: _____.

As you have seen, we often discuss descriptive chemistry according to the oxidation states that the elements of a group display. In Questions 15 through 43, tell the oxidation state of any noble gas or halogen in the element, compound, or ion.

15. Xe _____   25. $MnCl_2$ _____   35. $HClO_4$ _____
16. Cl_2 _____   26. HClO _____   36. $NaBrO_3$ _____
17. PtF_6 _____   27. NaClO _____   37. HOF _____
18. XeF_2 _____   28. Br^- _____   38. H_5IO_6 _____
19. XeF_4 _____   29. BrO_3^- _____   39. OCl^- _____
20. XeF_6 _____   30. Br_2 _____   40. ClO_2^- _____
21. $RbXeF_7$ _____   31. IO_3^- _____   41. $Zn(ClO_4)_2$ _____
22. AgI _____   32. SOF_2 _____   42. $NaBrO_4$ _____
23. $MgBr_2$ _____   33. HCl _____   43. $NaClO_3$ _____
24. KF _____   34. $CuCl_2$ _____

44. The names and symbols for the elements of Group 6A are _____, _____, _____, _____, and the quite rare element _____.

45. Because it does not have available d orbitals in its outermost shell, oxygen can bond covalently to a maximum of ____ other atoms.

46. Because of the availability of d orbitals in their outer shells, the heavier 6A elements sulfur, selenium, and tellurium, can bond covalently to as many as ____ other atoms.

47. The outermost electronic configuration of all Group 6A elements is _____.

48. Of the Group 6A elements heavier than oxygen, _____ is by far the most abundant.

49. All of the Group 6A elements form covalent hydrides having the formula _____, in which the 6A element is represented as E.

50. The major natural source of selenium and tellurium is in _____ ores.

51. The oxidation state of sulfur in the sulfides is ____ .

52. The dioxides of sulfur, selenium, and tellurium, arranged according to increasing covalent (decreasing ionic) character, are _____ (write the formulas).

53. Sulfur dioxide is the anhydride of the acid _____ (name, formula), whereas sulfur trioxide is the anhydride of the acid _____ _____ (name, formula).

54. The name of the ion S^{2-} is _____.

55. The name of the compound SO_2 is _____.

56. The formula for sodium sulfite is _____.

57. The formula for sodium hydrogen sulfate is _____.

58. The formula for sodium sulfate is _____.

59. The formula for sulfurous acid is _____.

60. The formula for sulfuric acid is _____.

61. The formula for the sulfate ion is _____.

62. The three nonmetallic (or predominantly nonmetallic) elements of Group 5A are _____, _____, and _____. The two more metallic elements of this group are _____ and _____.

63. Each of the Group 5A elements can exhibit at least some of the oxidation states ranging from _____ to _____.

64. The equation for the autoionization reaction of liquid ammonia is _____ _____.

65. The organic compounds in which one or more of the hydrogen atoms of ammonia have been replaced with organic groups are called _____; like ammonia, these compounds are all _____ (strong or weak?) bases.

66. The name of the major commercial process for producing nitric acid is the _____ process.

67. The two oxosalts of nitrogen that are sometimes used as meat additives to retard oxidation are _____ (name, formula) and _____ (name, formula).

68. Nitric acid is prepared by dissolving _____ (name, formula) in water; nitric acid is a _____ (strong or weak?) acid and a _____ (strong or weak?) oxidizing agent.

69. The existence of elements in several different forms in the same physical state, as exemplified by phosphorus and arsenic, is called _____.

In each of the following Questions, 70 through 84, supply either the name or the formula of the compound or ion, and tell the oxidation state of nitrogen in that compound or ion.

	Formula	Name	Oxidation State of Nitrogen
70.	NH_3	_____	_____
71.	HNO_3	_____	_____
72.	_____	nitrogen oxide	_____
73.	_____	nitrogen	_____

74. _____ nitrous acid _____

75. N_2O _____ _____

Formula	Name	Oxidation State of Nitrogen
76. _____	nitrogen dioxide	_____
77. _____	dinitrogen tetroxide	_____
78. NO_3^-	_____	_____
79. _____	ammonium nitrate	_____
80. NH_4^+	_____	_____
81. $Ca(NO_3)_2$	_____	_____
82. _____	potassium nitrite	_____
83. _____	potassium nitrate	_____
84. _____	lead(II) nitrate	_____

In each of the following Questions, 85 through 92, supply either the name or the formula of the compound or ion, and tell the oxidation state of phosphorus or arsenic in that compound or ion.

Formula	Name	Oxidation State of Phosphorus/Arsenic
85. $Ca_3(PO_4)_2$	_____	_____
86. P_4	_____	_____
87. _____	calcium dihydrogen phosphate	_____
88. _____	tetraphosphorus trisulfide	_____
89. H_3PO_3	_____	_____
90. H_3PO_4	_____	_____
91. H_3AsO_4	_____	_____
92. _____	phosphate ion	_____

93. The element that is second to oxygen in abundance in the earth's crust is _____.

94. Some gems and semiprecious stones are crystals of the compound _____ in the form of _____, but containing colored impurities.

95. The class of minerals in which aluminum atoms replace some silicon atoms in the silica structure is called _____.

96. The glass used in bottles and window panes is a fused mixture of _____ _____ (name, formula) and _____ _____ (name, formula).

97. Elements with properties intermediate between those of metals and those of nonmetals are called _____.

Multiple Choice

_____ 1. Which one of the halogens is conveniently purified by sublimation?
    (a) F_2 (b) Cl_2 (c) Br_2 (d) I_2

_____ 2. Which of the halogens has the lowest melting point?
    (a) F_2 (b) Cl_2 (c) Br_2 (d) I_2

_____ 3. Which of the halogens has the strongest X–X bond in the diatomic element?
    (a) F_2 (b) Cl_2 (c) Br_2 (d) I_2

_____ 4. Which of the halogens has the shortest X–X bond?
    (a) F_2 (b) Cl_2 (c) Br_2 (d) I_2

_____ 5. Which one of the following is *not* a common use of fluorine or one of its compounds?
    (a) additive to toothpaste   (b) refrigerant
    (c) manufacture of plastics   (d) oxidizing agent
    (e) dietary supplement to aid in thyroid function

_____ 6. Which one of the following is a likely method for preparing a free halogen?
    (a) $Cl_2 + 2Br^- \rightarrow 2Cl^- + Br_2$
    (b) $Br_2 + 2Cl^- \rightarrow 2Br^- + Cl_2$
    (c) $I_2 + 2F^- \rightarrow 2I^- + F_2$
    (d) $2HF \rightarrow H_2 + F_2$
    (e) $2KBr + I_2 \rightarrow 2KI + Br_2$

_____ 7. Which one of the following is *not* described in the chapter as a common use of some compound of bromine?
    (a) photographic film   (b) sedative
    (c) soil fumigant   (d) manufacture of plastics

_____ 8. Compounds commonly used as bleaching agents contain which one of the halogens?
    (a) fluorine   (b) chlorine   (c) bromine   (d) iodine

_____ 9. Which one of the following is a principal use of hydrofluoric acid and *not* of any other hydrohalic acid?
    (a) disinfectant   (b) bleaching agent   (c) dietary supplement
    (d) agent for etching glass   (e) oxidizing agent

_____ 10. What is the name of $NaClO_2$?
    (a) sodium hypochlorite   (b) sodium chlorite
    (c) sodium chlorate   (d) sodium perchlorate
    (e) sodium paraperchlorate

_____ 11. What is the formula for sodium iodate?
    (a) NaI (b) $NaIO$ (c) $NaIO_2$ (d) $NaIO_3$ (e) $NaIO_4$

_____ 12. What is the formula of the only oxoacid that contains fluorine?
(a) HF   (b) HOF   (c) HFO_2 (d) HFO_3 (e) HFO_4

_____ 13. Which one of the elements studied in this chapter is necessary to aid thyroid function?
(a) fluorine   (b) chlorine   (c) bromine   (d) iodine   (e) xenon

_____ 14. Which of the following kinds of salts are used as bleaching agents?
(a) halides   (b) halites   (c) hypohalites   (d) halates   (e) xenates

_____ 15. Which one of the Group 6A elements is *not* a solid at room temperature and atmospheric pressure?
(a) oxygen   (b) sulfur   (c) selenium   (d) tellurium   (e) polonium

_____ 16. Which of the following Group 6A elements has the highest first ionization energy?
(a) O   (b) S   (c) Se   (d) Te

_____ 17. Which one of the following is a major method for obtaining sulfur from deposits of the free element?
(a) strip mining   (b) treatment in a blast furnace
(c) deep shaft mining   (d) electrolysis
(e) pumping out the sulfur after melting with very hot water

_____ 18. Which one of the following is *not* mentioned in the chapter as a use of selenium?
(a) as a coloring agent in glass   (b) in photocopying machines
(c) in solar cells   (d) as a component of gunpowder

_____ 19. Which of the following Group 6A hydrides has the lowest melting point?
(a) H_2O (b) H_2S (c) H_2Se (d) H_2Te

_____ 20. Sulfur trioxide is environmentally significant because
(a) it is depleting the ozone layer.
(b) it helps protect streams from excessive contamination by phosphates.
(c) it dissolves in atmospheric water to produce "acid rain."
(d) it helps to absorb ultraviolet light that could attack the ozone layer.
(e) it is formed by the rotting of eggs.

_____ 21. The process in which sulfur dioxide is catalytically oxidized to sulfur trioxide is called the _____ process.
(a) contact   (b) Frasch   (c) Hall-Héroult   (d) Downs   (e) Ostwald

_____ 22. The Group 5A element with the highest melting point is
(a) nitrogen.   (b) phosphorus.   (c) arsenic.   (d) antimony.   (e) bismuth.

_____ 23. Nitrogen is made available to living organisms primarily by
(a) bacterial action in the nitrogen cycle.   (b) the Haber process.
(c) the Ostwald process.   (d) the use of detergents.
(e) the oxidizing action of nitric acid.

_____ 24. The multistep procedure by which nitric acid is produced commercially from ammonia is called
(a) the Haber process.   (b) the Ostwald process.   (c) respiration.
(d) eutrophication.   (e) the Frasch process.

_____ 25. The most important oxoacid of nitrogen in commercial uses is
(a) HNO_3.   (b) HNO_2.   (c) $H_2N_2O_2$.   (d) H_3NO_4.   (e) HN_3.

_____ 26. The largest single use of phosphorus compounds is in
(a) the fertilizer industry.   (b) the detergent industry.
(c) rocket propellants.   (d) insecticides and other poisons.
(e) smog preventive agents.

_____ 27. Clay minerals generally have _____ structures with _____ surface areas.
    (a) sheet-like, large   (b) sheet-like, small   (c) globular, large
    (d) globular, small   (e) irregular, small

Answers to Some Important Terms in This Chapter

1. nonmetals
2. noble gases
3. halogens
4. ionization energy
5. bond energy
6. displacement reaction
7. ternary acids

8. oxoacids
9. Frasch process
10. acidic anhydride
11. contact process
12. nitrogen cycle
13. Haber process

14. photochemical smog
15. PANs
16. photochemical oxidants
17. Ostwald process
18. silica
19. silicate

Answers to Preliminary Test

True-False

1. False. The heavier ones have been shown to form compounds.
2. True. They are to the right and above the metalloids.
3. False. All have quite low boiling and melting points, so that they are gases at room temperature and pressure, but they all will condense and solidify by cooling and compressing.
4. False. Almost all of the compounds of the noble gases involve d orbitals of that element, and thus deviate from the octet rule by having 10, 12, or 14 electrons in the valence shell of the noble gas atom.
5. False. They react only with very strong oxidizing agents, like fluorine (F_2).
6. True. This is consistent with the decrease in their ionization energies.
7. False. All are diatomic, but bromine is a liquid, and iodine is a solid.
8. False. Their properties, especially their chemical properties, show more similarity within this group than in most other groups of the periodic table.
9. True. See the discussion of the relative polarizability of fluoride and iodide in Section 28-3.
10. True. However, their properties become less nonmetallic—more metallic—from the top of the column to the bottom, such that astatine is generally considered a metalloid.
11. False. While it is true that all halogens form many stable halides (−1), all of them except fluorine commonly exhibit oxidation states of +1, +3, +5, and +7. (Fluorine seems to show +1 in a few compounds, like hypofluorous acid, HOF, but apparently the oxidation states are +1, 0, and -1, respectively.) These higher oxidation states are discussed, for instance, in Section 28-7 for the oxoacids and oxyanions of the halogens.
12. False. It is because of its strength as an oxidizing agent that no other chemical agent can directly oxidize fluoride ions to fluorine gas, so it is usually prepared electrochemically. Section 28-4 describes a chemical synthesis that was recently discovered.
13. True

14. False. The compounds range from quite ionic, when halogens combine with active metals, to quite covalent, when they combine with other nonmetals.

15. False. The differences in the properties within a group become more pronounced as the group being examined switches from near the ends of the periodic table to those nearer the middle.

16. True. For example: H_2O, H_2S, H_2Se, and H_2Te. However, they can have other oxidation states.

17. False. Oxygen never exhibits an oxidation state as high as +6. Can you tell why? The higher oxidation states +2, +4, and +6 are quite common for the heavier members of Group 6A.

18. False. H_2O is odorless. H_2S, H_2Se, and H_2Te are poisonous, bad-smelling gases. For instance, H_2S is the smell of rotten eggs.

19. True. As discussed in Section 28-10, the strength of these substances as acids increases as we descend within the group.

20. True. Remember that the trend throughout the periodic table is that metallic properties increase going down within a group.

21. False. There are, for instance, large deposits of free sulfur, predominantly as S_8 molecules, along the U.S. Gulf Coast. See also the photo on page 1065.

22. True. The major sulfide ores are galena (PbS), iron pyrites (FeS_2), and cinnabar (HgS).

23. False. At some temperatures, liquid sulfur is dark brown. Read in the first part of Section 28-8 how the properties of sulfur in the solid and liquid phases change with temperature.

24. False. Read the second part of Section 28-8.

25. False. All of these compounds, except for water, are gases. Remember that water has very unusual properties because of its ability to form hydrogen bonds.

26. True. Its boiling point at 760 torr is –10°C.

27. False. It is a liquid (boiling point 44.8°C).

28. True. The formula is H_2SO_3.

29. False. Its properties are described in Section 28-12.

30. False. Nitrogen is very stable and unreactive.

31. False. It is diamagnetic, all electrons being paired. You may wish to review the discussion of bonding in O_2 and N_2 in terms of molecular orbitals, Chapter 9.

32. True. It exhibits all oxidation states from –3 to +5.

33. True. Amines can be thought of as being derived from NH_3 by the replacement of one or more of its hydrogen atoms with organic groups, represented by R, R′, or R″. Amines that have one hydrogen atom replaced by an organic group can be represented by RNH_2, those with two replaced hydrogen atoms by RNHR′, and those with three by NRR′R″.

34. True. Any molecule that contains unpaired electrons is paramagnetic. NO, which contains an odd number of electrons (7 + 8), must have an unpaired electron.

35. False. If you think about it, this statement just makes no sense, whether or not you know anything about oxides of nitrogen. In order to act as an oxidizing agent, the substance must already be in a higher oxidation state and be able to be reduced.

36. True. The reaction is: $3NO_2 + H_2O \rightarrow 2HNO_3 + NO$.
    See the Chemistry In Use essay, page 1075.
    Determine the oxidation state of nitrogen in each compound.

37. False. There are no d orbitals in the valence shell ($n = 2$) of nitrogen.

38. True. This is discussed in Section 28-17.

39. False. It is reactive with regard to oxidation by atmospheric oxygen, but it does not react with water, so it is stored under water to protect it from air.

40. False. This element does not occur free in nature. About 87% of the silicon in the earth's crust is in the form of silica (SiO_2) or the silicate minerals.

41. False. It has a polymeric structure, occurring in several solid forms. SiO_2 does not exist as discrete molecules.

42. True. Not many such compounds actually form. One of the few stable such species is the hexafluorosilicate ion, SiF_6^{2-}.

Short Answer

1. helium. It has 93% of the lifting power of hydrogen, which is flammable.
2. xenon. More than 200 compounds of krypton, xenon and radon have been prepared. Xenon compounds include XeF_2, XeF_4, XeF_6, $XeOF_2$, $XeOF_4$, XeO_2F_2, XeO_3F_2, XeO_2F_4, XeO_3, and XeO_4.
3. F > Cl > Br > I. Electronegativity generally decreases going down any column on the periodic table.
4. positive. This means that the elements are easily reduced and that they are good oxidizing agents.
5. $I_2 < Br_2 < Cl_2 < F_2$, going up the column on the periodic table.
6. ns^2np^5. The total ($2 + 5 = 7$) is consistent with the group number, 7A.
7. F < Cl < Br < I. The sizes of the halogens increases doing down the column.
8. fluorine; astatine. Metallic character increases going down any group in the periodic table.
9. greater; greater. Study Section 28-3.
10. hydrogen halides
11. +1; +3; +5; +7. Be aware of which halogens commonly exhibit which oxidation states. See Table 28-5.
12. HF « HCl = HBr = HI. Water is such a strong base that it does not distinguish among the acid strengths of HCl, HBr, and HI. Review the leveling effect, Section 10-7.
13. NaXO or NaOX (X = Cl, Br, I); bleaching agents (especially with Cl)
14. HClO < $HClO_2$ < $HClO_3$ < $HClO_4$. You learned in Chapter 6 (and saw again in Chapters 10 and 18) that perchloric acid, $HClO_4$, is one of the few common strong acids. For any element, acidity of the oxoacids increases with increasing oxidation state of the central atom.

In answering Questions 15 through 43, it may help for you to remember the following: In free elements, the oxidation state is always zero. You may need to review the determination of oxidation states, Section 5-7.

15.	Xe	0	24.	F	−1	33.	Cl	−1
16.	Cl	0	25.	Cl	−1	34.	Cl	−1
17.	F	−1	26.	Cl	+1	35.	Cl	+7
18.	Xe +2;	F −1	27.	Cl	+1	36.	Br	+5
19.	Xe +4;	F −1	28.	Br	−1	37.	F	−1
20.	Xe +6;	F −1	29.	Br	+5	38.	I	+7
21.	Xe +6;	F −1	30.	Br	0	39.	Cl	+1
22.	I	−1	31.	I	+5	40.	Cl	+3
23.	Br	−1	32.	F	−1	41.	Cl	+7

42. Br +7   43. Cl +5

44. oxygen (O); sulfur (S); selenium (Se); tellurium (Te); polonium (Po)
45. 4
46. 6
47. ns^2np^4
48. sulfur
49. H_2E
50. sulfide. It should not surprise you that sulfur and selenium are found together in the same chemical state.
51. -2
52. tellurium dioxide (TeO_2) < selenium dioxide (SeO_2) < sulfur dioxide (SO_2).
    The electronegativities of these elements are: O = 3.5, S = 2.5, Se = 2.4, Te = 2.1. Thus, the differences in electronegativities for the bonds are: S–O = 1.0, Se–O = 1.1, Te–O = 1.4. The bonds become less covalent moving down the column.
53. sulfurous acid (H_2SO_3); sulfuric acid (H_2SO_4). Calculate the oxidation states for sulfur in SO_2 and H_2SO_3, SO_3 and H_2SO_4. Generally, the oxidation state of the central atom of an acid anhydride remains the same when water is added to produce the corresponding acid.
54. sulfide. Review naming of compounds in Sections 6-3 and 6-4.
55. sulfur dioxide
56. Na_2SO_3. Na^+ and SO_3^{2-}
57. $NaHSO_4$. Na^+ and HSO_4^-
58. Na_2SO_4. Na^+ and SO_4^{2-}
59. H_2SO_3
60. H_2SO_4
61. SO_4^{2-}
62. nitrogen, phosphorus, arsenic; antimony, bismuth
63. -3; $+5$
64. $2NH_3(\ell) \rightleftharpoons NH_4^+ + NH_2^-$. This reaction takes place to a much smaller extent than the corresponding one for water ($K_{H_2O} = 10^{-14}$, $K_{NH_3} = 10^{-35}$), so ammonia is a more basic solvent than is water. Liquid ammonia is a very useful solvent for some reaction systems.
65. amines; weak
66. Ostwald
67. sodium nitrite ($NaNO_2$); sodium nitrate ($NaNO_3$). There is, however, considerable concern about the formation of carcinogenic (cancer-causing) compounds from these additives.
68. nitrogen dioxide (NO_2); strong; strong. Dissolving NO_2 in water is part of the Ostwald process. You should be aware that this is an oxidation-reduction process, so NO_2 is not to be considered the "anhydride" of nitric acid in the usual sense. Although nitric acid can be prepared by dissolving N_2O_5 (the oxide that already has nitrogen in the same oxidation state as HNO_3) in water, this is not the major method of preparation.
69. allotropism. The various forms are called allotropes or allotropic forms. (See Section 2-1.)
70. ammonia, -3. Review oxidation states (Section 5-7).
71. nitric acid, $+5$

72. NO, +2
73. N_2, 0. Sometimes this is called "dinitrogen," though not in this text.
74. HNO_2, +3
75. dinitrogen oxide, +1
76. NO_2, +4
77. N_2O_4, +4
78. nitrate ion, +5
79. NH_4NO_3, −3 in the ammonium ion, +5 in the nitrate ion
80. ammonium ion, −3
81. calcium nitrate, +5
82. KNO_2, +3
83. KNO_3, +5
84. $Pb(NO_3)_2$, +5
85. calcium phosphate, +5
86. white phosphorus, 0
87. $Ca(H_2PO_4)_2$, +5
88. P_4S_3, +1.5; $4(x) + 3(−2) = 0$
89. phosphorous acid, +3
90. phosphoric acid, +5
91. arsenic acid, +5
92. PO_4^{3-}, +5
93. silicon
94. SiO_2; quartz
95. aluminosilicates. These are mentioned in Section 28-18, with a few examples.
96. sodium silicate (Na_2SiO_3); calcium silicate ($CaSiO_3$). The chemistry of glass is quite complicated, but the major features are indicated in Section 28-18.
97. metalloids

Multiple Choice

1. (d). The melting point of iodine is 114°C.
2. (a). The melting point of fluorine is −220°C.
3. (b). This is one of the exceptions to the very regular trends in properties of the elements of Group 7A. See the bond energies tabulated in Table 28-4.
4. (a). Review atomic radii, Section 5-2.
5. (e). Iodine is the dietary supplement used to aid in thyroid function.
6. (a). Remember that in the displacement reactions, the lighter halogens replace the heavier ones from halides. This is consistent with our observation that the lighter the halogen, the greater its oxidizing power. Review displacement reactions, Section 6-8.
7. (d). See Section 28-4, part 3.
8. (b). See Section 28-4, part 2.
9. (d). See Section 28-6.
10. (b)
11. (d). This is one of the principal naturally occurring sources of iodine.
12. (b)

13. (d). See Section 28-4, part 4.
14. (c). See Section 28-7.
15. (a). The boiling point of oxygen is $-183°C$.
16. (a). This trend of decreasing first ionization energy going down the group prevails in all representative groups of the periodic table. Review Section 5-3.
17. (e). This is called the Frasch process, described in Section 28-8.
18. (d). See Section 28-8, part 2.
19. (b). The anomalously high melting point of water is due to its extensive hydrogen bonding.
20. (c). We have discussed acid rain several times in this text.
21. (a). Can you identify the processes mentioned in the other answers?
22. (c). The melting points of the elements of Group 5A reach a maximum in the middle of the group, $813°C$.
23. (a). This is sometimes accomplished with the aid of an enzyme called nitrogenase.
24. (b). The Ostwald process is described in Section 28-16, part 2.
25. (a). This is nitric acid. HN_3 is not an oxoacid and there is no such acid as H_3NO_4.
26. (a). Phosphate fertilizers are essential because the low solubility of most natural phosphates makes the phosphorus cycle very slow.
27. (a). These minerals, which represent an important class of the silicates and aluminosilicates, are discussed in Section 28-18.